Developments in Precambrian Geology 1

ARCHAEAN GEOCHEMISTRY

DEVELOPMENTS IN PRECAMBRIAN GEOLOGY 1
Advisory Editor B.F. Windley

ARCHAEAN GEOCHEMISTRY

Proceedings of the Symposium on Archaean Geochemistry:
The Origin and Evolution of Archaean Continental Crust,
held in Hyderabad, India, November 15–19, 1977

Edited by

B. F. WINDLEY

Department of Geology, The University, Leicester, Great Britain

and

S. M. NAQVI

National Geophysical Research Institute, Uppal Road, Hyderabad, India

ELSEVIER SCIENTIFIC PUBLISHING COMPANY

Amsterdam — Oxford — New York 1978

ELSEVIER SCIENTIFIC PUBLISHING COMPANY
335 Jan van Galenstraat
P.O. Box 211, 1000 AE Amsterdam, The Netherlands

Distributors for the United States and Canada:

ELSEVIER NORTH-HOLLAND INC.
52, Vanderbilt Avenue
New York, N.Y. 10017

Library of Congress Cataloging in Publication Data

Symposium on Archaean Geochemistry: the Origin and Evolution of Archaean Continental Crust, Hyderabad, India, 1977.
Archaean geochemistry.

(Developments in Precambrian geology ; 1)
Bibliography: p.
1. Geology, Stratigraphic--Archaean--Congresses.
2. Petrology--Congresses. 3. Geochemistry--Congresses.
I. Windley, Brian F. II. Naqvi, S. M. III. Title.
IV. Series.
QE653.S88 1977 551.7'12 78-17268
ISBN 0-444-41718-4 (v. 1)

ISBN 0-444-41718-4 (Vol. 1)
ISBN 0-444-41719-2 (Series)

Printed in The Netherlands

PREFACE

The geochemistry of Archaean rocks is currently a subject of much interest and activity, largely due to recent technological advances which have made available, readily although not so cheaply, equipment which allows rapid determination of radiometric age dates, and rock, mineral and REE analyses. A result of this technological revolution is an explosion in the number of research papers published in this field. Therefore the recent formation of the International Geological Correlation Project (IGCP) on "Archaean Geochemistry" is a timely event. The aim of the project is to stimulate and integrate some, and hopefully much, of the relevant research with a view to improving communication between, and experience of, the scientists concerned and so to assist in the cross-fertilisation of ideas and techniques which is essential for the healthy development of the subject. This book is the first to result from this IGCP project. Most of the papers were presented at an international meeting in Hyderabad, India, in late November 1977. To those instrumental in the organisation of the conference that gave rise to this volume we offer our acknowledgements and thanks: H. Narain, M.V.N. Murthy, T.V. Viswanathan, V.K.S. Vardan, G.R. Udas, C. Rama Swamy and S. Balakrishna.

The selected papers range across the common Archaean rock groups and vary from reviews to research reports. Several are by geochemists from India, reflecting the considerable activity in that sub-continent. It is worth noting that there is more known at the moment about the geochemistry of the Archaean rocks of India than of most other Archaean regions in the world, including, for example, Greenland and Labrador.

India contains one of the few Archaean cratons in which there is a marked unconformity between rocks of high and low metamorphic grade. In several places in southern India the greenschist grade Dharwar group (Bababudan-Chitradurga-Shimoga belts) rests with a basal oligomictic conglomerate unconformably on the amphibolite grade Sargur group (Sargur-Holenarasipur-Nuggihalli belts). The reader will see from papers in this volume that there is considerable dispute about the origin of some of these meta-supracrustal belts. On the one hand, the Sargur meta-sediments, meta-volcanic amphibolites and anorthositic intrusions, that occur in strips in tonalitic gneisses, are thought to be comparable with the high-grade rocks of Greenland and Labrador (Janardhan et al.); on the other, the Holenarasipur meta-supracrustals (similar in type to the Sargurs) are considered to belong to a highly metamorphosed greenstone belt (Naqvi et al.), the constituent anorthosites being interpreted as basaltic anorthositic flows (Drury et al.).

These differences in opinion, based on well-documented examples, reflect an important debate that echoes through the contemporary literature.

Knowledge of Archaean rocks, especially their geochemistry, is essential for an understanding of the physical-chemical conditions that operated during the early stages of earth history. These geochemical data give us a fingerprint of the composition and evolution of the early continents and oceans. The papers in this volume provide some of the latest information on this subject.

B.F. Windley and S.M. Naqvi

March 1978

CONTENTS

GEOTECTONICS AND CRUSTAL COMPOSITION

BASIC FACTORS IN ARCHAEAN GEOTECTONICS

S. A. Drury
Dept. of Earth Sciences, Open University,
Milton Keynes, Bucks., U.K.

Abstract

Consideration of heat flow parameters and possible lithospheric transformation and deformation during the Archaean enables constraints to be placed on the geotectonic framework of the early earth. They imply a transformation from primary magmagenesis at random centres, which reflect small-scale convective jets and impact-induced thermal anomalies, to linear ridge systems. Patterns of lithosphere consumption would change in concert with this transformation. The difficulty of eclogite formation, due to steep early geothermal gradients, implies the dominance of shallow-dipping zones of lithosphere consumption. This carries the consequence of secondary magmatism and formation of protocontinental material over broad areas of the early crust. The inter-relation between shallow-dipping imbrication of supracrustals and protosialic material, and calc-alkaline magmatism at the sites of lithosphere consumption helps explain many general features of high-grade Archaean terrains. Hypothetical buckling structures associated with horizontal deformation of sial-capped slabs of thin lithosphere help explain features of low grade terrains. Secular decline in overall heat flow implies, at some point in Precambrian time, the sudden appearance of eclogite-capped, descending lithosphere, steep-dipping subduction with narrow volcanic arcs and decreasing amplitude/wavelength ratios of lithospheric deformation.

INTRODUCTION

In recent years it has become popular to strip the plate tectonic concept of modern lithosphere of its revolutionary content of a planet in continual motion and change by combining it with the conservative and mechanistic philosophy of uniformitarianism and imposing them on some geological and geochemical similarities between ancient and modern rocks. This tendency reached its acme in Moorbath's (1975) conclusion that "... the only really major non-uniformitarian geological event that we can discern with any certainty is the formation of the Earth itself."

To penetrate beyond mere impressions about the early earth it is necessary to replace this conclusion with the premise 'Past events are the material basis for present processes'. By combining this materialistic approach with the wealth of knowledge about

(a) the observed characteristics of ancient rocks,

(b) the effects of varying P, T and activities of various fluid phases upon different actual and hypothetical earth materials in terms of melting, crystallization and rheological properties,

(c) the accretion and early fractionation of the earth,

(d) theoretical and indirect information about the earth's thermal evolution,

it should be possible to develop models concerned with the essence of the earth's lithospheric evolution. One vital constraint is that what we term modern plate tectonics is an end point containing some vestiges of past activity and towards which all past events have led.

The purpose of this paper is to review the available information about (b) and (d) until 2.5 b.y. ago, and to suggest a framework for geotectonic evolution in this period.

ARCHAEAN MODELS

Most hypotheses about Archaean lithospheric behaviour have focussed on the volcanic-sedimentary assemblages of the so-called greenstone belts, in particular on the volcanic rocks. Characterization of the tectonic setting of modern and recent lavas by analysis of trace element abundances and patterns has reached a level of considerable sophistication (e.g. Jakes and White, 1972; Pearce and Cann, 1972). On the basis of lithostratigraphic and geochemical similarities between Archaean and modern volcanics the following syllogistic hypotheses have been erected.

1) The calc-alkaline component of greenstone belts indicates that greenstone belts represent Archaean island arcs (e.g. Wilson, et al., 1965; Folinsbee et al., 1968; Hart et al., 1970; White et al., 1971; Goodwin, 1973; Jahn et al., 1974).

2) Basic-ultrabasic lavas are Archaean equivalents of Alpine ophiolites or modern ocean crust (Engel, 1968; Glikson, 1971; Anhaeusser, 1973; Jahn and Shih, 1974; Rivalenti, 1976).

Such actualistic inferences have led Anhaeusser (1973), Talbot (1973) and Burke et al. (1976) to propose fairly detailed plate tectonic models for the Archaean involving lateral accretion of island arcs and continent-continent collision.

Partial similarity between Cretaceous and Caledonide marginal basins and some greenstone belts in terms of structure, stratigraphy and geochemistry have encouraged Tarney et al., (1976) to infer the 'reasonableness' of such modern tectonic features having been important in the Archaean.

Anhaeusser et al. (1969), Windley (1973) and Hunter (1974), in response to the ensialic situation of some greenstone belts and the passing resemblance of their basic volcanics to ocean tholeiites have developed theories of incipient continental rifting and graben filling as the basis of Archaean models.

Very few geotectonic models have been proposed for Archaean high-grade terrains. Windley and Smith (1976), on the basis of similarities between the rock associations in amphibolite-granulite facies gneiss terrains and modern calc-alkaline batholith complexes of active plate margins, proposed a direct analogy between the two environments. They argued that the tonalitic component of high-grade terrains were products of crystal fractionation from mantle-derived basic magma at active continental margins. This approach has been carried further by Tarney and Windley (in prep.). Other hypotheses to explain the calc-alkaline gneisses with interleaved sediments, volcanics and layered igneous complexes involve: calc-alkaline volcanism (Sheraton, 1970; Bowes et al., 1971), crustal underplating by calc-alkaline plutonic magmatism (Holland and Lambert, 1975) and control of calc-alkaline magmatism by inter-thrusting of pre-existing crust (Bridgewater et al., 1974).

In existing literature there are few papers in which a decisive break is made with uniformitarianism. Glikson (1971) suggested that warping or loading of primitive basic-ultrabasic crust and concomitant partial melting could explain many of the features of early Archaean granite-greenstone terrains without invoking plate tectonic mechanisms. In view of the near certainty that the earth could not have escaped the mega-impacting of 4.2-3.8 b.y. recognised on the moon, Green (1972), Glikson (1976a) and Goodwin (1976) have developed models for major meteorite-induced thermal and mechanical disruption at this stage which formed the basis for terrestrial equivalents of lunar maria basalts (basic-ultrabasic greenstones) and diachronous trondjhemitic diapirs, now represented by the earliest greenstone-granite terrains. Lambert (1976) has made a detailed contribution on geothermal conditions for the period 3.6-2.6 b.y. based on the heat generation of simple models of the outer earth. On this theoretical basis he has erected a progressive evolutionary model for the earth in which plate tectonics is restricted to the last 1.0 b.y. Shaw (1972, 1976) has taken the bold step of attempting to outline the geochemical evolution of the early earth, rightly starting with the earth's accretion, and using geothermal and geochemical assumptions to produce a model for the sequential fractionation of sialic crust from the mantle. In his model convection-dominated random lithospheric tectonics passed into a plate-tectonic situation after 2.6 b.y.

ARCHAEAN GEOTECTONICS

The general behaviour of modern lithosphere appears to be connected with three processes:

(a) Emplacement of primitive, mantle-derived basaltic magmas at constructive margins (Lliboutry, 1969).

(b) Sinking of cool, basalt-capped lithosphere at destructive margins, aided by the density increase during the transformation of basalt to eclogite at depth (Elsasser, 1967).

(c) Gravitational sliding of lithosphere above the sloping upper surface of the asthenosphere near constructive margins (Hales, 1969).

The forces involved in these mechanisms contribute to lithospheric motion, but none can supply sufficient energy to maintain it; some form of convective energy transfer from the deep earth is the only possible driving force (MacKenzie, 1969). The relationship between deep convection and near surface motions is essentially dialectical, rather than mechanical, in the sense that while convection determines surface motions its form will be affected by lithospheric behaviour through feedback loops. This stems from the fact that all parts of the earth are in thermal and mechanical contact, e.g. sinking of large volumes of cool lithosphere can control the form of convection in the asthenosphere (MacKenzie, 1969). It is this fundamental relationship that is reflected in the 'observable' features of modern rocks. Subtly different conditions for the relationship in the past may have been reflected by considerable differences in the form taken by near-surface processes.

Due to the inadequacy of conductive and radiative energy transfer in possible mantle materials, the continual energy release by radioactive decay and the enormous initial energy content of the earth after accretion and core-formation, the only stable regime in the earth can be one of mantle-wide convection (Mackenzie and Weiss,

1975) in which the surface heat flow at any time equals the rate at which heat is generated. In its unlikely absence, rapid temperature rise would result in partial melting, when convection would be established.

Once radiative cooling at the surface enabled a wholly solid lithosphere to develop, it would be subject to penetration of mantle-derived magma, the motion of whatever lay beneath it and the associated thermal inhomogeneities, irrespective of the composition of this lithosphere. Up to the terminal major bombardment of the lunar surface, 4.1-3.9 b.y. ago, it seems unlikely that the earth's primary lithosphere escaped impact by large planetismals. Smith (1976) has estimated that, between 4.5 and 4.0 b.y., the earth may have been hit by 10^3 to 10^4 projectiles 10 to 100 km in radius. A variety of evidence suggests that a significant proportion of the earth's outer part was added by accretion of mixed-temperature meteoritic material after rapid core formation, and its admixture with more fractionated earlier material (Gast, 1960; Anders, 1968; Turekian and Clarke, 1969; Fanale, 1971; Murthy and Hall, 1972; Ganapathy and Anders, 1974; Hutchison, 1974, 1976; Morgan et al., 1976). Murthy (1976) has presented a series of constraints suggesting that the earliest crust was basic-ultrabasic with a minor anorthositic component. Such a regime can hardly be said to comply with uniformitarian concepts, though contained within it is the essence of the earth's subsequent development and of modern lithospheric motion.

To assess the form of lithospheric behaviour in the period 4.5 to 3.9 b.y. is no easy matter. Several important parameters need to be assessed, including:

(1) temperature variation with depth
(2) thermal effects of major impacts
(3) depths to melting of basaltic and pyrolite-peridotite compositions and thickness of lithosphere and underlying asthenosphere.

(1) It can be assumed that heat transfer in solid lithosphere is by conduction. In partially melted asthenosphere heat transfer will be dominated by convection. Convection gives more rapid transfer of thermal energy than conduction, therefore once convection is stabilised temperature gradients in the asthenosphere will be less than those nearer the surface and will be essentially adiabatic. The temperature at the upper surface of the asthenosphere will be on the solidus for the composition and pH_2O of mantle material. The

lowest melting temperature of ultrabasic rocks is greater than 1000 °C, and solidi for ultrabasic rocks slowly increase in temperature with pressure. Thus if the temperature gradient in the lithosphere decreased with time, the lithosphere must have increased in thickness in response to whole-earth cooling.

Green et al. (1975) showed that 3.4. b.y. peridotitic lavas were erupted in the Barberton greenstone belt at temperatures of 1650 °C. Green (1975) showed that the rare-earth element patterns of these komatiites, which require garnet fractionation from the melt, are compatible with initial melting at 200 km on a geotherm as shown in Fig. 1. In this the temperature gradient in the 3.4 b.y. lithosphere was about 25 °C km^{-1}.

Taking into account increased radiogenic heat production 3.6 b.y. ago, and assuming conduction alone, Lambert (1976) examined the thermal gradients in a variety of single- and multi-layered models of the upper mantle. Models with (a) a homogeneous mantle containing all heat producing elements, and (b) a 30 km crust containing 33% of heat production above a homogeneous mantle, gave very steep temperature gradients. Under these conditions mantle melting and convection would rapidly fractionate heat-producing elements towards the surface. The only stable configuration involved concentrations of heat production in the upper 100 km of the mantle which gave a model geotherm very close to that favoured by Green (1975).

Direct determination of Archaean temperature gradients from the mineralogy of metamorphic rocks are sparse, largely concentrated on high-grade sialic terrains younger than 3.0 b.y. and complicated by imprecision of mineralogical P/T indicators, in particular for the Al_2SiO_5 polymorphs. The presence of large tracts of Archaean cordierite-bearing gneisses has led to suggestions of temperature gradients as high as 55-60 °C km^{-1} at about 3.0 b.y.; however, Hensen and Green (1973) have deduced a gradient of around 35 °C km^{-1} for such rocks in Antarctica. Wells (1976) has demonstrated a gradient of 15-30 °C km^{-1} for 2.8 b.y. granulite facies metamorphism in West Greenland. Dougan (1977) has shown that the Archaean cordierite granulites of the Imataca Complex in Venezuela formed under a geothermal gradient of around 35 °C km^{-1}.

The thermochemical deductions of pressure and temperature by these authors are plotted on Fig. 2. It should be noted that all the fields are at higher temperatures than would be present at commensurate depths under normal, modern geothermal conditions.

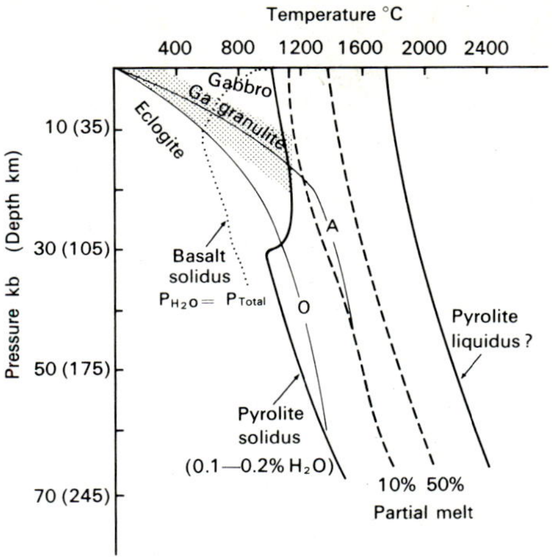

Fig. 1. Effect of modern oceanic and possible Archaean geotherms on melting of basalt and pyrolite and on mineralogy of basaltic compositions. Modern oceanic geotherm = O; Archaean geotherm = A; stability field of garnet granulite - stippled. After Green (1975).

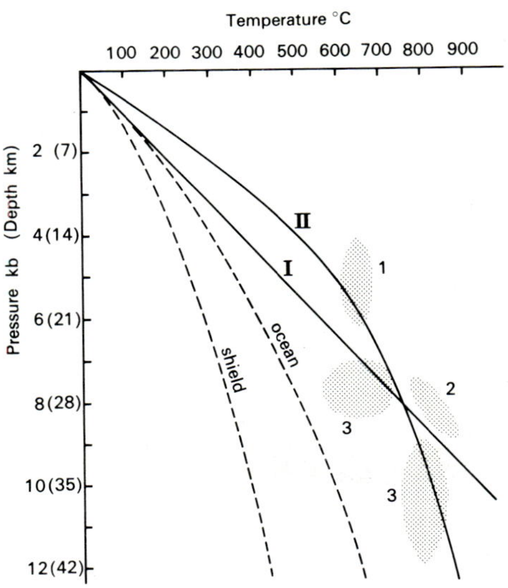

Fig. 2. Pressure and temperature fields for 3.0-2.8 by high-grade terrains. 1 = Venezuela, 3.0 by (Dougan, 1977); 2 = Antarctica, Archaean (Hensen and Green, 1973); 3 = West Greenland, 2.85 by (Wells, 1976). Modern steady-state geotherms = dashed curves; Archaean geotherm with no crustal heat production and mantle flow of 2 h.f.u. = I; Archaean geotherms with uniform crustal heat production = II.

Wells (1976) suggested that amphibolite and granulite facies gneisses from West Greenland were formed by a crustal geothermal gradient similar to that associated with average modern heat flow in near surface regions. Fig. 2 shows that this could only occur to depths of 35 km if the crust had no heat production and mantle heat flow was about 2 h.f.u. (cf. modern values of 0.8 h.f.u.).

The rocks on Fig. 2, however, lie on a model geotherm for <u>uniform</u> Archaean crustal heat production involving moderate mantle heat flow but a very high near-surface geothermal gradient (c. 43 ^{o}C km^{-1}). For an exponential decrease of heat-producing elements with depth in the crust (e.g. Hawkesworth, 1974; Glikson and Lambert, 1976, Fig. 6), the fields on Fig. 2 would form on a geotherm with an even higher near-surface geothermal gradient. This small amount of data suggests a surface heat flow at c. 3.0 b.y. between 2 and 3 times modern values, but indicates the urgency of a major programme of geothermometry and geothermal modelling for high- and low-grade Archaean terrains up to 3.8 b.y. old. If rapid core-formation and an initially hot semi-molten earth is accepted, it follows that the lithosphere has grown in thickness with progressive loss of heat through convection and surface radiation. If Green's (1975) near-surface temperature gradient of 25 ^{o}C km^{-1} for 3.4 b.y. is correct, then earlier lithospheric geotherms must have been steeper.

The work of Green and Ringwood (1967) on the basalt-to-eclogite transformation showed that eclogite is unlikely to form when the near-surface geothermal gradient is greater than 15 ^{o}C km^{-1}. In such a case, basalt-capped lithosphere is transformed to granulite facies assemblages that cannot sink into the underlying ultrabasic asthenosphere since their density (3.0-3.2) is less than that of mantle peridotite (3.3). Steep-angled subduction, with its associated magmagenetic features, which seems to be dependent on eclogite formation (Elsasser, 1967), was therefore an unlikely feature of Archaean geotectonics. The complete absence from the Archaean record of eclogites and their common companions, high pressure-low temperature 'blueschists', verifies this conclusion to some extent.

McKenzie and Weiss (1975) have suggested that convection in the asthenosphere, driven by internal heating and heating from below, takes a dual form: a small-scale flow superimposed on larger wavelength convection. In media of temperature-controlled viscosity, such as silicates, with heating from below, the small-scale flow probably has the surface expression of rising jets of hot material. Under early Archaean geothermal conditions the small-scale flow was probably dominant, large-scale movements probably being much slower (McKenzie and Weiss, 1975).

(2) Impact of large asteroid-sized planetismals between 4.5 and 3.9

b.y. would induce profound modifications of the outer earth's thermal structure through heating by seismic waves (Safronov, 1964; Levin, 1972). Such thermal anomalies would extend to depths of several hundred kilometers, depending on the size of planetismals. Impact structures up to 1000 km diameter might have characterised the surface, by analogy with lunar features such as Mare Imbrium. Green (1972) has suggested that large negative gravity anomalies would result over impacted areas. Levin (1972) and Goodwin (1976) postulated that the regions of major impacts would be characterised by mantle diapirism and that the earth's thermal inertia would ensure the longevity of thermal anomalies caused in this way, perhaps for periods of the order of 500 m.y. Goodwin (1976) further proposed that impact-induced mantle plumes, distributed in an asymmetric fashion, could have been responsible for the distribution of Archaean sialic material on the Pangaea continental reconstruction.

It can be stated with some confidence that, because of impacting phenomena and interference between large numbers of small-scale convection cells in the asthenosphere, the early Archaean surface heat flow was dominated by irregularly distributed positive thermal anomalies. Areas of low heat flow would form a polygonal network between such anomalies.

(3) Given early Archaean geotherms of the form shown in Fig. 1, it is clear that partial melting of wet pyrolite would occur over a wider depth range than at present and that partial melting of wet basalt systems would be at a much shallower depth than today. For Green's model geotherm III (1975) the lithosphere would be only 50 km thick and melting of wet basalt would be initiated at depths greater than 25 km (Fig. 1).

In the asthenosphere, the modern oceanic geotherm is probably sub-parallel to the peridotite solidus and with a low dT/dP because of the efficiency of convective heat transfer. A similar situation applying in the early Archaean would therefore be expected to increase considerably the depth to the base of the asthenosphere, presently at about 600 km below oceans. However, this lower boundary is more likely to represent an increase in activation energy for Nabarro-Herring creep, largely independent of melting and probably at a similar depth in the early Archaean (McKenzie and Weiss, 1975).

The tectonic implications of these unfortunately ill-defined, but definitely indicated, parameters are interesting and decidedly

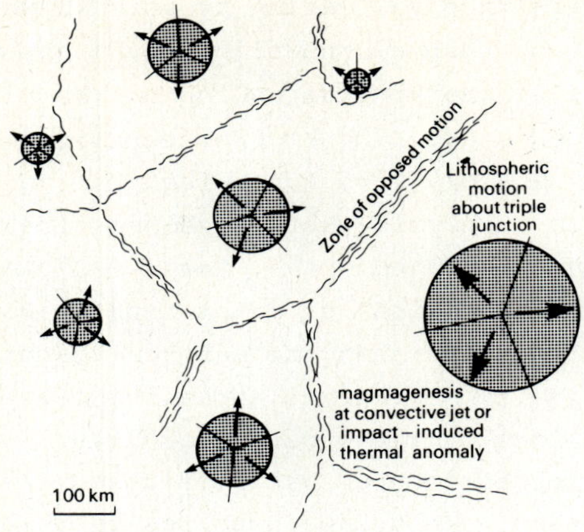

Fig. 3. Magmatism, lithospheric motion and polygonal interference pattern for early thermal regime dominated by convective jets and impact-induced thermal anomalies.

non-uniformitarian. While convection in the asthenosphere was dominated by small-scale flow in the form of upwelling hot jets and polygonal zones of downflow, modified by impact-induced thermal anomalies, additions to the crust in the form of primitive basic-ultra-basic magmas would be at irregularly distributed centres (Fig. 3). Such centres would be characterized by crustal thickening, depression of the base of the lithosphere and radial flow of low viscosity, high temperature lavas. Continual supply of magma would induce radial crustal growth about triple junctions away from the centres, perhaps aided by small-scale convection beneath. A polygonal network of opposed lithospheric motions would develop from such a system, producing linear belts in which the lithosphere would deform and thicken, either by buckling or by brittle failure and thrusting. Such belts could coincide with the zones of small-scale convective downflow.

This primitive regime would be transitory, for two reasons. Firstly, interference between adjacent centres and the formation of geoisothermal 'saddles' would lead ultimately to their coalescence and the development of linear zones of magmagenesis (Fig. 4). Secondly, the regime would be disrupted continually by impacts, which would brecciate and remix the magmatic products of the processes involved and induce the production of further mantle-

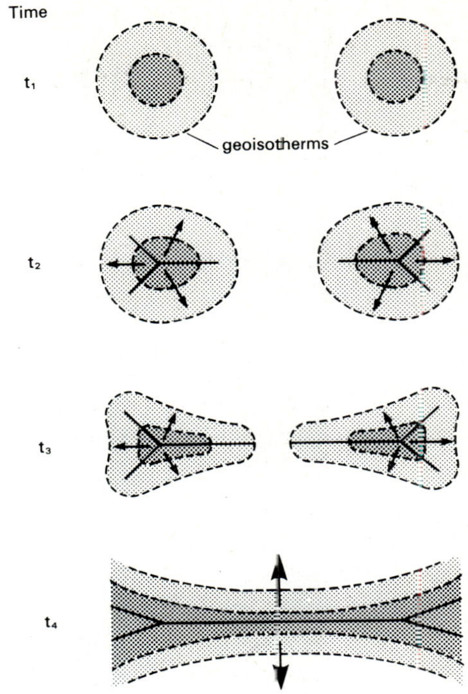

Fig. 4. Coalescence of thermal centres to form linear zones of primary magmagenesis.

derived magmas. Glikson (1976b) has suggested that the oldest komatiitic lavas of Archaean terrains represent terrestrial analogues of lunar mare basalts. Since identical magmas would be produced as a consequence of the earth's own geothermal development, the two different modes of production cannot be distinguished on geochemical grounds alone. The lunar basaltic rocks, in contrast, can only be explained by impact-induced melting.

Fracture of this basic-ultrabasic lithosphere at zones of opposed motion leads ultimately to one slab overriding the other. With geotherms which do not enter the field of eclogite stability such interactions cannot produce steeply descending lithospheric slabs. Rather, the low density of basalt-capped lithosphere would ensure that the lower slab was driven at a shallow angle, directly beneath the overriding slab (Fig. 5c). Consequent magmatic activity would then occur beneath a zone of great width relative to the length of the active boundary (Windley, 1977, p.318), the width depending on the rate of approach of the two expanding slabs and the duration of activity. This configuration is markedly different to modern Benioff zones and island arcs, where the width/length ratio is small and width is determined by the angle of dip on the descending slab, and largely independent of the rate of approach and duration of motion. Since the P-T boundary between the fields of eclogite and garnet granulite stability has been shown to be sharp and to pass through the origin of a graph of pressure against temperature (Green and Ringwood, 1967), as do geotherms, a small change in the geotherm close to this boundary would suddenly enable eclogite to form from basaltic materials. This suggests that the transformation from shallow-angled to steep subduction, together with all the endogenetic connotations, must have been sudden and not an evolutionary progression. This geotectonic revolution would coincide with a particular stage in the cooling history of the earth when local near-surface geothermal gradients fell below 15 $^{\circ}C\ km^{-1}$ for the first time in the earth's history.

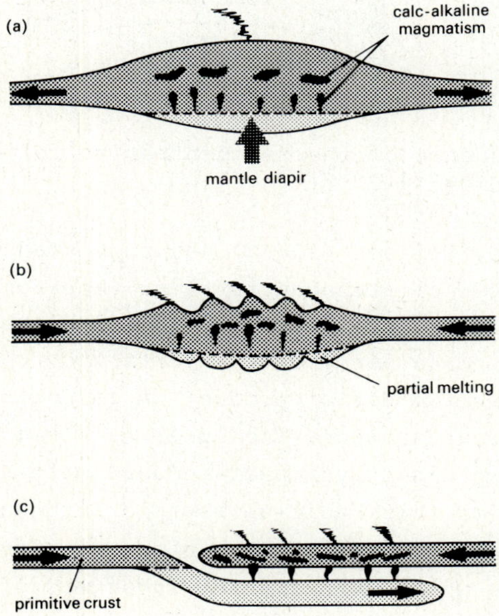

Fig. 5. Secondary magmatism related to crustal thickening by: a) burial in volcanic edifice b) buckling at zones of opposed lithospheric motion, c) shallow-angled thrusting at zones of opposed lithospheric motion.

Basalt-capped lithosphere is essentially transient in that at moderate rates of primary magma generation its residence time at the earth's surface is less than 10^8 years. Consequently, the details of its structural form are trivial in the context of this discussion. However, low density segments of lithosphere, capped by silicic materials generated by various processes (Fig. 5), probably remain at the surface from the date of their formation and consequently their response to stresses is of considerable importance. What follows owes much to Rutland's (1973) suggestion of lithospheric deformation in the Archaean, but incorporates analysis of the form taken by this deformation.

The outer part of the earth can be regarded as a two-layer system for the purpose of geotectonic analysis: a thin, high viscosity lithosphere above a thicker less viscous asthenosphere. The form taken by the common boundary during compression is governed by three important factors:

(a) the magnitude of the principal compressive stress

(b) the relative thicknesses of lithosphere and asthenosphere

(c) the viscosity contrast between the two layers.

During experimental compression of the interface between two materials of different viscosity, an originally sinusoidal form, whose wavelength depends partly on the viscosity contrast (Ramberg, 1964), develops to a series of sharp cusps of low viscosity material penetrating between broad lobes of high viscosity material (Ramsay, 1967 p. 383).

The wavelength and amplitude of such structures are also directly related to the thickness of the buckled layer. Thus a thin Archaean lithosphere, underlain by an asthenosphere containing a higher proportion of melt than today and consequently with a lower viscosity, could have developed such lobe and cusp structures at its base during lateral compression. The cusps would form linear arrays perpendicular to the maximum compressive stress (Fig. 6).

Since the cusps would be composed of high density mantle material displaced upwards, there are two important consequences. Firstly, they would induce positive gravity anomalies, downsagging of the surface being necessary to restore isostatic balance. Secondly, adiabatic uplift of ultrabasic materials at or near solidus temperatures would induce further melting (see Fig. 1) and the production of basic magmas in the vicinity of the cusps.

Lithospheric deformation of this style produces both thickening in the form of downward lobes, and <u>also</u> zones of thinning and incipient weakness associated with the cusps. The thickness of modern lithosphere, and the more viscous state of the modern asthenosphere, with lower degrees of partial melting, precludes prominent buckling of stressed lithospheric plates today; this is a basic assumption of plate theory.

The implications of this possibility is that at some stage during the Archaean, the earth's surface was characterised by relatively closely spaced linear basins with volcanic activity, between and interfering with zones of lithospheric consumption. Such features would only survive in sial-capped segments of the lithosphere, which could not be consumed (Fig. 6). Such intense 'intraplate' activity would wane as declining heat production allowed the lithosphere to thicken beyond the limits of lobe and cusp tectonics, when only broad swells and basins could form. The decoupling of mutually opposed lithospheric motions, by the development of steep subduction of eclogite-bearing lithosphere, would dramatically reduce the intensity of transmitted compressive stresses in the lithosphere, and hence the chances of intense deformation within plates.

DISCUSSION

The geotectonic scheme outlined in the previous section has petrogenetic implications which are closely matched by many general features of both Archaean high- and low-grade terrains. Of course, there are many local peculiarities which require more specific explanation based on detailed work.

(1) The lobe and cusp configuration of Fig. 6 implies that tectonically aligned elongate basins should characterise Archaean supracrustal activity. Many authors have noted the common sub-parallel arrangement of greenstone belts in the Superior, Yilgarn and Dharwar provinces. The main greenstone belts in these provinces are demonstrably younger than local tonalitic-trondjhemitic basement. Individual greenstone belts often show isotropic structural patterns related to both downsagging and vertical movements of synorogenic plutons. Symmetrical supply of sediment has been demonstrated (McGlynn and Henderson, 1972). They lack evidence of structural directionality and paired belts of metamorphism that characterise modern island arcs. Finally, there is no evidence of age zoning in these provinces that would result from successive accretion of island arcs or back-arc basins.

(2) A secular increase in lithospheric thickness implies that the

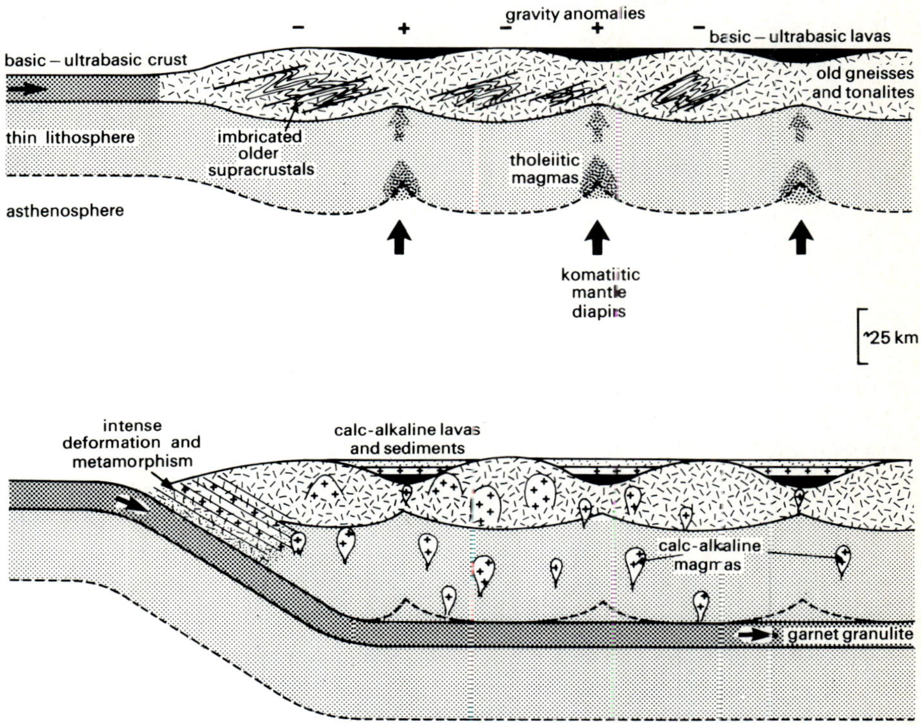

Fig. 6. Hypothetical form of compressed, thin Archaean lithosphere capped by protocontinental material, showing shallow-angled subduction of basalt-capped lithosphere and possible zones of magmagenesis.

wavelength of lobe-cusp structures showed increase. Most ensialic greenstone belts only span the 3.0-2.6 b.y. period which, together with their irregularity and small numbers, precludes the statistical analysis to detect such a secular change. However comparison between narrow Archaean greenstone belts and broad Proterozoic sedimentary basins suggests a decreasing tendency for deformation of sial-capped lithosphere.

(3) If sialic crust is the product of multi-stage reworking of primitive basic-ultrabasic crust, remnants of this initial crust

and ensimatic supracrustal sequences may be preserved in sialic crust. Glikson and Lambert (1976) summarise evidence for pre-sial supracrustals in Western Australia. However striking evidence comes from the low-grade terrain of the Dharwar craton of South India where an early set of supracrustal rocks (Radhakrishna, 1975) is intruded by the c. 3.4 b.y., tonalitic-trondjhemitic precursors of the Peninsular Gneisses (S.M. Naqvi, pers. comm., 1977). The latter are unconformably overlain by younger supracrustals in a series of large sub-parallel greenstone or schist belts which are older than 2.4 b.y. The older supracrustal rocks, in small synformal belts and enclaves within the Peninsular Gneisses, are dominated by basic-ultrabasic lavas and aluminous and iron-rich pelites. There is little evidence for sial-derived clastic sediments within these rocks (Naqvi, this volume). Drury et al. (this volume) describe the geochemistry of anorthositic basalts from one of these occurrences, whose REE patterns may indicate an origin as enclaves of earlier crust component enclosed in later ultrabasic lavas.

(4) The geotectonic scheme outlined in the previous section implies sequential emplacement of volcanics of different composition. Magmas derived from deep mantle melting (komatiites) during the period of compression of sialic slabs should be succeeded by magmas derived by partial melting of underriding basic-ultrabasic crustal slabs and hydrated peridotite in the overlying lithosphere. That is, by calc-alkaline and tholeiitic magmas. In general this is the observed case for greenstone belts ranging in age from 3.4 b.y. to 2.6 b.y., though the basal komatiite component of the volcanic cycles decreases in successively younger terrains. Glikson (1976b) summarises the details of this generalization.

(5) In the regime outlined by Fig. 6 and the accompanying text, the site of most intense deformation is that related to consumption of simatic crust at the edge of the sialic slab. The dominant deformation would amount to imbrication of sialic basement and supracrustals at shallow angles and intense flattening of the imbricated wedge. The contemporaneous production of calc-alkaline magmas below this wedge would lead to complex interference between deformation and intrusion. Crustal thickening in this dynamic regime would cause high-grade metamorphism in amphibolite to granulite facies. Descriptions of Archaean high-grade terrains on all continents conform in general to this pattern (e.g. Bridgwater et al., 1974). As discussed earlier, the isotopic and REE geochemistry of the silicic meta-igneous rocks in such complexes

suggest their derivation by mantle-depth partial melting of basic source materials leaving garnet-bearing residua.

(6) Shallow-angled consumption of basic crust implies the generation and upward emplacement of calc-alkaline magmas by low degree partial melting over very wide zones, rather than in narrow belts which characterize modern steep subduction (See Windley, 1977, p. 318). The dominant rocks of Archaean low-grade terrains are vast gregarious batholiths which separate and deform greenstone belts. Within individual provinces synorogenic plutons of this kind, with very close radiometric ages, occur over distances up to 1000 km perpendicular to the strike of greenstone belts. The most common characteristic of such plutons is the indication by their isotopic and REE geochemistry that they formed by partial melting of basic rocks at mantle depths.

The geotectonic regime outlined, like modern plate tectonics, is a diachronous system in that similar events may take place at different times in the earth's lithosphere. Not all the stages need manifest themselves, and some may be repeated with subtly different forms because of local peculiarities. I believe such a general, non-uniformitarian model can be shown to evolve in train with geothermal evolution into regimes that help explain the many different aspects of Proterozoic earth history. By the same token such a regime is the necessary prerequisite to the establishment of modern lithospheric motions and their surface expressions.

REFERENCES

Anders, E., 1968. Chemical processes in the early Solar system, as inferred from meteorites. Acc. Chem. Res., 1: 289-298.

Anhaeusser, C.R., 1973. The evolution of the early Precambrian crust of South Africa. Phil. Trans. R. Soc. London, A 273: 338-359.

Anhaeusser, C.R., Mason, R., Viljoen, M.J. and Viljoen, R.P., 1969. A reappraisal of some aspects of shield geology. Bull. geol. Soc. Am., 80: 2175-2200.

Bowes, D.R., Barooah, B.C. and Khoury, S.G., 1971. Original nature of Archaean rocks of North-west Scotland. Geol. Soc. Aust. Sp. Publ., 3: 77-92.

Bridgwater, D., McGregor, V.R. and Myers, J.S., 1974. A horizontal tectonic regime in the Archaean of Greenland and its implication for early crustal thickening. Precambrian Res., 1: 179-197.

Burke, K., Dewey, J.F. and Kidd, W.S.F., 1976. Dominance of horizontal movements, arc and microcontinental collisions during the later permobile regime. In: B.F. Windley (Editor), The Early History of the Earth, Wiley, London, 113-129.

Dougan, T.W., 1977. The Imatáca complex near Cerro Bolivar, Venezuela - a calc-alkaline Archaean protolith. Precambrian

Res., 4: 237-268.

Drury, S.A., Naqvi, S.M. and Hussein, S.M., in press. REE distributions in basaltic anorthosites from the Holenarasipur greenstone belt, Karnataka, South India. (This volume).

Elsasser, W.H., 1967. Convection and stress propagation in the upper mantle. Princeton University Technical Report 5.

Engel, A.E.G., 1968. The Barberton Mountain Land: clues to the differentiation of the Earth. Trans. Geol. Soc. S. Afr., 71 (annex): 255-270.

Fanale, F.P., 1971. A case for catastrophic early degassing of the earth. Chem. Geol., 8: 79-105.

Folinsbee, R.E., Baadsgaard, H., Cumming, G.L. and Green, D.C., 1968. A very ancient island arc. In: Knopoff, L., Drake, C.L., and Hart, P.J. (Editors), The crust and upper mantle of the Pacific area: AGU Geophysical Monograph, 12: 441-448.

Ganapathy, R. and Anders, E., 1974. Bulk composition of the moon and earth, estimated from meteorites. Geochim. Cosmochim. Acta. Suppl., 5: 1181-1206. Pergamon, New York.

Gast, P.W., 1960. Limitations on the composition of the upper mantle. J. geophys. Res., 65: 1287-1297.

Glikson, A.Y., 1971. Primitive Archaean element distribution patterns: chemical evidence and geotectonic significance. Earth Planet. Sci. Lett., 12: 309-320.

Glikson, A.Y., 1976a. Earliest Precambrian ultramafic-mafic volcanic rocks: Ancient oceanic crust or relic terrestrial maria. Geology, 4: 201-205.

Glikson, A.Y., 1976b. Stratigraphy and evolution of primary and secondary greenstones: Significance of data from shields of the Southern Hemisphere. In: B.F. Windley (Editor), The Early History of the Earth. Wiley, London, 257-277.

Glikson, A.Y., and Lambert, I.B., 1976. Vertical zonation and petrogenesis of the early Precambrian crust in Western Australia. Tectonophysics, 30: 55-89.

Goodwin, A.M., 1973. Plate tectonics and evolution of Precambrian crust. In: D.H. Tarling, and S.K. Runcorn, S.K. (Editors), Implications of Continental Drift to the Earth Sciences. Academic Press, London, 1047-1069.

Goodwin, A.M., 1976. Giant impacting and the development of continental crust. In: B.F. Windley (Editor), The Early History of the Earth. Wiley, London, 77-95.

Green, D.H., 1972. Archaean greenstone belts may include equivalents of lunar maria? Earth Planet. Sci. Lett., 15: 263-270.

Green, D.H., 1975. Genesis of Archaean peridotitic magmas and constraints on Archaean geothermal gradients and tectonics. Geology, 3: 15-18.

Green, D.H., Nichols, I.A., Viljoen, M. and Viljoen, R., 1975. Experimental demonstration of the existence of peridotitic liquids in earliest Archaean magmatism. Geology, 3: 11-14.

Green, D.H. and Ringwood, A.E., 1967. An experimental investigation of the gabbro to eclogite transformation and its petrological implications. Geochim. Cosmochim. Acta, 31: 767-833.

Hales, A.L., 1969. Gravitational sliding and continental drift. Earth Planet. Sci. Lett., 16: 31-34.

Hart, S.R., Brooks, C., Krogh, T.E., Davis, G.L., and Nava, D., 1970. Ancient and modern volcanic rocks: a trace element model. Earth Planet. Sci. Lett., 10: 17-28.

Hawkesworth, C.J., 1974. Vertical distribution of heat production in the basement of the Eastern Alps. Nature, 249: 435-436.

Hensen, B.J. and Green, D.H., 1973. Experimental study of the stability of cordierite and garnet in pelitic compositions at high pressures and temperatures. Contr. Mineral. Petrol. 38: 151-166.

Holland, J.G. and Lambert, R. St. J., 1975. The chemistry and origin of the Lewisian gneisses of the Scottish mainland: the Scourie and Inver assemblages and sub-crustal accretion. Precambrian Res., 2: 161-174.

Hunter, D.R., 1974. Crustal development of the Kaapvaal craton, 1. The Archaean. Precambrian Res., 1: 259-294.

Hutchison, R., 1974. The formation of the Earth. Nature, 250: 556-558.

Hutchison, R., 1976. Strontium and lead isotopic ratios, heterogeneous accretion of the earth, and mantle plumes. Geochim. Cosmochim. Acta., 40: 482-485.

Jahn, B. and Shih, C., 1974. On the age of the Onverwacht Group, Swaziland sequence, South Africa. Geochim. Cosmochim. Acta., 38: 873-885.

Jahn, B., Shih, C., and Murthy, V. Rama, 1974. Trace element geochemistry of Archaean volcanic rocks. Geochim. Cosmochim. Acta, 38: 611-627.

Jakes, P. and White, A.J.R., 1972. Major and trace element abundances in volcanic rocks of orogenic areas. Bull. geol. Soc. Am., 83: 29-40.

Lambert, R. St. J., 1976. Archaean thermal regimes, crustal and upper mantle temperatures, and a progressive evolutionary model for the Earth. In: B.F. Windley (Editor), The Early History of the Earth. Wiley, London, 363-373.

Levin, B.J., 1972. Origin of the Earth. In: A.R. Ritsema (Editor), The Upper Mantle. Tectonophysics, 13: 7-29.

Lliboutry, L., 1969. Sea-floor spreading, continental drift and lithosphere sinking with an asthenosphere at melting point. J. geophys. Res., 74: 6525-6540.

McGlynn, J.C. and Henderson, J.B., 1972. The Slave province. Geol. Ass. Can. Spec. Paper, 2: 505-526.

McKenzie, D.P., 1969. Speculations on the causes and consequences of plate motions. Geophys. J.R. astr. Soc., 18: 1-32.

McKenzie, D.P. and Weiss, N.O., 1975. Speculations on the thermal and tectonic history of the Earth. Geophys. J.R. astr. Soc., 42: 131-174.

Moorbath, S., 1975. The geological significance of early Precambrian rocks. Proc. Geol. Ass., 86: 259-279.

Morgan, J.W., Ganapathy, R., Higuchi, H. and Krahenbuhl, U., 1976. Volatile and siderophile trace elements in anorthositic rocks from Fiskenaesset, West Greenland: comparison with lunar and meteoritic analogues. Geochim. Cosmochim. Acta., 40: 861-887.

Murthy, V. Rama, 1976. Composition of the core and the early chemical history of the earth. In: B.F. Windley (Editor), The Early History of the Earth. Wiley, London, 21-31.

Murthy, V. Rama and Hall, H.T., 1972. The origin and composition of the Earth's core. Phys. Earth Planet. Interiors, 6: 123-130.

Naqvi, S.M. In press (this volume).

Pearce, J.A. and Cann, J.R., 1973. Tectonic setting of basic volcanic rocks determined using trace element analyses. Earth Planet. Sci. Lett., 19: 290-300.

Radhakrishna, B.P., 1975. The two greenstone groups in the Dharwar craton. Indian Mineralogist, 16: 12-15.

Ramberg, H., 1964b. Selective buckling of composite layers with contrasted rheological properties: a theory for simultaneous formation of several orders of folds. Tectonophysics, 1: 307-341.

Ramsay, J.G., 1967. Folding and fracturing of rocks. McGraw-Hill, New York, 568 pp.

Rivalenti, G., 1976. Geochemistry of metavolcanic amphibolites from South-west Greenland. In: B.F. Windley (Editor), The Early History of the Earth. Wiley, London, 213-223.

Rutland, R.W.R., 1973. Tectonic evolution of the continental crust of Australia. In: D.H. Tarling and S.K. Runcorn (Editors), Implications of Continental Drift to the Earth Sciences. Academic Press, London, 1011-1033.

Safronov, V.S., 1965. Sizes of the largest bodies fallen on planets in process of their formation (in Russian). Astron. Zh., 42: 1270-1276.

Shaw, D.M., 1972. Development of the early continental crust. Part 1. Use of trace element distribution coefficient models for the proto-Archaean crust. Can. J. Earth Sci., 9: 1577-1595.

Shaw, D.M., 1976. Development of the early continental crust. Part 2: Pre-Archaean, Proto-Archaean and later eras. In: B.F. Windley (Editor), The Early History of the Earth. Wiley, London, 33-53.

Sheraton, J.W., 1970. The origin of the Lewisian gneisses of North-west Scotland, with particular reference to the Drumbeg area, Sutherland. Earth. Planet. Sci. Lett. 8, 301-310.

Smith, J.V., 1976. Development of the Earth-Moon system with implications for the geology of the early Earth. In: B.F. Windley (Editor), The Early History of the Earth. Wiley, London, 3-19.

Talbot, C.J., 1973. A plate tectonic model for the Archaean crust. Phil. Trans. R. Soc. Lond., A 273: 413-427.

Tarney, J., Dalziel, I.W.D. and De Wit, M.J., 1976. Marginal basin 'rocas verdes' complex from S. Chile: a model for Archaean greenstone belt formation. In: B.F. Windley (Editor), The Early History of the Earth. Wiley, London, 131-146.

Tarney, J. and Windley, B.F. In prep.

Turekian, K.K. and Clark, S.P. Jr. 1969. Inhomogeneous accumulation of the Earth from the primitive solar nebula. Earth Planet. Sci. Lett., 6: 346-348.

Wells, P.R.A., 1976. Late Archaean metamorphism in the Buksefjorden region, Southwest Greenland. Contrib. Mineral. Petrol., 56: 229-242.

White, A.J.R., Jakes, P. and Christie, D.M., 1971. Composition of greenstones and the hypothesis of sea-floor spreading in the Archaean. Geol. Soc. Aust. Sp. Publ., 3: 47-56.

Wilson, H.D.B., Andrews, P., Moxham, R.L. and Ramlal, K., 1965. Archaean volcanism in the Canadian Shield. Can. J. Earth. Sci., 2: 161-175.

Windley, B.F., 1973. Crustal development in the Precambrian. Phil. Trans. R. Soc. Lond., A. 273: 321-341.

Windley, B.F., 1977. The Evolving Continents. Wiley, Lond., 385 pp.

Windley, B.F. and Smith, J.V., 1976. Archaean high-grade complexes and modern continental margins. Nature, 260: 671-675.

INFERRED COMPOSITION OF EARLY ARCHAEAN CRUST AND VARIATION IN CRUSTAL COMPOSITION THROUGH TIME

J.J.W. Rogers

Geology Department, University of North Carolina,
Mitchell Hall 029A, Chapel Hill, North Carolina 27514, USA

Abstract

Two shields formed in late Proterozoic time (Llano uplift of Texas and Red Sea Hills of Egypt) are characterised by large volumes of post-tectonic, potassic granite. These granites were formed as the final event in the stabilisation of the shields. The potassic granites are absent, or very rare, in presently exposed Archaean terranes, thus causing younger shields to have greater average abundances of lithophilic elements than older shields.

Weighted average compostions of exposed surface areas can be plotted for three shields of different age (2500 my, Archaean rocks of the Canadian shield; 1750 my, Proterozoic rocks of the Canadian shield; and 1000 my, Llano uplift of Texas). Extrapolation of these trends to 3500 my yields the following inferred composition of early Archaean crust: SiO_2, 4.0%; TiO_2, 0.5%; Al_2O_3, 17.0%; total Fe as Fe_2O_3, 5.4%; MgO, 3.1%; CaO, 3.9%; Na_2O, 4.6%; K_2O, 1.5%.

This inferred composition is sufficiently poor in K_2O that it is unlikely that significant quantities of potassium feldspar occurred in rocks of this age. The absence of potassium feldspar (and true granites) correlates with the absence of granitic debris in sedimentary rocks formed by the erosion of early Archaean crust. This observation also indicates that early Archaean crust was actually impoverished in potassium, and granitic materials in general, and that the absence of potassic granites from presently exposed older shields is not simply the result of removal of high-level granites by deep erosion of Archaean terranes.

INTRODUCTION

Controversies about the evolution of the earth's continental crust have centred around two problems. One problem is the rate of evolution of sial, with principal theories fitting into three broad categories: (1) early Archaean segregation of nearly all of the sialic material in the earth and production of continental masses that have simply been reworked by younger geologic events; (2) gradual growth of continents during the Archaean, with the process essentially complete about 2500 my ago and crustal reworking taking place since then; and (3) continual evolution of new sial throughout geologic time. A second problem is the possibility of variation in the composition of continental crust through time, with theories ranging from an essentially constant composition of the crust during all of geologic time to a progressive increase in lithophilic components from older to younger crusts. This short paper cannot review all of the pertinent literature. Reviews are provided by Condie (1976) and numerous papers in the NATO Advanced Study Institute Proceedings edited by Windley (1976); some of the more recent papers are those of Hargraves (1976), Lowman (1976), Kröner (1977), Naqvi (1977), Rogers (1977) and Taylor (1977).

This paper results from an effort to obtain compositional information on two comparatively young shields. The Llano uplift of central Texas, U.S.A., shows ages in the range 1000-1100 my (Zartman, 1964, 1965), with the last major event being the emplacement of a suite of potassic granites (Town Mountain Granites) c. 1000 my ago. The composition of the entire exposed area of the uplift has been estimated by Johnson et al. (1976). The Red Sea Hills of eastern Egypt contain rocks primarily showing ages in the range 500-1200 my (Hashad et al., 1972; El Shazly et al., 1973; Rogers et al., 1978). The last major event in the Red Sea Hills was the emplacement of a suite of potassic granites (Younger Granites) c. 600 my ago.

The term 'age of a shield' must be defined. In this paper it refers to the last major igneous or metamorphic event known to have affected a shield area. In many areas this event is recorded by the K/Ar dates of a variety of rock types, even where older dates are recognisable by Rb/Sr or U/Pb dating (e.g. see Stockwell, 1968, for delineation of age provinces in Canada). In a number of shields (e.g. the Llano uplift and Red Sea Hills), this final event coincides with the emplacement of post-tectonic granites. The final event also stabilises (cratonises) the shield and allows extensive development of platform sediments over the crystalline basement.

This paper: (1) discusses the possible variation in composition of shield areas with their age; (2) postulates a composition of the continental crust 3500 my ago; (3) discusses the question of whether the variation in composition results from variation in the composition of typical rock types (such as gneisses) or from variation in the proportions of different rock types; (4) postulates that younger shields contain significantly greater amounts of potassic granite than older shields; and (5) proposes that the greater abundance of potassic granites is not related to a shallower depth of erosion in younger shields but represents true differences in the bulk composition of continental crust segregated at different times in earth history.

VARIATION IN SHIELD COMPOSITION WITH TIME

The pioneering work on the composition of broad areas of shields was done by Eade and Fahrig (1971) on the Canadian shield. Estimates of large areas of the

Baltic and Ukrainian shields were published by Ronov and Migdisov (1971). The composition of the relatively small (c. 9000 km^2) Llano uplift has been estimated by Johnson et al. (1976).

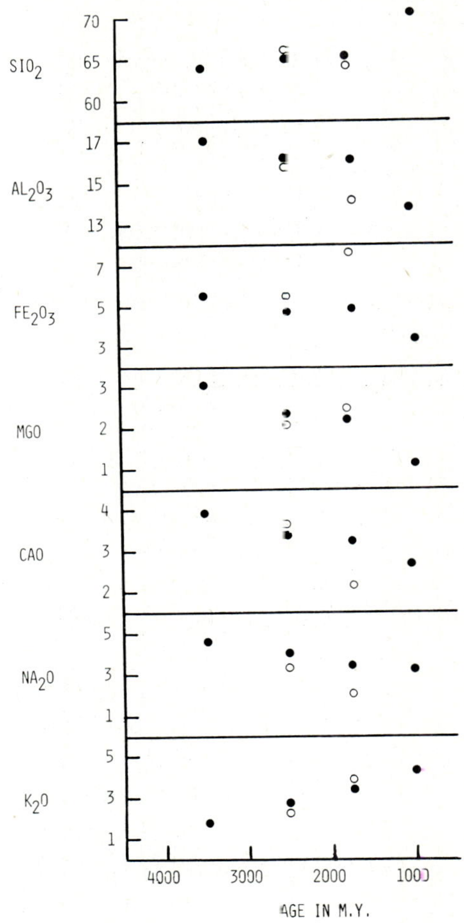

Fig. 1. Compositions of shields of various ages. The values plotted at 1000 my are for the Llano uplift (Johnson et al., 1976). The closed circles plotted at 1750 and 2500 my are from Eade and Fahrig (1971) for the Proterozoic and Archaean rocks of the Canadian shield. The open circles plotted at 1750 and 2500 my are from Ronov and Migdisov (1971) for the Proterozoic and Archaean rocks of the Baltic-Ukrainian shields. The values shown at 3500 my are extrapolated from the other closed circles (Llano and Canadian shields) and represent the inferred composition of the crust 3500 my ago.

Eade and Fahrig (1971) provided separate estimates of Archaean and Proterozoic areas of the Canadian shield, as did Ronov and Migdisov (1971) for the Baltic and Ukrainian shields. Fig. 1 shows diagrammatically the variation in composition with time for these two ages in the Canadian shield, two ages in the combined Baltic-Ukrainian shield, and the Llano uplift. The age plotted for the Archaean shields is 2500 my, the Kenoran event at the end of the Archaean in the Canadian shield. The age of the Proterozoic shields is rather arbitrarily chosen as 1750 my, the time of the Hudsonian orogeny in the Churchill province of Canada. The age of the Llano uplift is 1000 my, approximately the age of emplacement of the youngest granites.

Eade and Fahrig (1971) indicated that Proterozoic shields contain higher concentrations of lithophilic elements, such as potassium, and correspondingly lower concentrations of mafic elements than Archaean shields. Fig. 1 shows that the composition of the Llano uplift follows the same trend, with several of the variations being surprisingly linear.

POSTULATED COMPOSITION OF EARLY ARCHAEAN CRUST

If the compositions plotted in Fig. 1 are at all representative of the compositions of large volumes of continental crust having the ages shown, then it is possible to hypothesise sequential variation in the composition of the crust with age. Thus, the composition of an early Archaean crust at 3500 my age can be inferred by extrapolation, and these points are shown in Fig. 1. This inferred crust has the following composition: SiO_2, 64.0%; TiO_2, 0.5%; Al_2O_3, 17.0%; total Fe as Fe_2O_3, 5.4%; MgO, 3.1%; CaO, 3.9%; Na_2O, 4.6%; K_2O, 1.5%.

The rock type or types that constitute a crust of this composition are uncertain. A mixture of mafic and somewhat more silicic (trondhjemitic-tonalitic) rocks could yield a similar composition; Naqvi et al. (1978) have proposed that the crust that provided sedimentary debris for the Dharwar geosynclinal piles contained mafic volcanic rocks and trondhjemitic-tonalitic plutonic rocks. The inferred composition is also very similar to that of a granodiorite-tonalite. If the rock is essentially granodiorite, the modal mineralogy should include quartz, andesine, hornblende and biotite; it seems likely that the K_2O would be completely contained in biotite, and the probable absence of potassium feldspar is important in assessing the source rocks of Archaean sediments (discussed below).

REASONS FOR COMPOSITIONAL VARIATIONS

Differences in the composition of different shields can result from: (a) differences in the composition of individual rocks types (gneiss, granite, etc.) that are major components of all shields; (b) differences in the proportions of the major rock types that constitute the shields; or (c) some combination of (a) and (b). In order to investigate the principal reasons for the variations postulated in Fig. 1, the major rock types of shields have been subdivided in the following five suites: (1) granulite facies rocks; (2) metavolcanic rocks of geosynclinal, including greenschist, belts; (3) metasedimentary rocks of geosynclinal, including greenschist, belts; (4) gneiss-migmatite-granite, including trondhjemite-tonalite, terranes; and (5) post-tectonic potassic granites. These rock types have been chosen because they represent readily recognisable suites, and most of the mapped rock types of shield areas can be placed in one of them.

The preceding classification contains many inherent difficulties and flaws. Different geologists working in different areas (or even the same area) use different classifications of rock types for mapping and sampling purposes. Granulite facies rocks form a readily identifiable suite, but the diversity of rock types is so great that no satisfactory average composition has been obtained for any area. Geosynclinal metavolcanic and metasedimentary rocks are also fairly easily recognisable, but many maps of Precambrian terranes group both metasedimentary and metavolcanic rocks in one map unit identified by such terms as 'greenstone belt' or 'schist'. The gneiss-migmatite-granite terranes are very complex and particularly difficult to subdivide. Many geologists have recognised and mapped suites of calcalkaline 'granites' (some as mafic as tonalite and granodiorite) that are intrusive into older gneiss. In some areas, however, the degree of migmatisation is so intense that no adequate subdivision of gneiss, migmatite and granite can be made, particularly on maps of large areas. For this reason, investigation of compositions of rocks in this suite has been restricted to gneisses, and no effort has been made to determine an average composition of the calcalkaline intrusive rocks. The term 'schist' on some geologic maps also includes some rocks that should be included in the gneiss-migmatite-granite terrane. The post-tectonic potassic granites form one of the most easily identifiable suites in shield areas.

It is interesting to note that shields of different ages contain the various rock suites identified above in more or less the same stratigraphic order regardless of the general age of the shield. In the Canadian Archaean (e.g. Goodwin et al., 1972), the oldest rocks are gneiss-migmatite terranes and greenschist belts, with age relationships between the two suites being uncertain. The gneisses are intruded by calcalkaline batholiths, and the youngest crystalline rocks are high-level potassic granites. Similar relationships are found in shields as young as the late Proterozoic/early Palaeozoic area of eastern Egypt (El Ramly, 1972; discussions in Said, 1972), where gneisses and greenschist assemblages of probable mid-Proterozoic age were intruded by Older Granites (calcalkaline) c. 750 my ago and finally intruded by Younger Granites (potassic) c. 600 my ago. The fact that rocks in younger shields show chronologic sequences similar to those of older shields indicates that the young shields are not simply older shields whose clocks have been reset by a metamorphic event, which should yield identical ages for all rock types rather than a sequence of ages.

TABLE 1: Comparison of compositions of Precambrian gneisses

	1	2	3	4	5	6
SiO_2	70.4	64.8	65.4	71.5	70.2	69.2
TiO_2	0.4	0.9	0.5	0.3	0.3	0.5
Al_2O_3	13.7	15.3	16.0	14.8	14.2	15.9
ΣFe_2O_3	3.9	5.5	4.9	1.6	3.3	3.4
MgO	1.2	1.7	2.1	0.9	2.8	1.5
CaO	2.0	3.2	3.4	3.3	1.7	3.2
Na_2O	3.2	4.3	3.9	4.8	3.7	4.2
K_2O	4.8	2.1	2.8	1.6	2.7	2.3

1: Valley Spring Gneiss, central Texas, USA (Johnson et al., 1976); 1120 my age.
2: Gneiss of the Blue Ridge area of North Carolina, USA (Bryant and Reed, 1970); Grenville age (c. 1000 my).
3: Banded gneisses, migmatites, granitic gneisses and minor amphibolite inclusions of Canadian shield (Unit 7 of Table 21 in Eade and Fahrig, 1971); Archaean and Proterozoic.
4: Ancient Gneiss Complex of Kaapvaal craton, Swaziland (average of 5 samples of tonalitic gneiss from Hunter, 1970); Archaean.
5: Peninsular Gneiss of southern India (unweighted average of columns 50, 51, 53-55 of Naqvi et al., 1974); Archaean.
6: Lewisian gray gneiss of northern Scotland (Tarney, 1976); late Archaean.

TABLE 2: Comparison of compositions of Precambrian metabasalts

	1	2	3	4	5	6
SiO_2	49.6	49.9	51.2	47.2	52.1	51.0
TiO_2	1.3	0.7	0.9	1.1	1.0	0.7
Al_2O_3	14.2	14.3	13.7	16.5	15.2	15.5
ΣFe_2O_3	14.6	11.5	13.1	12.3	13.2	10.6
MgO	4.8	7.3	8.0	9.0	6.8	5.8
CaO	13.2	10.7	10.4	10.8	9.9	12.6
Na_2O	1.7	2.5	1.7	2.2	2.1	2.3
K_2O	0.22	0.16	0.15	0.3	0.32	0.56

1: Massive and banded amphibolites of Packsaddle Formation, central Texas, USA (Billings, 1962; calculated by Johnson et al., 1976); 1000 my metamorphism age.
2: Mafic, pillowed volcanic rocks of the Hoogenoeg Formation of the Onverwacht Group, Barberton Mountain Land, South Africa (Viljoen and Viljoen, 1969a); $>$3000 my age.
3: Metabasalts of the Coolgardie-Kurrawang belt of western Australia (Glikson, 1970); $>$ 2700 my age.
4: Basalts of the Kolar belt, peninsular India (average of hornblendic rocks from Divikara Rao et al., 1976); early Archaean.
5: Mafic volcanic rocks of the Yellowknife area, Canada (Baragar and Goodwin, 1969); Archaean.
6: Pillow lavas of Chitaldrug schist belt, India (Naqvi and Hussain, 1973); late Archaean.

NOTE: Values are as reported in original work without recalculation to 100%
ΣFe_2O_3 is total iron calculated as Fe_2O_3.

TABLE 3: Comparison of compositions of Precambrian greywackes

	1	2	3	4	5	6
SiO_2	64.5	70.9	62.8	69.9	63.3	66.2
TiO_2	-	0.3	0.7	0.4	0.8	0.5
Al_2O_3	12.5	12.2	15.2	12.3	13.0	10.2
ΣFe_2O_3	4.4	5.8	7.5	5.0	6.8	7.0
MgO	2.4	2.3	3.3	3.3	4.2	4.5
CaO	7.5	1.5	4.1	2.7	4.5	2.0
Na_2O	3.0	3.7	3.0	2.8	2.1	1.8
K_2O	2.3	2.9	2.4	2.6	2.0	1.6

1: Packsaddle metasediments, central Texas, USA (from Johnson et al., 1976); 1000 my metamorphism age.
2: Least-altered gneiss of northwest Adirondack Mts., New York (Engel and Engel, 1958); Grenville age (c. 1000 my).
3: Unweighted average of paragneisses and paraschists (Unit 5) from 5 suites in the Canadian shield (Ft Enterprise, Snowbird Lake, northern district of Keewatin, and Kasmere Lake from Table 19 of Eade and Fahrig, 1971; Red Lake-Lansdowne House area from Reilly and Shaw, 1967); Archaean and Proterozoic.
4: Unweighted average of 8 analyses of 'metamorphites' in Ancient Gneiss Complex of Swaziland (Table 2 in Hunter, 1974); Archaean, possibly 3500 my age.
5: Unweighted average of 9 metasedimentary unites from Chitaldrug schist belt of Dharwars, southern India (average of actinolite-chlorite-quartz schist, matrix of Talya conglomerate, micaceous schist, sericitic phyllite, matrix of Aimangala conglomerate, greywackes, matrix of Kurmerdikere conglomerate, chlorite schist and phyllite, and sericitic ferruginous phyllites; from Naqvi and Hussain, 1972); Archaean.
6: Sheba Formation of Barberton Mountain Land, S. Africa (Condie et al., 1970); early Archaean.

TABLE 4: Comparison of compositions of Precambrian potassic granites

	1	2	3	4	5
SiO_2	72.4	70.4	71.2	66.4	74.8
TiO_2	0.4	0.4	0.3	0.6	0.1
Al_2O_3	13.7	14.8	13.7	15.8	12.7
ΣFe_2O_3	3.1	2.4	3.2	4.8	1.0
MgO	0.4	0.9	0.8	1.6	0.1
CaO	1.3	1.8	1.8	2.8	0.6
Na_2O	3.4	4.7	3.4	3.8	4.1
K_2O	5.1	3.6	5.1	3.7	4.5

1: Average of post-tectonic Town Mountain Granites of central Texas, USA (Johnson et al., 1976); 1030 my age.
2: Young granite plutons of the Barberton region, Swaziland (average of F and G, 'older plutons' among the younger suite, Table VII, Viljoen and Viljoen, 1969b); 2800-2900 my age.
3: Young granite plutons of the Barberton region, Swaziland (average of H, 'younger plutons' among the younger suite, Table VII, Viljoen and Viljoen, 1969b); 2500-2600 my age.
4: High-level granite of Canadian shield (Unit 10 of Eade and Fahrig, 1971); Archaean and Proterozoic.
5: Younger Granites of Egypt (Rogers et al., 1978); 600 my age.

NOTE: Values are as reported in original work without recalculation to 100%. ΣFe_2O_3 is total iron calculated as Fe_2O_3.

Tables 1-4 show examples of the compositions of individual rock types from shields of different ages. The degree to which the cited analyses are representative of broad areas of the shields is highly variable. In Tables 1-4, the following values represent homogeneous suites or areally weighted averages of units based on grid sampling, point-counting of maps, stratigraphic sectioning, etc.: Table 1 - column 1 (Valley Spring Gneiss) and column 3 (gneisses of Canadian shield); Table 2 - most suites are sufficiently chemically homogeneous that the averages are satisfactory; Table 3 - column 1 (Packsaddle metasedimentary units), column 3 (paragneisses and paraschists of Canadian shield) and column 6 (Sheba Formation); Table 4 - most suites are sufficiently chemically homogeneous that the averages are satisfactory. Other values in the tables are unweighted averages, commonly of small numbers of samples.

Because of the sampling and mapping problems mentioned above, plus questions about the extent to which analyses are representative of rock types, it is difficult to draw specific conclusions from the data shown in Tables 1-4. It is clear that, within each suite, rocks of similar composition have been formed in shields of different ages. This conclusion does not prove, however, that the average composition of a particular rock type (gneiss, metavolcanic rock, etc.) does not change from one shield to another; sampling is simply not adequate for such a general conclusion. There is, however, no evidence for a variation in composition of individual rock types with age. That is, Tables 1-4 do not show progressive changes in the composition of gneisses, metavolcanic and metasedimentary rocks, and potassic granites related to their age of formation. Thus, it seems reasonable to investigate the possibility that variation in the bulk compositions of shields is related to changes in the proportions of rock types found in the shields rather than to changes in composition of individual rock types.

Table 5 shows the abundances of rocks in four shield areas of greatly different age. Uncertainties in this table result from the same problems of mapping and classification of rock types discussed above. In contrast to Tables 1-4, in the four areas shown in Table 5 it has been possible to separate calcalkaline, late-tectonic, granite plutons from the general gneiss-migmatite-granite terranes; greenschist-type metasedimentary and metavolcanic rocks, however, have been grouped as one unit. Despite uncertainties, the differences shown in Table 5 are sufficiently large that it is probably reasonable to draw three conclusions:

1. Granulite facies rocks are not found in young shields. This conclusion does not apply to those areas of young dates that are clearly formed by remetamorphism of older rocks. For example, the Grenville area of Canada (e.g. Wynne-Edwards, 1972) is a 1000 my terrane that contains abundant evidence of older, including Archaean, dates and structures superimposed by a younger, including granulite facies, event. Neither the Llano uplift nor the Red Sea Hills contain such evidence of older crust.

TABLE 5: Abundances of rock types in shields of different ages

	1	2	3	4
young, potassic granite	0.4	-	30	13
calcalkaline granite	24	8	-	28
metasedimentary and metavolcanic rocks	8	18	17	34
gneiss	52	65	35	8
granulite	14	8	-	-

1: New Quebec area (Superior province), Canada. The terminology used in the table correlates as follows with the units of Eade and Fahrig (1971):

young, potassic granite - Unit 10 (higher level granite, quartz monzonite, to monzonite)
calcalkaline granite - Unit 9 (massive to slightly foliated deeper level granitic rocks)
metasedimentary and metavolcanic rocks - Unit 5 (paragneiss, paraschist and minor amphibolite), and Unit 1 (volcanic and meta-volcanic rocks)
gneiss - Unit 7 (banded gneiss, migmatites, granitic gneisses and minor amphibolitic inclusions)
granulite - Unit 8 (pyroxene-bearing granite to granodiorite gneisses and granulite).

Percentage outcrop areas are the average for all areas in the New Quebec map region, as shown in Table 2 of Eade and Fahrig (1971). Rocks of the Superior province are Archaean, with the last major event c. 2500 my ago.

2: Peninsular India. Stratigraphy is generalised from the 1949 Geological Map of India (in Wadia, 1961) and summaries of Indian Precambrian geology (e.g. Pichamuthu, 1967; Naqvi et al., 1974; Rogers, 1974). Percentages of outcrop areas are determined by grid counting by the present writer of the 1949 Geological Map of India with the following assumptions: (a) charnockites and khondalites (on the map) are listed as granulite (in the table); (b) unclassified crystalline rocks are listed as gneiss; (c) Dharwarian system is listed as meta-sedimentary and metavolcanic rocks; and (d) granites are listed as calcalkaline granite. Only the portion of India south of the Indo-Gangetic plain has been included in the study. All rocks are Archaean.

3: Llano uplift, central Texas, USA (Johnson et al., 1976). The correlation of local terminology with the table is:

young, potassic granite - Town Mt. Granites; metasedimentary and metavolcanic rocks - Packsaddle Schist; gneiss - Valley Spring Gneiss.

Some 'fine-grained granites' are similar in composition to the Town Mt. Granites but are omitted from the tabulation. Ages of all rocks are in the range 1000-1100 my (Zartman, 1964, 1965).

4: Red Sea Hills, eastern Egypt. Percentages are calculated by the present writer from point counting of El Ramly's (1972) map. Correlation is as follows:

young, potassic granite - Younger Granites; calcalkaline granite - Older Granite; metasedimentary and metavolcanic rocks - metasediments and metavolcanics; gneiss - gneiss.

Rocks listed as gabbro-diorite, Dokhan volcanics and serpentine are omitted from the tabulation. Ages of all rocks are in the range 500-1200 my (Hashad et al., 1972; El Shazly et al., 1973). The younger Granites form a consistent group with an age of c. 600 my (Rogers et al., 1978).

2. Gneiss (including gneiss-migmatite-granite complexes) is probably more abundant in older shields than in younger ones.

3. Potassic granites are much more common in younger shields. As in the case of the granulites, this conclusion also does not apply to obviously reset terranes such as the Grenville, which contains no potassic granite plutons of the type described here.

Probably the most significant variation with regard to average compositions of shields is the increase in abundance of potassic granites toward younger shields. This increase in abundance is capable of causing virtually all of the apparent changes in composition, including the increase in abundances of lithophilic elements and decrease in abundances of mafic elements toward younger areas. Present information does not prove that this variation is the only signficant one, but it is clearly very important.

DISCUSSION

The fact that young shields contain an abundance of post-tectonic potassic granites raises two very interesting questions. First, are the young shields simply metamorphically and anatectically reworked portions of older, possibly Archaean shields? Second, do young shields appear to contain more potassic granite than old shields merely because the young shields have not been as deeply eroded? That is, did Archaean shields initially contain as high a proportion of potassic granite as the late Proterozoic shields do now, with the potassic granites having been removed by erosion from Archaean terranes? The possibilities are shown in Fig. 2.

Several lines of evidence tend to favour the concept of anatectic reworking of older shields. One is the clear evidence of tectonic overprinting in areas such as the 1000 my Grenville province (Wynne-Edwards, 1972) and the 500-600 my Pan-African belts of southern Africa (Kröner, 1977). These specific areas, however, do not contain significant amounts of post-tectonic, potassic granites. A second line of evidence is the presumption that the potassic granites should occupy the upper levels of crystalline terranes because of rising of the low-density magmas to shallow depth. On the basis of this concept, areas such as the Llano uplift and Red Sea Hills are the upper levels of older shields reactivated in late Proterozoic time. Continued erosion should presumably remove the zone of potassic granites and expose terranes similar to those of present Archaean shields (Fig. 2A).

Although the concept discussed in the preceding paragraph to the effect that apparently younger shields are merely upper levels of reworked older shields seems plausible, it does not fit at least four observations, some of which have been mentioned previously:

1. The various rock types in the Llano uplift and Red Sea Hills have not

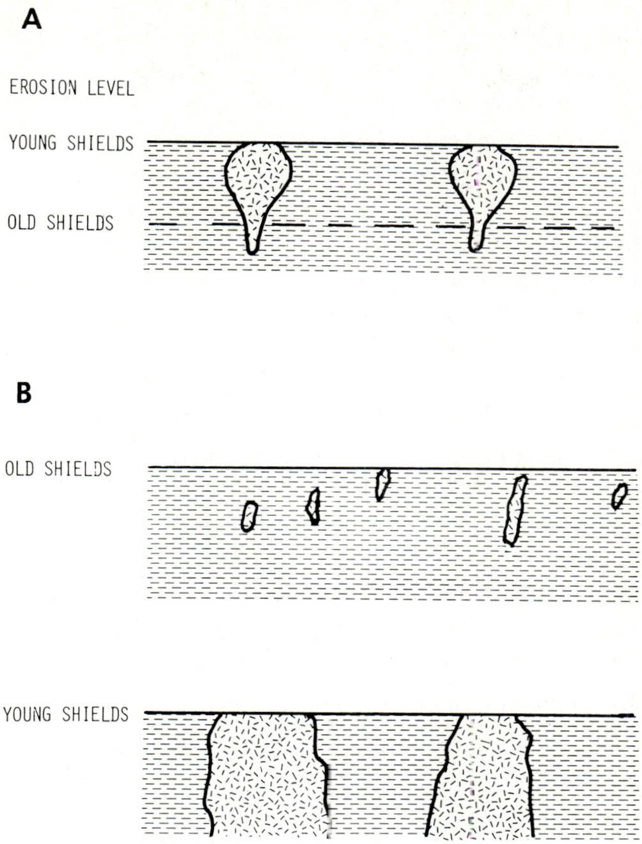

Fig. 2. Two possible explanations for the differences in currently measured abundances of potassic granites in shields of different ages. (Potassic granites are shown with a typical granite pattern of randomly oriented short lines; other shield rocks are shown by a pattern of parallel short lines.) A: Explanation based on difference in erosion level of Archaean and younger shields. Abundance of potassic granites at shallow levels permits their removal by erosion of Archaean terranes to deep crustal levels. B: Explanation based on differences in compostion of shields as initially formed, with older shields containing only very minor amounts of potassic granite. As discussed in the text, the absence of granitic detris from Archaean sedimentary rocks favours interpretation B.

simply been reset to the same age, as would be expected from a metamorphic and anatectic event, but show a sequence of ages similar to the sequence of ages shown by comparable rock types in Archaean shields.

2. Areas that are clearly formed by reworking of older shield material, such as the Grenville province and the Pan-African belts of southern Africa, do not contain appreciable quantities of post-tectonic potassic granite.

3. Limited strontium isotopic data for the potassic granites of eastern Egypt (not available for the Llano uplift) indicate initial Sr^{87}/Sr^{86} whole-rock ratios in the range 0.702-0.706. These ratios are clearly too low for material derived from average continental crust. Either the potassic granites have been obtained by partial melting of the mantle or of crustal materials with Rb/Sr ratios nearly as low as those of the mantle.

4. The erosion of the upper levels of Archaean shields does not appear to have produced the same kind of sedimentary debris that is now being formed by the erosion of the Llano uplift and Red Sea Hills. In particular, these younger shields are now providing large quantities of potassium feldspar for incorporation into sediments. A universal observation, however, of greywackes and other sedimentary rocks of Archaean greenschist belts is the virtual absence of potassium feldspar (e.g. Henderson, 1972). Naqvi et al. (1978) have shown that Archaean source terranes for Dharwar conglomerates were also devoid of true granites.

TABLE 6: Comparison of predicted composition of early Archaean crust with compositions of Archaean greywackes

	1	2	3	4
SiO_2	64.0	66.2	61.7	61.6
TiO_2	0.5	0.6	0.4	0.5
Al_2O_3	17.0	15.3	14.6	14.8
Fe_2O_3	5.4	5.7	5.1	6.6
MgO	3.1	2.7	3.5	2.5
CaO	3.9	1.7	5.7	3.9
Na_2O	4.6	3.1	4.1	4.2
K_2O	1.5	1.9	1.9	1.9

1. Predicted composition of early Archaean crust (from Fig. 1)
2. Greywackes from Yellowknife area, Canada (Henderson, 1972)
3. Greywackes from Chitaldrug schist belt, India (Naqvi, 1977)
4. Knife Lake Slates (mostly greywacke) from Minnesota, USA (Rogers and McKay, 1972)

Confirmation of the absence of granitic rocks from the exposed source terranes for Archaean greenschist belts may be found in Table 6. In Table 6 the estimated composition of 3500 my old crust (from Fig. 1) is compared with measured compositions of greywackes from three Archaean greenschist belts. The similarity of the composition of the greywackes to the inferred composition of the source area strongly suggests that the inferred composition of 3500 my old crust is approximately correct.

In summary, the preponderance of evidence indicates that the composition of continental crust has changed through time. Younger shield areas have greater amounts of potassic granite, higher concentrations of potassium (and presumably other lithophilic elements), and lower contents of mafic elements than older crusts (Fig. 2B). Thus, there appears to be progressive fractionation of lithophilic elements upward out of the mantle through time. Some of this fractionation may cause enrichment of older crusts in lithophilic elements, and some may result in the formation of wholly new continental sial, but the relative proportions of these two mechanisms are unknown.

In this regard, it is interesting to note that the composition of more recent eugeosynclinal sedimentary debris is very similar to the composition of Archaean crust and sediments derived from it (Rogers and McKay, 1972). In neither case has there been a significant contribution from granitic source rocks (containing potassium feldspar), despite the proximity of many Phanerozoic eugeosynclines to continental, sialic cratons. Modern eugeosynclines, therefore, are apparently compositionally similar to ancient Archaean crust and may be an effective chemical model for that crust.

REFERENCES

Baragar, W.R.A. and Goodwin, A.M., 1969. Archaean volcanism of the Canadian shield. In: A.R. McBirney (Editor), Proc. Andesite Conf., Bull. Oregon Dept. Geol. Min. Indust., 65: 121-142.

Billings, G.K., 1962. A geochemical investigation of the Valley Spring Gneiss and Packsaddle Schist, Llano uplift, Texas. Texas J. Sci., 14: 328-351.

Bryant, B. and Reed Jr., J.C., 1970. Geology of the Grandfather Mountain window and vicinity, North Carolina and Tennessee. US geol. Surv. Prof. Pap., 615, 190 pp.

Condie, K.C., 1976. Plate Tectonics and Crustal Evolution. Pergamon, New York, 288 pp.

Condie, K.C., Macke, J.E. and Reimer, T.C., 1970. Petrology and geochemistry of early Precambrian greywackes from the Fig Tree Group, South Africa. Bull. geol. Soc. Am., 81: 2759-2775.

Divakara Rao, V., Venkatanarayana, B. and Ramana, Y.V., 1976. Geochemistry of subsurface hornblendic rocks from Kolar and the Archaean crustal composition of the Peninsular shield. J. geol. Soc. India, 17: 245-25[illegible].

Eade, K.E. and Fahrig, W.F., 1971. Geochemical evolutionary trends of continental plates; a preliminary study of the Canadian shield. Bull. geol. Surv. Canada, 179, 51 pp.

El Ramly, M.F., 1972. Geological map of the basement rocks in eastern Egypt. Geol. Surv. Egypt, Cairo.

El Shazly, E.M., Hashad, A.H., Sayyah, T.A. and Bassyuni, F.A., 1973. Geochronology of Abu Swayel area, South Eastern Desert. Egypt. J. Geol., 17: 1-18.

Engel, A.E.J. and Engel, C.G., 1958. Progressive metamorphism and granitisation of the major paragneiss, northwest Adirondack Mountains, New York. Bull. geol. Soc. Am., 69: 1369-1414.

Glikson, A.Y., 1970. Geosynclinal evolution and geochemical affinities of early Precambrian systems. Tectonophys., 9: 397-433.

Goodwin, A.M. et al., 1972. The Superior Province. In: R.A. Price and R.J.W. Douglas (Editors), Variations in Tectonic Styles in Canada. Geol. Assoc. Canada Sp. Pap., 11: 527-623.

Hargraves, R.B., 1976. Precambrian geologic history. Science, 193: 363-371.

Hashad, A.H., Sayyah, T.A., El Kholy, S.B. and Youssef, A., 1972. Rb/Sr isotopic age determination of some basement Egyptian granites. Egypt. J. Geol., 16: 269-281.

Henderson, J.B., 1972. Sedimentology of Archean turbidites at Yellowknife, Northwest Territories. Canad. J. Earth Sci., 9: 882-902.

Hunter, D.R., 1970. The Ancient Gneiss complex in Swaziland. Trans. geol. Soc. S.Afr., 73: 107-150.

Hunter, D.R., 1974. Crustal development in the Kaapvaal craton; pt I, the Archaean. Precamb. Res., 1: 259-294.

Johnson, L.A., Rogers, J.J.W. and Nagy, R.M., 1976. Composition of the Precambrian Llano uplift, central Texas, USA. Geochim. Cosmoch. Acta, 40: 1419-1420.

Kröner, A., 1977. Precambrian mobile belts of southern and eastern Africa; ancient sutures or sites of ensialic mobility? A case for evolution towards plate tectonics. Tectonophysics, 40: 101-135.

Lowman Jr., P.D., 1976. Crustal evolution in silicate planets: implications for the origin of continents. J. Geol., 84: 1-26.

Naqvi, S.M., 1977. Archaean sedimentation of the Dharwars in the central part of the Chitradurga schist belt, Karnataka, India. Bull. geophys. Res., 15: 17-30.

Naqvi, S.M. and Hussain, S.M., 1972. Petrochemistry of early Precambrian metasediments from the central part of the Chitaldrug schist belt, Mysore, India. Chem. Geol., 10: 109-135.

Naqvi, S.M. and Hussain, S.M., 1973. Geochemistry of Dharwar metavolcanics and composition of the primeval crust of peninsular India. Geochim. Cosmoch. Acta, 37: 159-164.

Naqvi, S.M., Divakara Rao, V. and Narain, H., 1974. The protocontinental growth of the Indian shield and the antiquity of its rift valleys. Precamb. Res., 1: 345-398.

Naqvi, S.M., Divakara Rao, V., Rogers, J.J.W., Satyanarayana, K., Hussain, S.M. and Narayana, B.L., 1978. The petrochemistry of the conglomerates from late Archaean-early Proterozoic geosynclinal piles of South India and their geotectonic implications. Canad. J. Earth Sci. (in press).

Pichamuthu, C.S., 1967. The Precambrian of India. In: K. Rankama (Editor), The Precambrian, No. 3. Wiley, New York, pp. 1-96.

Reilly, G.A. and Shaw, D.M., 1967. An estimate of the composition of part of the Canadian shield in northwestern Ontario. Canad. J. Earth Sci., 4: 725-739.

Rogers, J.J.W., 1974. Problems concerning the evolution of the Precambrian shield of peninsular India. Bull. geophys. Res., 12: 103-118.

Rogers, J.J.W., 1977. Three arguments for continual evolution of sial throughout geologic time. In: C. Ponnamperuma (Editor), Chemical Evolution of the Early Precambrian. Academic Press, New York, pp. 27-39.

Rogers, J.J.W. and McKay, S.M., 1972. Chemical evolution of geosynclinal material. In: B.R. Doe and D.K. Smith (Editors), Studies in Mineralogy and Precambrian Geology, Mem. geol. Soc. Am., 135: 3-28.

Rogers, J.J.W., Ghuma, M.A., Nagy, R.M., Greenberg, J.K. and Fullagar, P.D., 1978. Plutonism in Pan-African belts and the geologic evolution of northeastern Africa. Earth Planet. Sci. Lett. (in press).

Ronov, A.B. and Migdisov, A.A., 1971. Geochemical history of the crystalline basement and the sedimentary cover of the Russian and North American platforms. Sedimentology, 16: 137-185.

Said, R., 1972. Special issue on the basement complex of Egypt. Ann. geol. Surv. Egypt, 1, 280 pp.

Stockwell, C.H., 1968. Geochronology of stratified rocks of the Canadian shield. Canad. J. Earth Sci., 5: 696-698.

Tarney, J., 1976. Geochemistry of Archaean high-grade gneisses, with implications as to the origin and evolution of the Precambrian crust. In: B.F. Windley (Editor), The Early History of the Earth. Wiley, Lond., pp. 405-417.

Taylor, S.R., 1977. Island arc models and the composition of the continental crust. In: Island Arcs, Deep Sea Trenches, and Back-arc Basins. Am. geophys. Un. Maurice Ewing Ser., 1: 325-335.

Viljoen, R.P. and Viljoen, M.J., 1969a. The geological and geochemical significance of the upper formations of the Onverwacht Group. Geol. Soc. S.Afr. Sp. Publ., 2: 113-151.

Viljoen, M.J. and Viljoen, R.P., 1969b. The geochemical evolution of the granitic rocks of the Barberton region. Geol. Soc. S. Afr. Sp. Publ., 2: 189-219.

Wadia, D.N., 1961. The Geology of India. MacMillan, Lond., 531 pp.

Windley, B.F., 1976. The Early History of the Earth (Editor), Wiley, Lond., 619 pp.

Wynne-Edwards, H.R., 1972. The Grenville province. In: R.A. Price and R.J.W. Douglas (Editors), Variations in Tectonic Styles in Canada. Geol. Assoc. Canada Sp. Pap., 11: 263-334.

Zartman, R.E., 1964. A geochronologic study of the Lone Grove pluton from the Llano uplift, Texas. J. Petrol., 5: 359-408.

Zartman, R.E., 1965. Rubidium-strontium age of some metamorphic rocks from the Llano uplift, Texas. J. Petrol., 6: 28-36.

GEOCHRONOLOGY

APPLICATION OF U-PB ZIRCON AND OTHER ISOTOPIC STUDIES TO THE IDENTIFICATION OF ARCHAEAN ROCKS IN THERMALLY AND TECTONICALLY OVERPRINTED TERRANES: LEWISIAN COMPLEX OF SCOTLAND

D. R. Bowes
Department of Geology, University of Glasgow,
Glasgow G12 8QQ, Scotland

Abstract

Isotopic data are used to test the interpretation that the gneisses and associated amphibolite facies rocks, which make up a large proportion of the Lewisian complex, are the result of 'reworking' or 'reactivation' on a crustal scale. The resultant model is that of a late Archaean crustal pile, variously affected by thermal and tectonic overprinting during early-middle Proterozoic times, but largely retaining its major chemical, mineralogical and fabric features. Apart from younger igneous intrusions, some lower Proterozoic sediments and localized zones of intense overprinting, the characters of the Lewisian complex can be used as indications of the nature of materials and processes of late Archaean times.

INTRODUCTION

Determination of the extent to which a shield complex, formed in Archaean times, retains its original characteristics despite thermal and tectonic overprinting, or whether it largely represents the product of younger episodes resulting from extensive 'reworking' or 'reactivation' of a pre-existing high-grade gneiss complex (e.g. Watson, 1973) and/or the addition of new crustal material, is a necessary prerequisite for the erection of a model for crustal evolution as well as for the utilization of geochemical data from the rock units of the shield indicating the nature of products and processes of Archaean times. An important means of elucidating the extent of operation of the various processes is provided by the behaviour of isotopic systems, both whole-rock and mineral. In the case of the Lewisian complex of Scotland U-Pb zircon studies play an important role, particularly because of the low Rb proportions in many of the rocks (e.g. Bowes et al., 1971; Lyon and Bower, 1977) which considerably inhibits the use of the Rb-Sr whole-rock method.

On the basis of studies by Sutton and Watson (1951, 1962, 1969), Watson (1965) and others, the shield area of Scotland has been set up as an example of the operation of intense 'reactivation' over large crustal segments and the concept applied to the development of other shield areas (Watson, 1973), including Greenland (e.g. Pulvertaft, 1968). The interpretation proposed for the Lewisian complex of Scotland is that the gneisses and associated amphibolite facies rocks, comprising a large proportion of the complex (Fig. 1), were developed during the early-middle Proterozoic Laxfordian orogeny (c. 1950-1700 m.y.) as the result of massive 'reactivation' of high-grade gneisses and granulites that had been formed in late Archaean times (during the Scourian orogeny - c. 2800-2700 m.y.) together with the metamorphism and deformation of early Proterozoic doleritic minor intrusions ('Scourie dykes'). The model of crustal evolution presented involved the activity of migmatite fronts and the development of virtually new foliated rocks as the result of intense metamorphic and structural modification; this led to the presence of small relict masses which retained structural patterns and textures of Archaean times amongst a 'reactivated' or 'regenerated' complex (the 'Laxfordian complex') which showed rapid variations in tectonic styles (Sutton and Watson, 1951, Figs. 2, 13). The type area is in the vicinity of Loch Laxford (Fig. 1) with large areas from Loch Laxford to Durness, Gruinard Bay to Loch Torridon, the islands of Rona and Raasay in the N.W. Highlands, and much of the Outer Hebrides showing the effects of 'strong Laxfordian reworking' (Sutton and Watson, 1969, Fig. 1; Watson, 1973, Fig. 3). A key factor in this interpretation is the relationships of metamorphic and structural features of the gneisses and related rocks to basic minor intrusions which were assumed to belong to one single series ('Scourie dykes') and to be a stratigraphic time marker (Sutton and Watson, 1951, p. 292). Isotopic dating has been used as confirmation of the interpretation, but the dates are from Rb-Sr and K-Ar mineral studies (Sutton and Watson, 1969, pp. 119, 121). However, subsequent U-Pb zircon studies, together with Rb-Sr and Pb-Pb whole-rock studies, place severe restraints on the extent to which the concept of 'reactivation' can be applied to the Lewisian complex of Scotland and put into context the concept of the 'isotopic clock' being reset in places but not in others (cf. Watson, 1973, p. 446). These factors, in turn, bear on the interpretation of crustal evolution for shield areas bordering the eastern part of the North Atlantic Ocean, the utilization of geochemical data from rock suites of the complex, and regional correlations such as those that have been made between Scotland and Greenland.

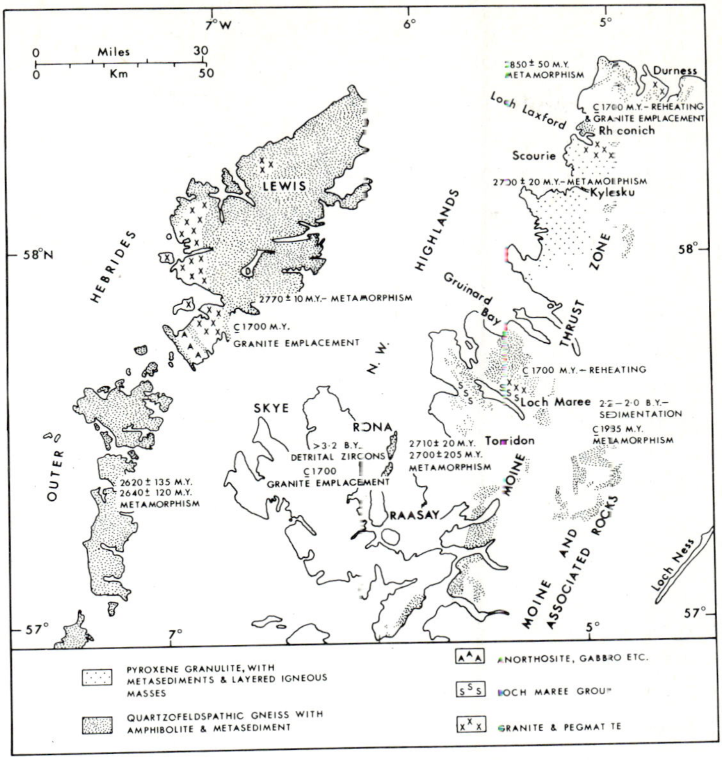

Fig. 1. Outline geological map of NW Scotland showing distribution of the Lewisian complex and age determinations.

EXTENT OF OVERPRINTING DURING LAXFORDIAN OROGENY

In the area immediately to the north of the type area for the Laxfordian orogeny (i.e. around Loch Laxford, Fig. 1), gneisses of the Rhiconich group (Dash, 1969) have been dated by the Rb-Sr whole-rock method at 1860 ± 185 m.y. (Lambert and Holland, 1972; age recalculated, like all other Rb-Sr ages given here, using $\lambda\ ^{87}Rb\ 1.42 \times 10^{-11} yr^{-1}$; uncertainly recalculated from published data using the computer program of McIntyre et al., 1966). This result, which is highly dependent upon one specimen of a biotite-hornblende rock with a much more favourable Rb-Sr ratio than the other analysed specimens, is concordant, within the errors, with that for gneisses from the Rhiconich group in the nearby Durness district (1695±130 m.y. – Rb-Sr whole-rock, Lyon and Bowes, 1977). The age is interpreted as the time of homogenization of rubidium and strontium isotopes over a scale greater than that of sample size (c. 8-10 cm^3). The age corresponds with that of granite and pegmatite emplacement (1720± 30 m.y. from U-Pb zircon studies) which is very abundant in this region and leads to the local development of injection complexes. This granite age is considerably younger than the 1935± 50 m.y. Rb-Sr whole-rock age for the dynamothermal metamorphism of cover rocks in an early phase of the Laxfordian orogenic episode (c. 1950-1700 m.y.) from elsewhere in the Lewisian complex (Bikerman et al., 1975). However it corresponds with that for pegmatite emplacement and abundant granite emplacement in other places in the complex (c. 1700 m.y. from Rb-Sr muscovite studies – Lyon et al., 1973; 1715±20 m.y. from U-Pb zircon studies – van Breemen et al., 1971). It also generally corresponds with the 1700±50 m.y. $^{40}K/^{40}Ar$ isochron data for hornblende from amphibolites in the Durness district which Lyon et al. (1977) interpret as the time at which hornblende passed through its blocking temperature (c.500-550°C according to Hanson and Gast, 1967) at the end of the orogenic episode or during the beginning of the epeirogenic uplift episode of the Laxfordian orogenic cycle. A younger c.1600 m.y. K-Ar biotite age for the gneiss is related to the time when the temperature passed through the c.300-350°C blocking temperature (Hanson and Gast, 1967), associated with uplift.

Hence the various lines of evidence point to the type area for the Laxfordian orogeny, and adjacent regions, having been subjected to intense metamorphism during early-middle Proterozoic times. On the basis of the Rb-Sr and K-Ar data, this could be interpreted as thermal metamorphism associated with extensive granitic injection towards the end of the Laxfordian orogenic cycle, followed by epeirogenic uplift and crustal cooling (Lyon and Bowes, 1977). Alternatively it could be interpreted as the result of dynamothermal metamorphism during the Laxfordian episode with gneiss formation (Lambert and Holland, 1972), particularly as the rocks of the district show evidence of a polyphase deformational sequence (e.g. Chowdhary and Bowes, 1972). However the U-Pb zircon isotopic systems provide additional evidence as near Durness (Fig. 1) systems pointing back to Archaean times are shown for zircon from both granite and gneiss. The existence of zircon xenocrysts at least 2730 m.y. old in a granitic vein point to the presence of Archaean rocks at the depth at which the magma was derived. In addition, an upper intersection age of 2850±50 m.y. (Fig. 2) for zircon, from the gneiss dated by Rb-Sr whole-rock methods at c. 1700 m.y., not only indicates an Archaean age for the gneiss, but also suggests that the c. 1700 m.y. metamorphic overprinting was not associated with intense penetrative deformation, as this would have resulted in the removal of lead from the zircons and a complete or nearly complete resetting of the isotopic systems (Lyon and Bowes, 1977). However, annealing of the zircons associated with the elevation of temperature in a largely passive mass of basement, at the time of the c.1950-1900 m.y. dynamothermal metamorphism shown by cover rocks elsewhere in the NW Highlands (Bikerman et al., 1975), would have made the lattice resistant to the movement of water, and so to lead loss (Mumpton and Roy, 1961), in the same manner as demonstrated for thermal aureoles of igneous intrusions (Zartman and Stern, 1967). Neither subsequent tectonic overprinting without penetrative mineral growth, as has been demonstrated for the region (Bowes, 1976a, p.370-1; Bowes and Hopgood, 1976, p.65-9), nor elevation of temperature associated with the abundant c.1700 m.y. granitic intrusions, would then have caused lead loss from the zircons. This interpretation is consistent with both the data available and the rehomogenization of the Rb-Sr whole-rock systems in the gneisses, indicating that diffusion of some elements was over distances of more than c.8-10 cm. The extent of diffusion in the amphibolites was probably considerably less, but the re-setting of mineral systems in both gneisses and amphibolites is not, by itself, evidence for 'reactivation' on the scale proposed by Sutton and Watson (1969).

Elsewhere in the Lewisian complex of the NW Highlands (Rona – Fig.1; Lyon et al., 1973), the retention of isotopic systems in gneisses that show the effects of polyphase tectonic overprinting during the Laxfordian episode (Hopgood and Bowes, 1972a), but only localized c.1700 m.y. acidic and basic minor intrusions, is indicated by the agreement of U-Pb zircon (2710±20 m.y. – Fig. 2) and Rb-Sr whole-rock (2700±205 m.y.) ages. The same is the case for the Outer Hebrides where U-Pb zircon (2770±10 m.y. – Fig. 1; Pidgeon and Aftalion, 1972), Rb-Sr whole-rock (2640±120 m.y.) and Pb-Pb whole-rock (2640±120 m.y. – Moorbath et al., 1975) ages have been determined for tectonically overprinted but generally not strongly thermally overprinted gneisses. This indicates that in those regions, rocks formed in the Scourian episode were not subsequently subjected to metamorphic and tectonic activity of sufficient intensity to significantly disturb whole-rock systems with regard to Rb, Sr, U and Pb and probably most elements, except for elements such as Ar. Most K-Ar mineral dates fall into the 1750-1500 m.y. range and this is consistent with them passing through their respective blocking temperatures during cooling and crustal uplift in the later stages of the Laxfordian cycle. However some K-Ar mineral ages extend to nearly 2100 m.y. (Holland and Lambert

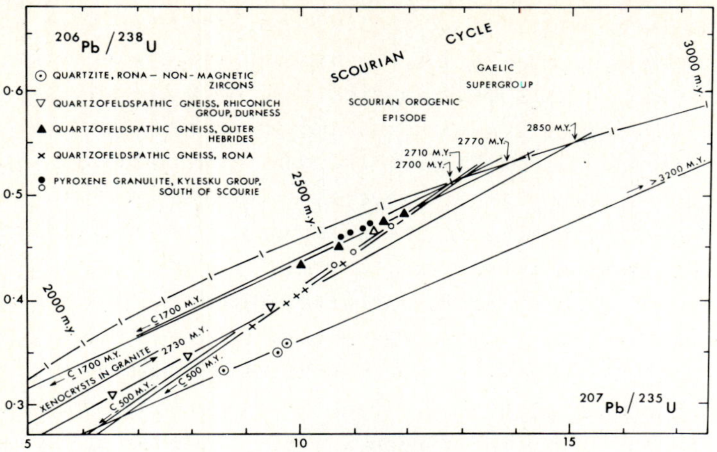

Fig. 2. U-Pb zircon data from the Lewisian complex; sources given in text.

1972; Moorbath and Park, 1972; Lyon et al., 1973; Moorbath et al., 1975). These are interpreted as representing incomplete overprinting of the isotopic systems of an earlier episode during the Laxfordian episode, although the possibility of the effects of introduction of excess ^{40}Ar, produced by degassing of deeper crustal rocks during the Laxfordian episode, cannot be entirely ruled out (Moorbath et al., 1975, p. 220).

Evidence for variable extents and times of rehomogenization of Rb-Sr whole-rock, Rb-Sr mineral and K-Ar mineral isotopic systems is given in the district from south of Gruinard Bay to Loch Maree (Fig.1). In a part where there are abundant granitic and pegmatitic veins emplaced late in the structural sequence, small masses of biotite-rich gneisses give a Rb-Sr whole-rock date of 1705$\pm$55 m.y., i.e. about the time of abundant emplacement of granites elsewhere in the complex. However while the widespread quartzofeldspathic gneiss, whose crustal history goes back to 2.8 to 2.7 b.y. ago, gives a computed data of 2.3$\pm$1.3 b.y. from Rb-Sr whole-rock data, inclusion of data from a micaceous variety of gneiss results in an apparent age of 1640$\pm$510 m.y. (Bikerman et al., 1975). Other gneisses, whose crustal history can also be deduced to go back to at least 2.7 b.y. but which have been brought into structural juxtaposition with a younger sedimentary assemblage (Loch Maree Group) whose deposition is unlikely to have been before 2.2 b.y. ago, give a Rb-Sr whole-rock age of 1925$\pm$100 m.y. This generally corresponds with the 1935$\pm$50 m.y. age for dynamothermal metamorphism of the metasediments during an early stage of the Laxfordian orogenic episode (Bikerman et al., 1975). The younger c.1700-1400 m.y. ranges of Rb-Sr and K-Ar mineral dates in the district (Giletti et al., 1961; Evans and Park, 1965; Moorbath and Park, 1972) are related to the closure of different mineral systems at different times during uneven crustal uplift. Here the controls were the fall of geo-isotherms, the composition and size of the mineral grains and the identity of the isotopic systems. Likewise the extent of the rehomogenization of the whole-rock systems was controlled by the mineralogical composition of the rocks and the disposition of the geo-isotherms resulting from crustal position and proximity to igneous masses. Here, as throughout most of the Lewisian complex, the gross nature of the products of the c.2.7 b.y. episode appears to have been retained with the isotopic systems indicating the dominance of the effects of c.1700 m.y. overprinting over penetrative deformation. Hence the evidence is strongly against the operation of wholesale 'reworking' or 'reactivation' over extensive crustal segments. Whether there was local 'reactivation' in particular crustal situations, e.g. the gneisses near Loch Maree dated at c.1900 m.y., requires further investigation.

The strongly foliated amphibolite masses which occur within the gneissose terranes played an important role in the formulation of the concept of 'reactivation' relating to the Lewisian complex. It can commonly be demonstrated that these lithological units show a community of structural elements with the c.2800-2700 m.y. old gneisses (e.g. Hopgood and Bowes, 1972a); accordingly they must be at least as old as the gneisses. In fact they more commonly exhibit the oldest recognized metamorphic fabric elements, developed in the early stages of the Scourian episode, than do the quartzofeldspathic gneisses (Chowdhary and Bowes, 1972, Fig.2, Photo 2). On these grounds they are interpreted as representing the products of late Archaean processes and having a bulk chemical composition representative of activity at that time: they cannot be the metamorphosed equivalents of early Proterozoic dolerite intrusions ('Scourie dykes'), a correlation which was a key factor in the interpretation of Sutton and Watson (1951) that much of the Lewisian complex was the result of strong 'regeneration' in Proterozoic times. Also correlated with the early Proterozoic dolerites were pyroxene-bearing basic minor intrusions which cross-cut foliated and lineated amphibolites (Bowes and Ghaly, 1964; Ghaly, 1966) and whose relationships to isotopically dated material indicate their emplacement relatively late in the Laxfordian episode (cf. Hopgood and Bowes, 1972b). Location of such relatively young pyroxene-bearing basic masses, which were not strongly affected by metamorphism and tectonism during the remaining part of the Laxfordian episode, corresponds with the location of what Sutton and Watson (1951, Fig.13) mark out as relict masses of an older gneiss complex amongst gneisses that were interpreted as the products of 'strong Laxfordian reworking'.

The availability of U-Pb zircon data, together with associated Rb-Sr and Pb-Pb whole-rock and Rb-Sr and K-Ar mineral isotopic data, now permits the erection of a crustal model in which the products of the c.2800-2700 m.y. old Scourian orogenic episode retain their essential characteristics. It also permits the unambiguous determination of time markers which allow the determination of a stratigraphy not beset by the previous confusion.

Provided the constraints imposed by the isotopic evidence are borne in mind, much of the rock assemblage, previously interpreted by some workers as representing the products of 'regeneration' during Proterozoic times, can be used as indicative of the processes and products of late Archaean times. Those igneous masses that cut the gneissose terrane, and whose time of emplacement can be established with reference to isotopically dated material, provide evidence for the nature of processes and products of their respective times of development (e.g. Hopgood, 1971b; Hopgood and Bowes, 1972b). U-Pb zircon studies now in progress should provide further time markers for the basic masses (cf. Kratz et al., 1976).

In addition to amphibolite facies gneissose terrane, the Lewisian complex contains granulite facies terrain, particularly in the district from Scourie to the south, in the NW Highlands. There localized narrow linear belts, in which there has been intense tectonism associated with retrogressive metamorphism, cut across the generally flat-lying granulites (Peach et al., 1907; Sutton and Watson, 1951; Bowes, 1969). The existence of two generations of these zones of 'reactivation' has been demonstrated by Khoury (1968) and Evans and Lambert (1974). One generation is associated with the c.2400-2200 m.y. old Inverian episode in which amphibolite facies gneisses were formed at the expense of pyroxene granulites; the other one is associated with the Laxfordian episode in which schists on the amphibolite-greenschist facies boundary were developed at the expense of the earlier-formed granulites and gneisses (pre-Torridonian crush zones of Peach et al., 1907). These localized belts acted as channelways for the uprise of connate and/or juvenile waters through deep parts of the basement during the respective episodes (Tarney, 1973; Beach, 1976). In the wide belt (c.2.5 km) which separates the granulitic terrane around Scourie from the gneissose terrane around Loch Laxford, late Archaean isotopic systems remain in at least some rocks with Rb-Sr whole-rock data on a muscovite-bearing gneiss indicating an age of 2745±270 m.y. Muscovite/whole-rock isochrons yield significantly younger ages (c.1750 m.y.) indicating mild redistribution of radiogenic strontium at that time (Lyon et al., 1975), which generally corresponds with that of granitic injection in the Lewisian complex. The interpretation of this zone (referred to by some authors as the 'Laxford front') as a gradational zone, associated with a migmatite front, between 'non-reactivated' Archaean granulites, together with early Proterozoic dykes to the south and 'reactivated' gneisses and amphibolites to the north (cf. Sutton and Watson, 1962, Fig. 3), is not supported by isotopic and structural evidence. Geophysical evidence suggests the juxtaposition of different levels of basement (Bott et al., 1972), while this and the corresponding zones have characteristics of zones of basement tectonics in other shield areas. The 'reactivation' they exhibit is localized and not regionally expressed, as indicated by the nature of U-Pb isotopic systems of zircons in the 2700±20 m.y. old granulites of the Kylesku group (Fig. 2; Pidgeon and Bowes, 1972). While some of these show slight effects of the Laxfordian episode (c.1700 m.y. lower chord-concordia intersection), in others corresponding effects are apparently lacking, with a lower intersection age of c.500 m.y. suggesting a Caledonian isotopic influence. This does not necessitate serious disturbance even of biotite Rb-Sr and K-Ar ages, but the movement of hydrothermal solutions through metamict zircons during a mild Caledonian reheating. Accordingly the nature of the granulites can be taken as representative of the products of the c.2700 m.y. Scourian episode (Sheraton, 1970; Bowes et al., 1971).

LENGTH OF ARCHAEAN HISTORY

For the Lewisian complex, not only has the concept of wholesale reactivation been proposed as being operative in the Laxfordian episode (a proposal not in accord with isotopic evidence - see above), but the concept has been proposed as having been operative in the Scourian episode, whose products Sutton and Dearnley (1964) suggested were derived from a 'Pre-Scourian complex'. Taken together these proposals imply a model of crustal evolution in which parts of the Lewisian complex had been twice 'reactivated'. They also imply a relatively long Archaean crustal history and have been used as a basis for correlations between Scotland and Greenland which imply an Archaean history back to c.3800 m.y. for the Lewisian complex. (cf. Bridgwater et al., 1973).

In these and related subsequent proposals, the time significance of basic minor intrusions has been an important factor in correlations. For example, the suggestion of Francis (1973, p. 161) of a very old cover to basement relationship that was first deformed in the Scourian orogeny, and then again in the Laxfordian orogeny, is dependent upon correlations, from place to place, of basic minor intrusions that were assumed to represent one generation of intrusions ('Scourie dykes') emplaced in early Proterozoic times (cf. Sutton and Watson, 1951, 1969). On this basis, the long sequence of events prior to the emplacement of what Hopgood (1971a) had previously shown to be much younger dykes (emplaced in middle Proterozoic times) was taken as representative of a long Archaean history, including that of an 'original gneiss complex' before the Scourian episode. Subsequently Davies (1975) proposed a pre-2800 m.y. history of the Lewisian complex which is directly comparable with that of part of West Greenland where the Archaean history goes back to at least c.3700 m.y. The key to this proposed correlation and lengthy Archaean history is the assumption that basic minor intrusions, now amphibolites, in NW Scotland correspond with the middle Archaean Ameralik dykes of Greenland.

Chapman and Moorbath (1977) have tested this hypothesis that the Lewisian complex in the Scourie-Loch Laxford district (Fig. 1), as described by Davies (1975), has a long Archaean history. On the basis of Pb-Pb whole-rock studies, they argue strongly against the presence of ancient sialic crust, like that in West Greenland, in this part of NW Scotland. A date of 2680$\pm$60 m.y. agrees with earlier work of Moorbath et al. (1969) relating to a more widely distributed collection of rocks from the Lewisian complex, as does the postulation that the presumed igneous rocks, from which the products of the Scourian episode were derived, separated from the upper mantle or basic lithosphere no earlier than c.2800-2900 m.y. ago.

This evidence from Pb-Pb isotopic studies that the Lewisian complex has only a relatively short Archaean history accords with other evidence from U-Pb zircon, Rb-Sr whole-rock and Pb-Pb whole-rock isotopic studies presented by Pidgeon and Bowes (1972), Lyon et al. (1973) and Moorbath et al. (1975). Hence on present evidence the various lithological units making up the crustal pile which was affected during the Scourian orogenic episode (Gaelic Supergroup — Bowes 1976b), can only have had a relatively short crustal history before intense metamorphism and deformation. Interpretations that invoke 'reactivation' of crust formed in early-to-middle Archaean times are unsubstantiated. The only material yet shown to have an age greater than c.2850-2900 m.y. is part of an inhomogeneous zircon population from a bedded quartzite within some of the c.2700 m.y. old gneisses (Bowes et al., 1976). $^{207}Pb/^{206}Pb$ data for round non-magnetic grains amongst this population indicate an age of at least 2840 m.y. and there is the possibility that they have a history going back to 3250 m.y. (Fig. 2). Their derivation must have been from crustal material in existence in the crustal segment before the Lewisian complex. However none of this material, such as that recognized in North Norway (Taylor, 1975) and West Greenland (Escher and Watt, 1976), has yet been recognized by isotopic means in the Precambrian rocks of Scotland.

DISCUSSION AND CONCLUSIONS

Isotopic evidence from U-Pb zircon and related studies point to the development of the Lewisian complex from c.2850 m.y. ago, with the dominant development being in the 2850-2700 m.y. period, i.e. in late Archaean times. Subsequent crustal additions, which were generally relatively small, occurred through early and middle Proterozoic times and the history of the Lewisian complex ended with the initiation of the Grenville cycle, products of which are now recognized in western and northwestern parts of the British Isles (van Breemen et al., 1978). Thermal and tectonic overprinting associated with igneous additions and tectonic activity during Proterozoic times have variously affected the late Archaean assemblage. However over most of the crustal segment the major chemical, mineralogical and fabric features have remained extant. Only in localized linear belts which cut the granulite facies terrane has there been intense 'reactivation'. Elsewhere, the amphibolite facies gneisses and related rocks which make up the bulk of the complex are not the products of extensive 'reworking' or 'reactivation' of earlier-formed crustal material as has previously been proposed. Accordingly, provided the extent of subsequent movement indicated by the isotopic systems is taken into account, a large proportion of this shield complex can be used as representative of the processes and products of late Archaean times.

The conflict of views concerning the extent of 'reworking' or 'regeneration' in the Lewisian complex is paralleled by a conflict of views concerning the extent to which crystalline rocks in West Greenland represent the 'reworking' of earlier-formed crustal material (cf. Chadwick et al., 1974; Moorbath, 1977, p. 166-171). In both instances the positive nature of the evidence from isotopic studies highlights the inadequacy of the geological criteria used as a basis for postulating 'reactivation' as a major contributing factor to shield evolution; in the case of the Lewisian complex this relates particularly to the use of basic minor intrusions as time-markers. These two examples also suggest that in shield areas where 'reactivation' has been proposed as a major factor, without constraints of isotopic studies, the use of geochemical and other data as representative of particular (and even presumed) time periods needs to be treated with caution. However, in the case of the Lewisian complex of Scotland, isotopic studies of U-Pb zircon and other isotopic systems have shown that there is a considerable body of data, particularly geochemical data, that can be used with some confidence as indicative of the nature of processes and products of late Archaean times. In this and other ways, the Lewisian complex is similar to the products of the Presvecokarelian episode in the Baltic Shield (e.g. in Finland – Bowes, 1975, 1976c) and the characteristics of its exposed parts are likely to play a significant role in the interpretation of seismic data relating to the deeper parts of the crust below NW Europe.

REFERENCES

Beach, A., 1976. The interrelations of fluid transport, deformation, geochemistry and heat flow in early Proterozoic shear zones in the Lewisian complex. Phil. Trans. R. Soc. Lond. A, 280: 569-604.

Bikerman, M., Bowes, D.R. and Van Breemen, O., 1975. Rb-Sr whole-rock isotopic studies of Lewisian metasediments and gneisses in the Loch Maree region, Ross-shire. J. geol. Soc. Lond., 131: 237-254.

Bott, M.H.P., Holland, J.G., Storry, P.G. and Watts, A.B., 1972. Geophysical evidence concerning the structure of the Lewisian of Scotland, N.W. Scotland. J. geol. Soc. Lond., 128: 599-612.

Bowes, D.R., 1969. The Lewisian of Northwest Highlands of Scotland. In: M. Kay (Editor), North Atlantic geology and continental drift: a symposium. Mem. Am. Assoc. Petrol. Geol., 12: 575-594.

Bowes, D.R., 1975. Scotland – Finland Precambrian correlations. Bull. geol. Soc. Finl., 47: 1-12.

Bowes, D.R., 1976a. Tectonics in the Baltic Shield in the period 2000-1500 m.y. ago. Acta geol. Polon., 26: 355-376.

Bowes, D.R., 1976b. Archaean crustal history in northwestern Britain. In: B.F. Windley (Editor), The Early History of the Earth. Wiley, London, 469-479.

Bowes, D.R., 1976c. Archaean crustal history in the Baltic Shield. In: B.F. Windley (Editor), The Early History of the Earth. Wiley, London, 481-488.

Bowes, D.R., Barooah, B.C. and Khoury, S.G., 1971. Original nature of Archaean rocks of North-West Scotland. Spec. Publ. geol. Soc. Aust., 3: 77-92.

Bowes, D.R. and Ghaly, T.S., 1964. Age relations of Lewisian basic rocks, south of Gairloch, Ross-shire. Geol. Mag., 101: 150-160.

Bowes, D.R. and Hopgood, A.M., 1976. Significance of structural trend in Precambrian terrain. Acta geol. Polon., 26: 57-82.

Bowes, D.R., Hopgood, A.M. and Pidgeon, R.T., 1976. Source ages of zircons in Archaean quartzite, Rona, Inner Hebrides, Scotland. Geol. Mag., 113: 545-552.

Bridgwater, D., Watson, J. and Windley, B.F., 1973. The Archaean craton of the North Atlantic region. Phil. Trans. R. Soc. Lond. A, 273: 493-512.

Chadwick, B., Coe, K., Gibbs, A.D., Sharp, M.R. and Wells, P.R.A., 1974. Field evidence relating to the origin of c. 3000/Myr gneisses in southern West Greenland. Nature Lond., 248: 136-137.

Chapman, H.J. and Moorbath, S., 1977. Lead isotope measurements from the oldest recognized Lewisian gneisses of north-west Scotland. Nature Lond., 268: 41-42.

Chowdhary, P.K. and Bowes, D.R., 1972. Structure of Lewisian rocks between Loch Inchard and Loch Laxford, Sutherland, Scotland. Krystalinikum, 9: 21-51.

Dash, B., 1969. Structure of the Lewisian rocks between Strath Dionard and Rhiconich, Sutherland, Scotland. Scott. J. Geol., 5: 347-374.

Davies, F.B. 1975. Origin and ancient history of gneisses older than 2,800 Myr in Lewisian Complex. Nature Lond., 258, 589-591.

Escher, A. and Watt, W.S. (Editors), 1976. Geology of Greenland. Grønlands geol. Unders., Copenhagen.

Evans, C.R. and Lambert, R.St.J., 1974. The Lewisian of Lochinver, Sutherland; the type area for the Inverian metamorphism. J. geol. Soc. Lond., 130: 125-150.

Evans, C.R. and Park, R.G., 1965. Potassium-argon age determinations from the Lewisian of Gairloch, Ross-shire. Nature Lond., 205: 350-352.

Francis, P.W., 1973. Scourian-Laxfordian relationships in the Barra Isles. J. geol. Soc. Lond., 129: 161-189.

Ghaly, T.S., 1966. The Lewisian geology of the area between Loch Shieldaig and Loch Braigh Horrisdale, Gairloch, Ross-shire. Scott. J. Geol., 2: 282-305.

Giletti, B.J., Moorbath, S. and Lambert, R.St.J., 1961. A geochronological study of the metamorphic complexes of the Scottish Highlands. Q.J. geol. Soc. Lond., 117: 233-272.

Hanson, G.N. and Gast, P.W., 1967. Kinetic studies in contact metamorphic zones. Geochim. Cosmochim. Acta, 31: 1119-1153.

Holland, J.G. and Lambert, R.St.J., 1972. Chemical petrology and K-Ar ages of the Lewisian gneisses of Rona, near Skye, Scotland. Geol. Mag. 109: 339-347.

Hopgood, A.M., 1971a. Structure and tectonic history of Lewisian Gneiss, Isle of Barra, Scotland. Krystalinikum, 7: 27-60.

Hopgood, A.M., 1971b. Correlation by tectonic sequence in Precambrian gneiss terrains. Spec. Publ. geol. Soc. Aust., 3: 367-376.

Hopgood, A.M. and Bowes, D.R., 1972a. Application of structural sequence to the correlation of Precambrian gneisses, Outer Hebrides, Scotland. Bull. geol. Soc. Am., 83: 107-128.

Hopgood, A.M. and Bowes, D.R., 1972b. Correlation by structural sequence in Precambrian gneisses of north-western British Isles. Rep. XXIV Int. geol. Congr., 1: 195-200.

Khoury, S.G., 1968. The structural geometry and geological history of the Lewisian rocks between Kylesku and Geisgeil, Sutherland, Scotland. Krystalinikum, 6: 41-78.

Kratz, K.O. et al., 1976. Age limits of the Jatulian complex of Karelia. Doklady U.S.S.R. Acad. Sci. (Geology Section), 231(5): 1191-1194.

Lambert, R.St.J. and Holland, J.G., 1972. A geochronological study of the Lewisian from Loch Laxford to Durness, Sutherland, Scotland. J. geol. Soc. Lond., 128: 3-19.

Lyon, T.D.B. and Bowes, D.R., 1977. Rb-Sr, U-Pb and K-Ar isotopic study of the Lewisian complex between Durness and Loch Laxford, Scotland. Krystalinikum, 13: 53-72.

Lyon, T.D.B., Gillen, C. and Bowes, D.R., 1975. Rb-Sr isotopic studies near the major Precambrian junction between Scourie and Loch Laxford, Northwest Scotland. Scott. J. Geol., 11: 333-337.

Lyon, T.D.B., Pidgeon, R.T., Bowes, D.R. and Hopgood, A.M., 1973. Geochronological investigation of the quartzofeldspathic rocks of the Lewisian of Rona, Inner Hebrides. J. geol. Soc. Lond., 129: 389-404.

McIntyre, G.A., Brooks, C., Compston, W. and Turek, A., 1966. The statistical assessment of Rb-Sr isochrons. J. geophys. Res., 71: 5459-5468.

Moorbath, S., 1977. Ages, isotopes and evolution of Precambrian continental crust. Chem. Geol., 20: 151-187.

Moorbath, S., Allaart, J.H., Bridgwater, D. and McGregor, V.R., 1977. Rb-Sr ages of early Archaean supracrustal rocks and Amitsoq gneisses at Isua. Nature Lond., 270: 43-45.

Moorbath, S. and Park, R.G., 1972. The Lewisian chronology of the southern region of the Scottish mainland. Scott. J. Geol., 8: 51-74.

Moorbath, S., Powell, J.L. and Taylor, P.N., 1975. Isotopic evidence for the age and origin of the "grey gneiss" complex of the southern Outer Hebrides, Scotland. J. geol. Soc. Lond., 131: 213-222.

Moorbath, S., Welke, H., and Gale, N.H., 1969. The significance of lead isotopic studies in ancient high-grade metamorphic basement complexes, as exemplified by the Lewisian rocks of Northwest Scotland. Earth Planet, Sci. Lett., 6: 245-256.

Mumpton, F.A. and Roy, R., 1961. Hydrothermal stability studies of the zircon-thorite group. Geochim. Cosmochim. Acta, 21: 217-238.

Peach, B.N. et al., 1907. The geological structure of the North-west Highlands of Scotland. Mem. geol. Surv. U.K.

Pidgeon, R.T. and Aftalion, M., 1972. The geochronological significance of discordant U-Pb ages of oval-shaped zircons from a Lewisian gneiss from Harris, Outer Hebrides. Earth Planet. Sci. Lett., 17: 269-274.

Pidgeon, R.T. and Bowes, D.R., 1972. Zircon U-Pb ages of granulites from the Central Region of the Lewisian, northwestern Scotland. Geol. Mag., 109: 247-258.

Pulvertaft, T.C.R., 1968. The Precambrian stratigraphy of western Greenland. Rep. XXIII Int. geol. Congr., 4: 89-107.

Sheraton, J.W., 1970. The origin of the Lewisian gneisses of northwest Scotland, with particular reference to the Drumbeg area, Sutherland. Earth Planet Sci. Lett., 8: 301-310.

Sutton, J. and Dearnley, R., 1964. In: Discussion of Bowes, D.R., Wright, A.E. and Park, R.G. Layered intrusive rocks in the Lewisian of the North-west Highlands of Scotland. Q.J. geol. Soc. Lond., 120: 188-189.

Sutton, J. and Watson, J., 1951. The pre-Torridonian metamorphic history of the Loch Torridon and Scourie areas in the North-west Highlands, and its bearing on the chronological classification of the Lewisian. Q. J. geol. Soc. Lond., 106: 241-307.

Sutton, J. and Watson, J., 1962. Further observations on the margin of the Laxfordian complex of the Lewisian near Loch Laxford, Sutherland. Trans. R. Soc. Edinb., 65: 89-106.

Sutton, J. and Watson, J., 1969. Scourian-Laxfordian relationships in the Lewisian of northwest Scotland. Geol. Assoc. Canada Spec. Pap., 5: 119-128.

Tarney, J., 1973. The Scourie dyke suite and the nature of the Inverian event in Assynt. In: R.G. Park and J. Tarney (Editors), The Early Precambrian of Scotland and related rocks of Greenland. University of Keele, Keele, 105-118.

Taylor, P.N., 1975. An early Precambrian age for migmatitic gneiss from Vikan i Bø, Vesteralen, North Norway. Earth Planet. Sci. Lett., 27: 35-42.

van Breemen, O., Aftalion, M. and Pidgeon, R.T., 1971. The age of the granitic injection complex of Harris, Outer Hebrides. Scott. J. Geol., 7: 139-152.

van Breemen, O., Halliday, A.N., Johnson, M.R.W. and Bowes, D.R., 1978. Crustal additions in late Precambrian times. In: D.R. Bowes and B.E. Leake (Editors), Crustal evolution in northwestern Britain and adjacent regions. Geol. J. Sp. Iss. 10 (in press).

Watson, J., 1965. Lewisian. In: G.Y. Craig (Editor), The Geology of Scotland. Oliver & Boyd, Edinburgh, 49-77.

Watson, J., 1973. Effects of reworking on high-grade gneiss complexes. Phil. Trans. R. Soc. Lond. A, 273: 443-455.

Zartman, R.E. and Stern, T.W., 1967. Isotopic age and geological relationships of the Little Elk Granite, Northern Black Hills, South Dakota. U.S. geol. Surv. Prof. Pap., 575-D: 157-163.

GEOCHRONOLOGY AND GEOCHEMISTRY OF ARCHAEAN TONALITIC GNEISSES AND GRANITES OF SOUTH KANARA DISTRICT, KARNATAKA STATE, INDIA

M. N. Balasubrahmanyan

Geological Survey of India, 29 Jawaharlal Nehru Road, Calcutta: 700016

Abstract

The migmatised ancient gneissic complex and the south Kanara granitic suite form the basement to the greenstone belts on the western seaboard of Karnataka, as evidenced by the varying structural disposition and the occurrence of an oligomictic conglomerate between them. The gneissic complex is tonalitic to granodioritic in composition and is inferred to be as old as 3200 Ma, with a possibility of being even older than 3500 Ma on the basis of K/Ar and Pb/Pb dates on minerals from cross-cutting pegmatites. The south Kanara granitic suite, with a porphyritic phase is meso— to katazonal in aspect, composite in nature with consistent planar foliae, has concordant and discordant relationships with the gneissic complex and has cross-cutting pegmatite and aplite dykes. In composition it is granodioritic with a fine-grained tonalitic variant.

Geochemically the transformation of the ancient gneisses to the 'granitic' suite was marked by decrease in Cr, and increase in Ba, La, K/Rb, K/Pb, Ba/Sr, Ca/Sr and Ti/Nb. The porphyritic granite and pegmatite belong to a later phase of the granitic activity. They are granodioritic to adamellitic in composition and, being geochemically alike, represent a two-stage crystallisation trend of a granitic melt from a porphyritic to a pegmatitic stage, with decrease in Ba, Sr, Cr, Zr, Pb and increase in Rb, La and Nb. The minimum K/Ar age of 2564 Ma of a biotite from a cross-cutting pegmatite is confirmed by a crude total rock Rb/Sr isochron age of 2669 ± 60 Ma. The high $^{87}Sr/^{86}Sr$ initial ratio of 0.7056 confirms the reworked nature of the suite. It is therefore postulated that the granite complex of South Kanara evolved due to the partial melting of a tonalitic gneissic complex about 3600 Ma old to give rise to the plutonic tonalitic-granodioritic-adamellitic suite at c. 2700 Ma, which is similar to granitic rocks in other granite—greenstone complexes. The similarity between the age of the granite (and some gneisses in the region) and that of the granulite facies metamorphism is suggestive of a mode of origin similar to that envisaged for the Rhodesian Craton-Limpopo belt.

INTRODUCTION

The western seaboard of South Kanara, bounded by the Western Ghats and the Arabian Sea, reveals a granodiorite—granite suite of batholithic dimensions (the Kanara Batholith) within a vast regional migmatitic gneissic complex (the Kanara Gneissic Complex) overlain by metamorphosed sediments and volcanics with an oligomictic quartz conglomerate at their base belonging to part of the Dharwar Supergroup (Balasubrahmanyan et al., 1976, Pl. 1). The structural trends of the gneissic complex and the Dharwar schists are discordant. The regional structure of the gneissic complex trends NNE-SSW with an antiform plunging NNE in the north, swinging to NW-SE in the south, with the batholith located along the anticlinal axis. The Dharwar schist belt has isoclinal folds trending NNW-SSE. The shear zones trending NNW-SSE in the gneisses adjoining the schist belt could be attributed to the reaction of the basement to the fold movements associated with the development of the schist belt.

Very small 'granitic' bodies are synkinematic and located in the fold cores of the gneisses. Late kinematic 'granites' with intrusive contacts have a wide extent; they vary in composition from granodiorite to granite. Xenoliths of amphibolite, hornblende schist, charnockite and sillimanite gneiss (with cordierite) in the pluton and clinopyroxene (hornblende)—plagioclase—magnetite and calc-silicate rocks in the gneissic complex illustrate the high grade of metamorphism. The first two types of xenoliths are abundant and commonly have a rim of biotite at their contact with the granite.

The following rock sequence is inferred :

Sub-recent to Recent	-	Sands, Tile clays
Tertiary	-	Laterite
Cretaceous-Tertiary	-	Rhyodacite-Granophyre
Proterozoic	-	Dolerite Dykes Dharwar Supergroup

ARCHAEAN	Kanara Batholith	Pegmatite, aplite, quartz veins
		Porphyritic biotite granite, biotite and hornblende granite - quartz monzonite, granodiorite, hybrid facies
	Kanara Gneissic Complex	Pegmatites Granite gneiss Streaky biotite gneiss Banded gneiss
	Metamorphites	Amphibolite, pyroxene - magnetite rock, calc silicate gneiss, sillimanite gneiss.

PETROGRAPHY

1. Gneissic Complex

(a) Banded Biotite Gneiss

This is the least migmatised unit with alternating biotite (often hornblende)-rich melanocratic layers with feldspar-quartz-rich leucocratic layers. The contrasting mineralogy of the bands is revealed by the alteration of the biotite (hornblende)-quartz association with rare micro-perthite and the microcline-plagioclase-quartz association. Rounded inclusions of zircon in biotite, calcic oligoclase, deformed quartz and microperthite characterise the former ; idiomorphic zircon, sodic albite, string and spindle microperthites the latter.

(b) Streaky biotite gneiss

This is a granodioritic biotite gneiss with nebulitic foliae with a tendency for the biotite to segregate into discontinuous streaks with drawn out feldspars often grading into augen gneiss and banded gneiss. The rock has a medium-grained texture with corroded atoll-like hornblendes and biotites in a matrix of clear microcline perthite associated with albite-oligoclase and deformed quartz.

(c) Granitic Gneiss

This is a noncoherent, diffuse type with a gneissic foliation, medium to coarse grained with microcline-perthite, altered oligoclase, deformed quartz and biotite. The perthite is often a replacement vein perthite enclosing grains of biotite, quartz and plagioclase. The plagioclase is sericitised and epidotised, whereas the biotite is in places completely altered to chlorite. This gneiss is distinct from the granitic rocks of the Kanara batholith in fabric, lack of primary flow texture, related primary joints, paragenesis and mineralogy.

2. Kanara Batholith

The Kanara Batholith consists of granodiorite, quartz monzonite (adamellite) and granite forming a composite intrusive with pegmatite and aplite veins. Special features of the batholith are a porphyritic facies in which planar foliation is paralleled by tabloids of feldspar phenocrysts, indicating a primary flow texture and a hybrid facies, which is confined to the contacts with a variety of rocks ranging from quartz diorite to granite. The nonporphyritic facies of the main body varies from a marginal type with a granodioritic composition with pronounced foliae, through a quartz monzonite (adamellite) to a granite *sensu stricto*.

(a) Hybrid facies

This is characterised by the abundance of basic inclusions of all sizes and shapes in various degrees of reaction with the granite and it is cut by granitic veins.

(b) Granodiorite

Abundant biotite with small feldspar crystals defining platy flow layers aligned parallel to the segregations characterise this type which is generally developed at the border of the batholith. A hypidiomorphic granular texture is composed of dominant plagioclase (An_{15-30}), moderate microcline-perthite, undulose quartz with streaks of biotite with sericitisation and replacement of the plagioclase by microcline, ubiquitous sodic albite at plagioclase-microcline boundaries indicating late stage alteration. Biotite is often replaced by muscovite. The approximate mode (vol. %) is : plagioclase 33.3, quartz 44.9, microcline-perthite 14.7, biotite 6.2 and others 0.9.

(c) Biotite and hornblende granite - quartz monzonite

This is the most common type with a medium- to coarse-grained texture with grey feldspars (rarely as phenocrysts) and quartz with a subresinous to greasy lustre. Biotite, plagioclase (An_{15-20}) coarse microperthite, antiperthite, microcline and quartz with accessory zircon, apatite, opaque ores, rare sphene and secondary epidote and calcite are the

Table 1. Mean chemical composition of gneisses and granites of South Kanara

Wt %	1	2	3	4	5	6	7	8
SiO_2	68.88	70.93	73.30	73.46	68.01	68.61	67.66	65.87
TiO_2	0.80	0.80	0.50	0.60	0.42	0.38	0.64	0.82
Al_2O_3	14.65	14.13	13.38	13.84	15.81	14.42	14.00	15.25
Fe_2O_3	1.64	1.28	1.22	1.26	0.90	1.13	0.50)	5.98
FeO	2.34	1.73	1.62	1.72	2.25	1.10	3.90)	
CaO	2.79	2.59	1.90	1.95	3.48	1.72	2.79	3.98
MgO	1.32	0.77	0.52	0.47	1.67	4.22	1.14	1.11
MnO	0.09	0.09	0.06	0.06	0.05	0.11	0.06	0.08
K_2O	2.24	2.19	3.12	1.86	1.80	2.41	3.62	2.38
Na_2O	4.39	4.64	4.05	4.05	4.77	4.67	4.56	4.22
P_2O_5	0.21	0.19	0.14	0.14	0.13	0.06	0.19	0.29

CIPW Norm	1	2	3	4
Q	24.4	28.8	33.8	37.4
Or	17.8	12.9	18.5	11.0
Ab	37.1	39.7	34.2	34.2
An	9.9	11.0	9.1	9.7
Co	-	-	-	1.6
Di	5.1	0.9	0.3	-
Hy	0.9	2.3	2.3	2.3
Mt	2.4	1.9	1.8	1.8
Il	1.5	1.5	0.9	1.1
Ap	0.5	0.3	-	-

Ratios	1	2	3	4
K/Rb	286.1	449.9	528.4	540.2
Ba/Rb	6.9	8.2	5.8	1.9
Ba/Sr	0.7	1.7	1.8	1.2
K/Pb	1437.7	1815.0	1726.0	1540.0
K/Ba	49.2	131.2	74.0	154.0
K/Sr	65.0	107.4	129.5	154,0
Ca/Sr	74.7	111.3	68.0	139.0
Ca/Y	153.0	185.0	135.0	139.0
Ti/Zr	34.5	31.0	20.0	12.0
Ti/Nb	480.0	510.0	300.0	240.0
K/Na	0.57	0.53	0.86	0.51
Rb/Sr	0.175	0.282	0.328	0.840
D.I.	79.3	81.4	86.5	82.6

1: Gneisses (South Kanara)
2: Granites (South Kanara)
3: Porphyritic granite (South Kanara)
4: Pegmatite (South Kanara) (Analyst - Dr N.R. Sen Gupta)
5: Gneiss (Barberton Mountain Land) (Viljoen and Viljoen, 1969)
6: Shimoga gneiss (Divakara Rao et al., 1974)
7: Uvak II gneiss (Labrador) (Collerson et al., 1975, p. 249)
8: Amîtsoq gneiss, Normal Series, 110857 (Lambert and Holland, 1976, p. 192)

constituent minerals. Myrmekite, quartz-muscovite intergrowths, replacement of microcline by muscovite are some paragenetic features. The modal composition (vol. %) is variable.

Microcline and microperthite	:	34.4 - 40.8
Plagioclase	:	30.5 - 21.8
Quartz	:	30.0 - 33.4
Biotite	:	4. 6 - 2.1
Hornblende	:	0 - 1.8
Others	:	0. 5 - 0.1

The perthites are partly replacement and partly exsolved types. Progressive potash (? enrichment) is evidenced by the replacement of hornblende by biotite, abundance of potash feldspar, reduction in plagioclase content, replacement of plagioclase by microcline and the development of myrmekite. Inversion to microcline was followed by late deuteric development of muscovite and calcite.

(d) Porphyritic biotite granite

Microcline - perthite phenocrysts are aligned parallel to flow layers, and set in a coarse matrix with biotite, oligoclase and quartz. Phenocrysts contain inclusions of biotite and plagioclase. Deformation has given rise to undulose quartz and perthitic texture.

MAJOR ELEMENT ABUNDANCES

Table 1 details the major element abundances, norm and trace element ratios given as a mean for the gneiss, 'granite' porphyritic granite and pegmatite. The analyses were kindly made by Dr. N.R. Sengupta, Chemist (Sr.), of GSI by orthodox wet chemical methods.

1. Gneissic Complex

The mean composition of the gneisses closely resembles that of the Uivak II Gneiss of Labrador (Collerson et al., 1975), the Barberton Mountain gneiss (Viljoen and Viljoen, 1969), and the Shimoga gneiss (Divakara Rao et al., 1974). Compared with the Amîtsoq Gneiss (Normal Series) the Kanara gneiss is poorer in iron and calcium, and it is poorer in magnesium with respect to the Shimoga gneiss. The K/Na ratio of 0.57 of the Kanara gneiss is comparable with the ancient tonalitic gneiss of the Barberton granite-greenstone province. The recent suggestion that the 'Peninsular Gneiss' may, in part at least, represent an ancient tonalitic gneiss basement (Radhakrishna, 1974) is aided by the occurrence of tonalitic pebbles with a 3250 Ma age in the Kaldurga conglomerate (Venkatasubramanian and Narayanaswamy, 1974). With regard to the K_2O : Na_2O ratio (Fig. 1) the mean composition is granodioritic, but individual

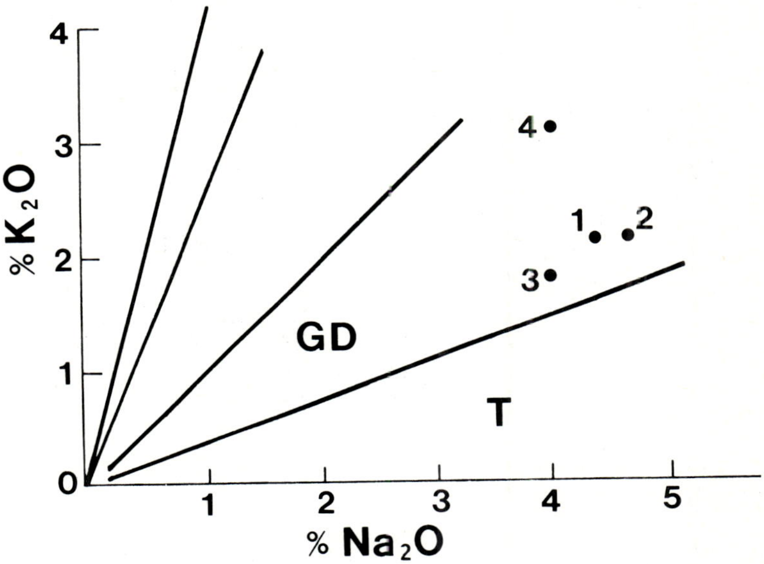

Fig. 1 : Mean Composition in relation to K_2O : Na_2O.
1. Gneiss ; 2. 'Granite'; 3. Pegmatite; 4. Porphyritic Granite.

compositions are tonalitic to granodioritic (not plotted in Fig.1). The K_2O : CaO ratios (Fig. 2) also indicate the same composition with the average nearer the tonalite/granodiorite boundary. The normative Q – Ab – Or values (Fig.3) for the two gneisses plot close to the Archaean amphibolite facies gneisses of Canada, Greenland and Scotland (Lambert et al., 1976, p. 380, Fig. 3). K_2O : Na_2O : CaO ratios (Fig. 3) are similar to Na_2O-rich gneisses (Lambert et al., 1976, p. 379, Fig.1), and contrast with K_2O-rich Archaean gneisses in the USSR. With regard to FeO – Alk – MgO (weight %) the Kanara gneiss plots closer to the Alk end than gneisses from the Canadian Shield, the Lewisian and E. Greenland (Lambert et al., 1976, p. 381, Fig. 4).

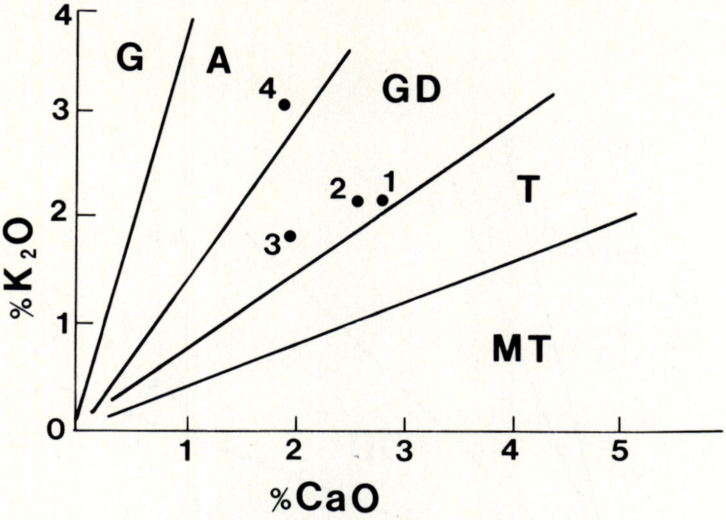

Fig. 2 : Mean Composition in relation to K_2O : CaO.
1. Gneiss ; 2. 'Granite'; 3. Pegmatite; 4. Porphyritic Granite.

2. Kanara Batholith

The average composition of the nonporphyritic facies of granodiorite and its fine-grained equivalents, which form the bulk of the area, plots near the tonalite/granodiorite boundary with respect to the K_2O : Na_2O (Fig.1) and K_2O : CaO ratios (Fig.2). The porphyritic facies is distinctly a granodiorite with an adamellite trend. The pegmatite dyke is also granodioritic.

TRACE ELEMENT CHEMISTRY

The Rb and Sr values were estimated by the author using pelletised whole-rock powder with a boric acid skirt in a Phillips Manual PW1540 X-ray Vacuum Spectrometer with the use of a single standard following the ratio method of Pankhurst (1969). Other trace elements were determined by H.B. Das, D.K. Indra and T. N. Das of GSI using an optical emission spectrograph.

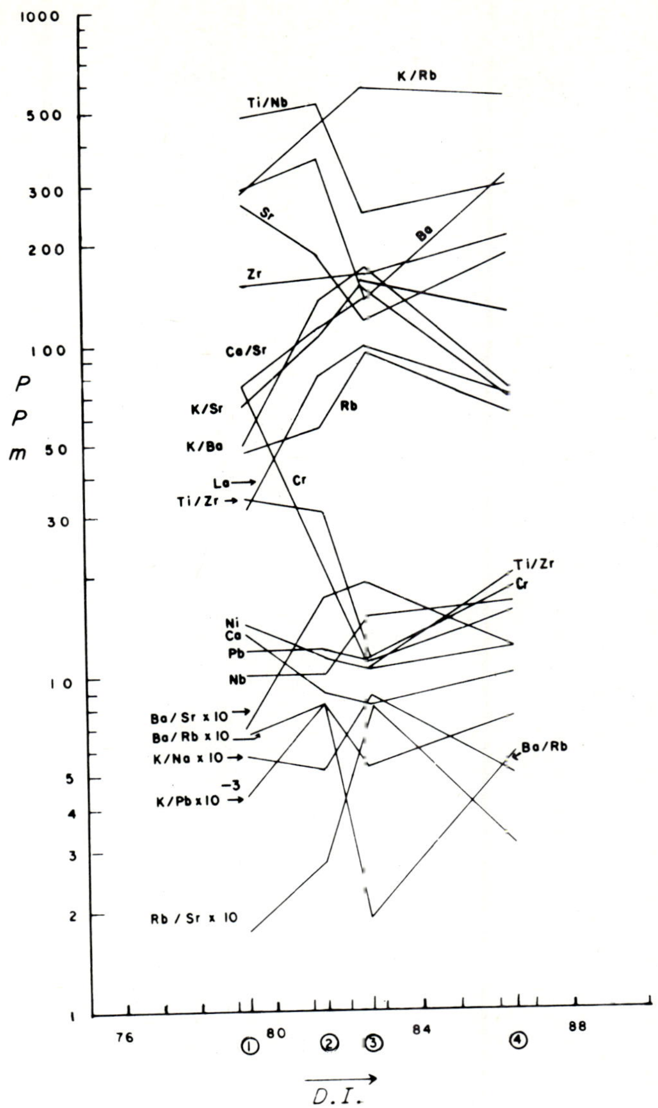

Fig. 3 : Variation diagram of trace elements and trace element ratios with respect to D. I. (i.e. normative Q + Ab + Or). 1. Gneiss ; 2. Granite; 3. Pegmatite; 4. Porphyritic Granite.

The variation in trace element characteristics is depicted in Fig.3, where the abundances as well as the relevant ratios are plotted against the normative Q + Ab + Or content. The relevant ratios are detailed in Table 1 and Fig.3.

TABLE 2: Trace element data on gneisses (ppm)

Rock type	Ba	Sr	Rb	Cr	Co	Zr	La	Y	Nb	Ni	Pb
1. Gneisses (South Kanara)	290.0	266.6	46.7	75.0	13.3	150.0	30.0	13.0	10.0	18.3	11.6
2. Gneisses (Shimoga)				9.1	5.0	121.0		38.0		10.7	
3. Gneisses (Sivaganga)				8.0	4.0	195.0		48.0		5.2	
4. Gneisses (Sira)				9.0	5.0					5.0	
5. Gneisses (Latahar)	300.0			-	10.0-22.0					10.0	20.0-80.0
6. Gneisses (Yellandlapad)					30.0-100.0					30.0	10.0
7. Granites and feldspathic gneisses (Mosabani)	15-720		50-356	1-39	10-24					18-76	
8. Champion gneiss (Crawford, 1969)		262.00	83.70								
9. Peninsular gneiss		494.03	133.56								
World Average (for calcic granites)		440.00									

Other data from Sankaran (1974)

TABLE 3: Trace element data on granitic rocks (ppm)

Rock type	Ba	Sr	Rb	Cr	Co	Zr	La	Y	Nb	Ni	Pb
1. Granite(S.Kanara)	360.0	187.5	52.9	23.8	9.0	155	73.8	10	10	11.3	11.9
2. Por.Gr.(S.Kanara)	316.6	183.3	60.2	18.3	10.0	210	70.0	10	11.5	12.0	16.6
3. Gr. pegmatite (S.Kanara)	133.3	116.6	98.0	11.6	8.3	160	96.6	10	15.0	11.6	10.0
4. Granite I (Singhbhum)	83-340	275-752	50-165	1-13	4-10	50-265		10-20		4-34	10-16
5. Granite II (Singhbhum)	130-720	350-1260	50-100	5-21	4-9	143-320		18-30		10-42	10-19
6. Granite III (Singhbhum)	50-1800	200-860	50-250	4-14	4-7	64-272		10-53		4-110	10-21
7. Granodiorite (Singhbhum)	130-570	520-1175	50-185	11-82	10-25	128-256		12-35		14-56	10-40
8. Granite(Singhbhum)	760-870	250-300	73-85	18-27	5-8	170-210				16-23	45-73
9. Soda granite	560-730	170-210	65-70	26-35	22-30	170-240				65-78	23-32
10. Granodiorite (Seraikaela)	26-1105	16-640		47-80	3-20	38-300		13-82		6-78	10-55
11. Grey granite (Closepet)		80-180	129-211								48-50
12. Pink granite (Closepet)		110-125	84-131								40-45
13. Granite(Chitaldrug)			60								40
14. Granite(Hosdurga)		119-196									
15. Granodiorite (Latehar)				-	15			500		50	100
16. Granite(Latehar) (Crawford, 1969)	100-400	750		-	10-32			40-66		10	55-80
Closepet granite		246.16	100.70								
Chitaldrug granite		17.325	406.42								
World average for high Ca granites		440.00									

Other data from Sankaran (1974)

The mean value for the Kanara gneiss is shown in Table 2, and for the Kanara batholith components in Table 3 compared with other occurrences in the Indian Shield as given by Sankaran (1974). Ba, Sr, Rb, Zr, Pb and Ni values compare well with the gneisses of Latehar, and of Mosabani (Cr and Co are exceptionally higher - nearly eight fold). Ba, Sr, Rb, Cr, Co, Zr, Y and Ni compare well with the Singhbhum granodiorite. In relation to the younger plutons of Karnataka (viz. Closepet, Chitaldurg, and Hosdurga) the older Kanara suite as a whole shows higher Sr, and lower Rb and Pb.

The K/Rb ratio of 286 for the gneisses is well within the crustal range (120 - 480, Taylor, 1965). The K/Rb values for the Kanara batholith component are moderately high (450, 528, 540) and they are higher than the normal trend. This may reflect the depth range in which the gneisses and batholith members crystallised.

Positive correlation of K/Rb, K/Sr, K/Ba, K/Pb, Ba/Sr ratios with increasing Q + Ab + Or from gneisses, through nonporphyritic and porphyritic granites, to pegmatite is indicative of a calc-alkali trend (Nockolds and Allen, 1953). The negative correlation indicated by Ti/Nb and Ti/Zr also appear to be normal. Ba/Rb variation is negative between gneiss, granite and porphyritic granite, and positive between porphyritic granite and pegmatite. Ca/Sr rises steeply between gneiss through granite to porphyritic granite and falls between granite/porphyritic granite and pegmatite.

1. Discussion

The close geochemical affinity between the granitic rocks and the gneisses is very apparent (Table 1). Debasification of the gneiss would result in the composition of the granitic rocks. With the available data it is postulated that partial melting of the tonalitic-granodioritic South Kanara gneiss resulted in the formation of the granite. Comparison of trace element chemistry also supports this hypothesis. South Kanara gneisses have lower Rb and higher Sr contents than the later granite. Similar geochemical relations exist between the Ancient Tonalite gneiss and late granites of the Barberton area (Viljoen and Viljoen, 1969). In the South Kanara area, the transformation of gneiss to granite was marked by decrease in Cr, Ni and Co and increase in Ba, La, K/Rb, K/Pb, K/Ba, Ba/Sr, K/Sr, Ca/Sr and Ti/Nb. It has already been observed that K/Ba, K/Pb and K/Rb and K/Sr rise with the K-feldspar phase. The moderately high K/Rb for the granites implies crystallisation at depth. On partial melting the felsic portion chiefly crystallised to give rise to anatectic granite. At P_{H_2O} = 2 Kb and a temperature of 700°C anatexis of gneisses sets in (Winkler, 1967, p. 195) at the high temperature range of the amphibolite facies of regional metamorphism. If the Q — Ab — Or values are plotted in a ternary diagram at 2 Kb P_{H_2O} and with Ab/An = 3.8 cotectic, the trend of partial melting and crystallisation is clear. Such a trend, since some alkali migration is indicated for the South Kanara granites, would imply some metasomatic alteration (Lambert and Holland, 1976, p. 195). The porphyritic granite and pegmatite possibly represent a late phase of granitic activity and perhaps represent two stages of crystallisation of a 'granite' melt. From the porphyritic to the pegmatite stage there is a decrease in Ba, Sr, Cr and Pb and an increase in Rb, La and Nb which is more or less common (Goldschmidt, 1954). Fall of Ba from porphyritic granite to pegmatite is compatible with its incorporation in K-bearing minerals at higher temperature. Variation in the other elements is a reflection of the change of environment, i.e. from an early magmatic to a late pegmatitic stage.

GEOCHRONOLOGY

The geochronology of the area was investigated by U/Pb, Rb/Sr and K/Ar methods. Rb/Sr ratios were estimated by XRF. A Thompson — CSF THN 206 30 cm 60° sector solid source mass spectrometer with chart recording and peak-switching device was used. K was estimated by the Li internal standard method using a Baird — Atomic Flame Photometer as well as by a Perkin — Elmer 503 Atomic Absorption Spectrophotometer. Argon isotope dilution analysis was carried out with a VG-Micromass 600 gas source mass spectrometer with 62 mm radius 90° sector tube evacuated by an ion pump (U.H.V. better than 1 x 10^{-9} torr) with 2 KV accelleration voltage and an electromagnetic scanning device with chart recorder output.

The system had an on-line all metal stainless steel spike system (Argon 38) and a fusion system in a resistance furnace. Extrapolation of the isotope ratios to zero time (to obviate the small memory) with necessary correction for fractionation

TABLE 4: K/Ar aga data

SL No.	Sample No.	Material	Host rock	K%	Vol. $^{*40}Ar$ (nl/gm)	Atm. $^{40}Ar\%$	Age Ma	Error ± Ma
1	193	Bi	Peg.	7.49	2353.05	0.65	3068	41
2	23	Bi	Peg.	7.18	2440.12	0.30	3184	38
3	23	Hb	Peg.	0.96	317.32	18.9	3180	54
4	109	Bi	Peg.	7.30	3157.17	0.4	3552	42
5	9	Bi	Peg.	7.78	2979.13	0.8	2564	48
6	40	Bi	Por.Gr.	7.12	2445.41	1.0	3200	54

Constants used: $\lambda_{\beta} = 4.963 . 10^{-10}\ a^{-1}$; $\lambda_{e} = 0.581 . 10^{-10}\ a^{-1}$;
$^{40}K = 0.01167$ atom %

TABLE 5: Rb/Sr analytical data

SL No.	Sample No.	Material	Rb (ppm)	Sr (ppm)	$^{87}Rb/^{86}Sr$	$^{87}Sr/^{86}Sr$	Age (Ma)
1	7	Aplite W.R.	102.1	297.2	0.913	0.744	Isochron
2	130	Grano-diorite W.R.	44.9	267.8	1 1.477	0.756	2669 ± 60
3	13	Porph. Gr. W.R.	153.8	235.9	1.961	0.785	

Constants used: $^{87}Rb = 1.42 . 10^{-11}\ a^{-1}$

was always carried out. Routine analysis of I.G.S. London standard biotite (A 89 and Mo 40) was carried out for spike calibration and use of calibrated air capsules for measuring the bias of the mass spectrometer.

The minimum age for the gneiss-granite complexes is given by the whole-rock K-Ar age of the youngest dolerite dyke at 2193 $\pm$ 45 Ma (Balasubrahmanyan, 1975) cutting the complex.

1. Gneissic Complex

The K/Ar ages of biotites from different suites of pegmatites cutting the gneissic complex indicate a probable minimum age of 3068 $\pm$ 41 Ma (Table 4, No.1). This is supported by a higher value of 3184 $\pm$ 38 Ma for the hornblende fraction (Table 4, No. 2) and 3180 $\pm$ 54 Ma for a biotite fraction (Table 4, No.3) from a pegmatite cutting the least migmatised tonalitic hornblende - biotite gneiss. That the gneisses could be even older than 3552 Ma is indicated by a biotite age of 3552 $\pm$ 42 Ma for another pegmatite (Table 4, No.4). This is supported by dating of an allanite from the same pegmatite which indicates 3200 Ma by the K/Ar method. The analytical data for allanite are as follows:

U = 313.6 ppm Pb = 2932 ppm.

$^{208}Pb/^{206}Pb = 12.251 \pm 0.015$

$^{207}Pb/^{206}Pb = 0.27315 \pm 0.00027$

$^{204}Pb/^{206}Pb = 0.002842 \pm 0.00004$

Assuming common lead model 3200 Ma with = 8.99, the ages will be.

$^{207}Pb/^{206}Pb$ = 3174 Ma

$^{207}Pb/^{235}U$ = 1355 Ma

$^{206}Pb/^{238}U$ = 514 Ma.

The discrepancy is due to lead-loss and if lead-loss is fairly recent, the $^{207}Pb/^{206}Pb$ minimum age of 3174 Ma may not be far from the truth. As such it is reasonable to assume that the K/Ar ages do not show excess argon in view of confirmation by the U/Pb minimum age. As such, it is reasonable to assume that the gneissic complex is at least 3200 Ma old and if the highest K/Ar age of 3552 Ma is not discarded could be even older than 3550 Ma.

2. Granodiorite-porphyritic granite – aplite suite

The whole-rock Rb–Sr isochron age of three components (Table 5 : Fig. 4) is 2669 $\pm$ 60 Ma with an initial $^{87}Sr/^{86}Sr$

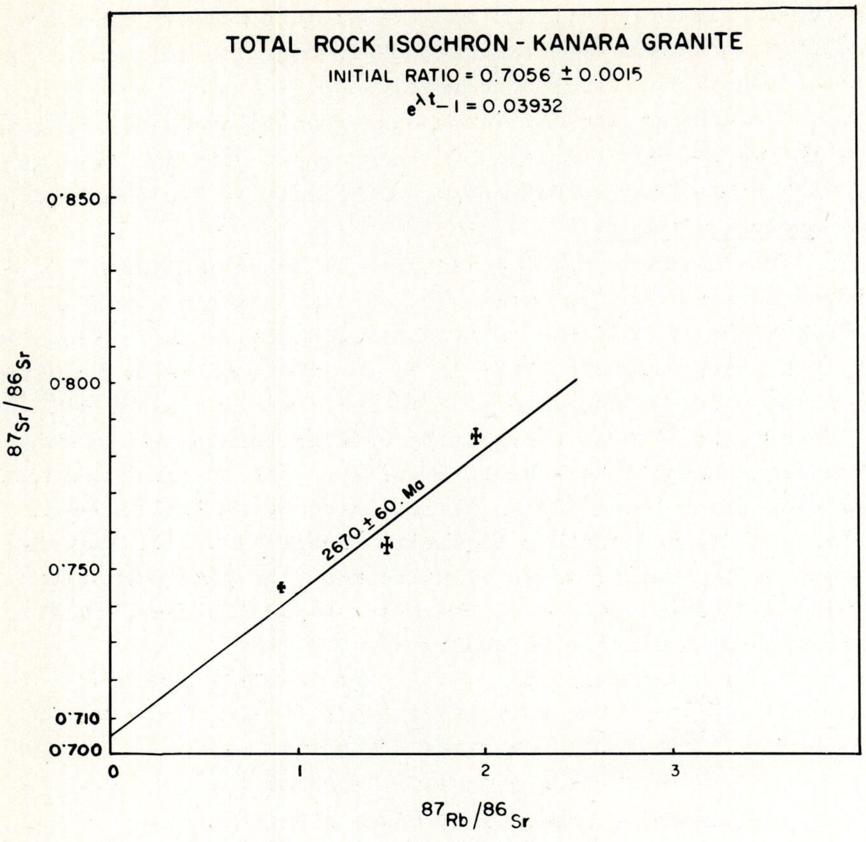

Fig. 4 : Total Rock Isochron - Kanara Granite.

ratio of 0.7056 ± 0.0015. This age is comparable with that of other granitic plutons in the Indian Shield such as the Arsikere and Banavara (2625 ± 90 Ma, Venkatasubramanian and Narayanaswamy, 1974), and J. N. Kote granite (2575 - 2620 Ma, Crawford, 1969). The initial ratio is not typical for mantle derivation of rocks of this age (Faure and Powell, 1972). The higher ratio is slightly lower than the value for continental crust if these were due to reworking of the 3200 Ma - 3600 Ma South Kanara gneissic complex (cf. Shaw, 1976, p. 46, Fig. 5). It does not corroborate recent thinking that such reworking is impossible in view of low initial ratios as advocated by Moorbath et al. (1975).

With regard to the age of charnockites in the region this age of 2700 Ma is interesting in relation to the probable processes that operated locally at the time. The granulite facies metamorphism is invoked for explaining the age and initial ratio (Crawford, 1969). This part of the shield was originally included by Fermor (1936) in the charnockitic province and the presence of sillimanite gneiss, calc-silicate rock, pyroxene-magnetite - quartz rock and charnockite confirms such a relationship. This does not conflict with the age of the charnockites in as much as the age of the nearest charnockitic suite to the South Kanara area pertains to the granulite facies metamorphism of the Kushalnagar charnockite at 2618 ± 46 Ma (Spooner and Faibairn, 1970), which may be of an older igneous source rock such as the 2900 - 3100 Ma old diorite (Balasundaram and Balasubrahmanyan, 1973) in the 3010 Ma old khondalitic terrain (Perraju, pers. comm.). It is pertinent to record here that Fyfe (1973, p. 16) suggested that "in other parts of Africa where extensive granulite terrains of similar age are exposed, I would suggest that these could represent the base of the crust as might occur beneath the Rhodesian batholith complex". The granitic plutons and the younger Peninsular gneisses, 2600 - 2700 Ma old in this part of the Indian Shield, appear to be similar to those of the Rhodesian Archaean batholith pattern as envisaged by A.M. Macgregor.

The similar age of granulites, 'granites' and some gneisses in the region indicates probability of partial melting producing these rock types with steep thermal gradients "in the more radioactive Archaean" (Fyfe, 1973, p. 13). The mechanism of such fusion being triggered by basic magmas has been suggested by Fyfe (1973, p.16), evidence for which seems to rest on probable gabbroic complexes below the partially molten crust, which can be confirmed in this region only after the absolute dates for these are known.

CONCLUSIONS

The model for the evolution of the gneiss-granite complexes of South Kanara concerns the ? partial melting of a tonalitic gneissic complex, c. 3600 Ma in age to give rise to a plutonic tonalitic - granodioritic granite suite c. 2700 Ma ago; this is similar to granulite-granite complexes in other Archaean terrains. It would be useful to test this tenet which is based on the Sr initial ratio by an extensive study of the area by Rb/Sr methods. The age and origin of the gneissic complex is still unknown ; U/Pb dating of zircons could be useful to reveal its ancient age. The study of REE patterns would also help in a better understanding of the geological processes in such an Archaean terrain.

ACKNOWLEDGEMENTS

The author thanks Dr. S.V.P. Iyengar for his guidance and Sri A. Sarkar for discussions.

Dr. Snelling and Mr. C.C. Rundle of IGS., London, set up and calibrated the all-metal Argon Micromass mass spectrometer in GSI., Calcutta which enabled cross checking of some dates. Dr. I.G. Swainbank of IGS., London helped the author in the allanite dating which is gratefully acknowledged. To the chemists who carried out the major and trace element analysis, the author is grateful. Shri R. K. Sundaram's support in the development of the Unit is thanked. Shri V.K.S. Varadan, Director General, evinced keen interest in this study and permitted the presentation of this paper for which the author is grateful.

REFERENCES

Balasubrahmanyan, M.N., 1975. The Age of the Dykes of South Kanara, Mysore State. Misc. Pub. Geol. Surv. India, 23 (1); 236 - 239.

Balasubrahmanyan, M.N., Ram Mohan Rao, G. and Venkataraman, S., 1976. The Gneissic Complex and Kanara Batholith, Mysore State. Misc. Pub. Geol. Surv. India. 23 (2), 363 - 374.

Balasundaram, M.S. and Balasubrahmanyan, M.N., 1973. Geochronology of the Indian Precambrian. Bull. Geol. Soc. Malaysia, 6; 213 - 226.

Collerson, K. D., Jesseau, C.W. and Bridgewater, D., 1976. Crustal development of the Archaean Gneiss Complex : Eastern Labrador In: B. F. Windley (Editor), The Early History of the Earth, Wiley, Lond. pp. 237 - 256.

Crawford, A.R., 1969. Reconnaissance Rb-Sr dating of the Precambrian rocks of southern Peninsular India. J. geol. Soc. India, 10: 117 - 166.

Divakara Rao V., Naqvi S.M., Satyanarayana K. and Hussain, S.M., 1974. Geochemistry and Origin of the Peninsular gneisses, Karnataka, India. J. geol. Soc. India, 15: 270 - 277.

Faure, G. and Powell, J.L., 1972. Strontium Isotope Geology, Springer-Verlag Berlin, 188 pp.

Fermor, L.L., 1936. An attempt at correlation of the ancient schistose rocks of Peninsular India, Mem. Geol. Surv. India, 70 pp.

Fyfe, W.S., 1973. Granites past and present In:L.A. Lister (Editor), Symposia on granites, Gneisses and Related Rocks. Spec. Pub. 3, Geol. Soc. S. Afr., pp. 13 - 16.

Goldschmidt, V. M., 1954. Geochemistry, Clarendon Press, Oxford.

Lambert, R. St. J., Chamberlain, V.E. and Holland, J.G. 1976. The Geochemistry of Archaean Rocks In:B.F. Windley (Editor), The Early History of the Earth, Wiley, Lond. pp. 377 - 388.

Lambert, R. St. J. and Holland, J.G., 1976. Amitsoq gneiss geochemistry : preliminary observations. In:B.F. Windley (Editor), The Early History of the Earth. Wiley, Lond. pp. 191 - 202.

Moorbath, S., Powell, J.L. and Taylor, P.N., 1975. Isotopic evidence for the age and origin of the "grey gneiss" complex of the southern outer. Hebrides, Scotland., J. geol. Soc. Lond., 131; 213 - 222.

Nockolds, S. R. and Allen, R., 1953. The geochemistry of some igneous rock series. Geochim. Gosmochim. Acta., 44; 105 - 142.

Pankhurst, R.J., 1969. Strontium Isotope Studies related to Petrogenesis in the Caledonian Basic Igneous Province of N.E. Scotland. J. Petrol., 10 (1); 115 - 143.

Radhakrishna, B. P., 1974. Peninsular Gneissic Complex of the Dharwar Craton - a suggested model for its evolution. J. geol. Soc. India, 15 (4); 439 - 454.

Sankaran, A.V., 1974. A review of geochemical studies on some Precambrian plutonic complexes of India. J. geol. Soc. India., 15 (4); 407 - 412.

Spooner C.M. and Fairbairn H.W., 1970. Strontium 87/ Strontium 86. Initial Ratios in Pyroxene Granulite Terrains. J. geophys. Res., 75 (32): 6706 - 6713.

Taylor, S.R., 1965. The applications of trace element data to problems of petrology. In:Phys. and Chem. of the Earth, Pergamon Press, I Edn. 6, 133 - 213.

Venkatasubramanian, V.S. and Narayanaswamy, R., 1974. The age of some gneissic pebbles in Kaldurga conglomerate, Karnataka, S. India. J. geol. Soc. India. 15; 318 - 319.

Viljoen, M. J. and Viljoen, R.P., 1969. The Geochemical evolution of the granitic rocks of the Barberton Region. Spec. Pub. geol. Soc. S. Afr. 2: 189 - 218.

Shaw, D. M., 1976. Development of the early continental crust. Part 2 : pre-archaean, protoarchaean and later eras. In:B.F. Windley (Editor), The Early History of the Earth, Wiley, Lond. pp. 33 - 53.

Watson, J., 1976. Mineralization in Archaean Provinces, In:B.F. Windley (Editor), The Early History of the Earth, Wiley, Lond. pp. 443-454.

Winkler, H.G.F., 1967. Petrogenesis of Metamorphic Rocks, Springer - Verlag, 237 pp.

Rb-Sr GEOCHRONOLOGY OF ROCKS FROM THE KOLAR SCHIST BELT, SOUTH INDIA

N.S. Bhalla, J.M. Gupta, Tikam Chabria and S.G. Vasudeva
Atomic Minerals Division, Atomic Energy Dept, Hyderabad-500016, India

Abstract

This paper presents the results of whole-rock age measurements based on the Rb-Sr method of one specimen of the "Champion gneiss" and two specimens of both the "Patna granite" and biotite-granites from Bisanattam, bordering the main Kolar schist belt. The analysed specimens define an isochron of 2866 ± 96 my with an initial $^{87}Sr/^{86}Sr$ ratio of 0.702 ± 0.002.

The ages obtained are much older than those reported by Ramamurthy and Sadashivaiah (1967) and Crawford (1969) for the "Champion gneiss". They conform more to the dates given by Venkatasubramanian et al. (1971) and confirm that the emplacement of the "Champion gneiss" and the "Patna granite" was a very early geological event in the evolution of the Peninsular gneissic complex.

INTRODUCTION

The greenstone-gneissic complex of the central part of the southern Indian Peninsular Shield consists mostly of large belts of greenschists bounded by gneissic rocks. These gneissic rocks, generally referred to as the "Peninsular gneiss", appear to have been emplaced over a long period of geological history. This is indicated by the limited available geochronological data.

An attempt has been made in this paper to date the granitic rocks associated with one of the most important greenstone belts, the "Kolar schist belt". This includes the so-called "Champion gneiss", whose age and origin is very much in dispute, and the granitic rocks bordering the western margin of the Kolar belt which are considered to be part of the Peninsular gneiss. The age data presented here confirm the current view of the polyorogenic evolution of the Peninsular gneissic complex as a whole.

GEOLOGY

The Kolar schist belt, about 100 km east of Bangalore in the state of Karnataka, is traceable for c. 80 km in the NS direction with a width of 4-6 km. It has been recognised as the Keewatin-type and shows close similarity to the classical greenstone belts of the world. Basic volcanics, occurring as amphibolites and hornblende schists with minor acid volcanics, graphitic schists, conglomerates, ironstones and cherts, are the lithological units associated with this formation. The imprint of several periods of deformation has given rise to tight folds with

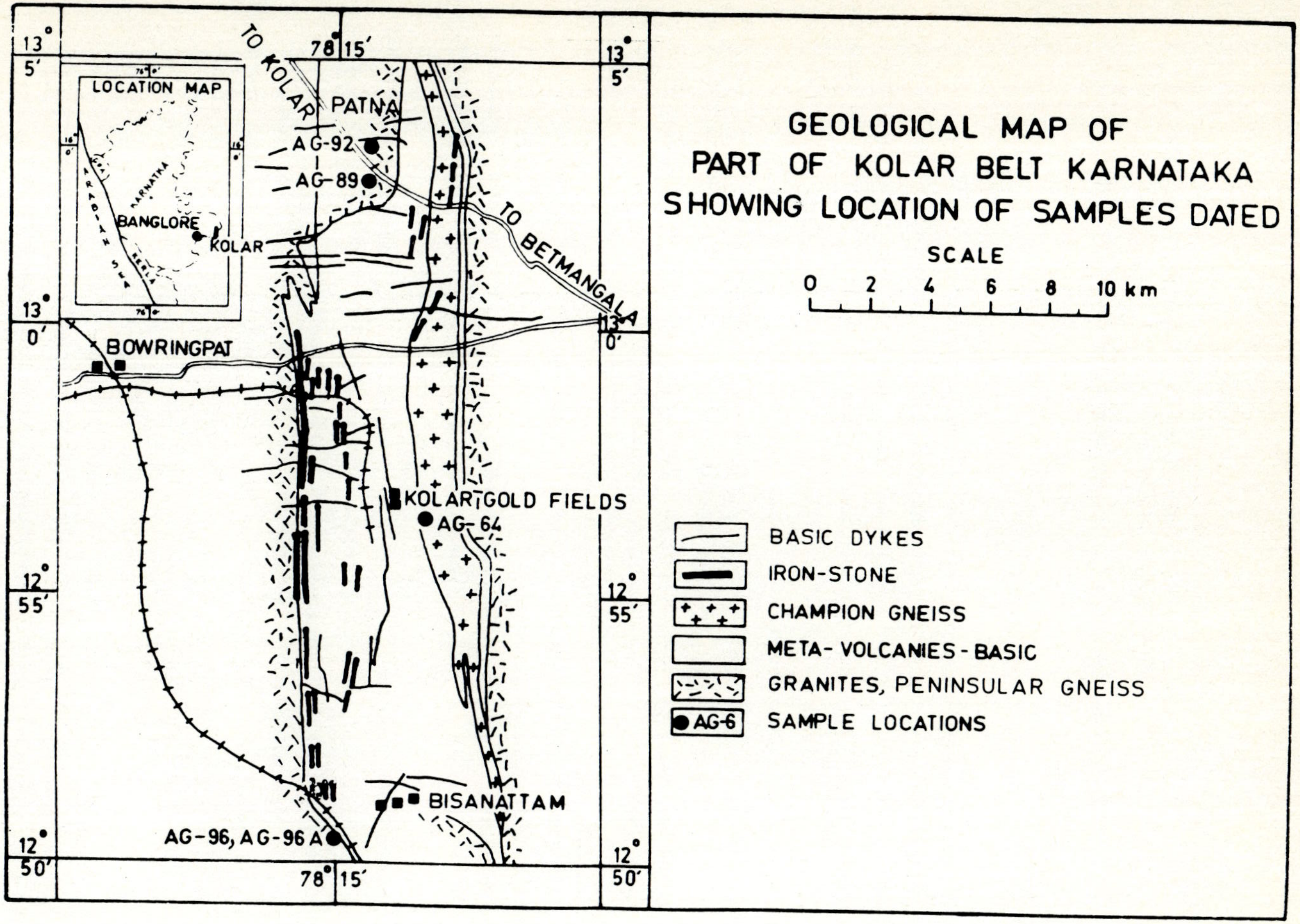

Fig. 1. Geological map of part of Kolar Belt, Karnataka, showing location of samples dated.

N-S axes and several faults. The rocks show considerable evidence of invasion by later tonalites, resulting in the widespread development of migmatitic gneiss (Radhakrishna, 1974).

There are two distinct granitic bodies associated with the Kolar schist belt: one borders both sides of the schist belt and the other occurs as a small linear body, the "Champion gneiss". The latter has long been of special interest on account of its association with gold mineralisation. Recently, however, it has been proposed that it is a group of metamorphosed rhyolites, rhyodacites, tuffs and agglomerates (Ziauddin and Narayanaswamy, 1976). The granites bordering the schist belt are classified as part of the Peninsular gneiss and Ramakrishna et al. (1974) suggest that they are the products of granitisation of the pre-existing rocks, although there is clear field evidence to indicate that they are later than the greenschists.

LOCATION OF THE SAMPLES

Rb-Sr ages have been determined for five granites from three localities. One represents the Champion gneiss from the type area in the Kolar gold fields (AG-64) (Lat. 12°56' N, Long. 78°17' E) and the other four samples are from two localities, one near Patna village on the Kolar-Betmangala road (AG-89 and 92) (Lat. 13°4' N, Long. 78°16' E), and the other from the area of the Bisanattam railway station (AG-96, 96A) (Lat. 12°50' N, Long. 78°15' E). The sample locations are shown in Fig. 1.

PETROGRAPHY

The Champion gneiss (AG-64) from the vicinity of the Kolar gold fields is a fine-grained porphyritic microgranite transformed into augen gneiss. Porphyroblasts of blue opalescent quartz in a fine-grained mesostasis of quartz and altered feldspars (oligoclase and microcline) are the major constituents. Muscovite, garnet, biotite, chlorite, calcite, apatite, opaques and zircon are the accessories. The feldspars are sericitised and the biotite shows alteration to chlorite and calcite (cf. Rama Rao, 1940, 1962).

The granites from the vicinity of Patna village (AG-89, 92) and those from near the Bisanattam railway station (AG-96, 96A) are identical. They are medium grained, and major constituents are strained quartz, sericitised orthoclase, albite and microcline perthite. Blue-green hornblende, biotite, chlorite, epidote and sphene are the accessories. Hornblende is altered to biotite and chlorite.

ANALYTICAL TECHNIQUE

The Rb-Sr isotopic data have been obtained on a Nier-type mass spectrometer with 60° deflection magnetic analyser of 6" radius of curvature. The equipment, constructed earlier, employs a thermal ionisation source using rhenium filaments for the production of positive ions. Rubidium and strontium for the analyses have been used in chloride and nitrate forms respectively, obtained by chemical extraction from the whole-rock samples. The standard isotope dilution technique has been employed for the quantitative determination of Rb and Sr using enriched ^{87}Rb and ^{84}Sr tracers. The separation of the respective elements was carried out on an ion exchange column with Dowex 50W resin. The procedure followed is similar to that described by Compston et al. (1965).

DISCUSSION

The Rb-Sr data are given in Table 1.

Table 1. Rb-Sr data

Sample No.	Rb (ppm)	Sr (ppm)	$^{87}Rb/^{86}Sr$	$^{87}Sr/^{86}Sr$
AG-64	132.5 ± 1.5	224.4 ± 2.2	1.72 ± 0.03	0.775 ± 0.002
AG-89	173.9 ± 1.1	349.4 ± 3.0	1.45 ± 0.02	0.764 ± 0.003
AG-92	180.8 ± 1.1	329.1 ± 3.3	1.60 ± 0.02	0.768 ± 0.001
AG-96	135.4 ± 0.7	500.5 ± 4.7	0.785 ± 0.009	0.733 ± 0.002
AG-96A	118.5 ± 0.7	629.0 ± 5.5	0.546 ± 0.007	0.725 ± 0.001

Observed strontium isotopic ratios have been normalised taking $^{86}Sr/^{88}Sr = 0.1194$ for any mass discrimination in the run. The errors quoted with each analysis are the standard errors. Age calculations are based on an ^{87}Rb decay constant of $1.42 \times 10^{-11}\ y^{-1}$, as recommended by the Subcommission on Geochronology (Steiger and Jager, 1977). The isochron for the samples has been drawn by the method of least squares fitting of a straight line following the treatment of Williamson (1968). The method takes into account the statistical errors in both the co-ordinates and minimises the weighted sum of the squared residuals. The errors associated with the age and initial $^{87}Sr/^{86}Sr$ have been computed from the variance obtained for the slope and intercept respectively.

Fig. 2 shows the plotted data points and the best-fit line through them. The isochron defines an age of 2866 ± 96 my with an initial ratio of 0.702 ± 0.002. This agrees, within the limits of experimental error, with the age of 2820 ± 100 my determined by Venkatasubramanian et al. (1971) for the Peninsular gneisses occurring on the eastern and western borders of the Kolar schist belt.

One sample of the analysed Champion gneiss also falls on the same isochron, suggesting that it is of the same age as the bordering granites. This age is not in agreement with that reported by Ramamurthy and Sadashivaish (1967) at 2480 my and Crawford (1969) at 2650 my. In spite of the discrepancy, however, the indicated age suggests that the Champion gneiss is no older than the surrounding granites. This also implies that both the Champion gneiss and the bordering granitic bodies, which are part of the Peninsular gneisses in the area, belong to the same epoch, though petrologically they appear different.

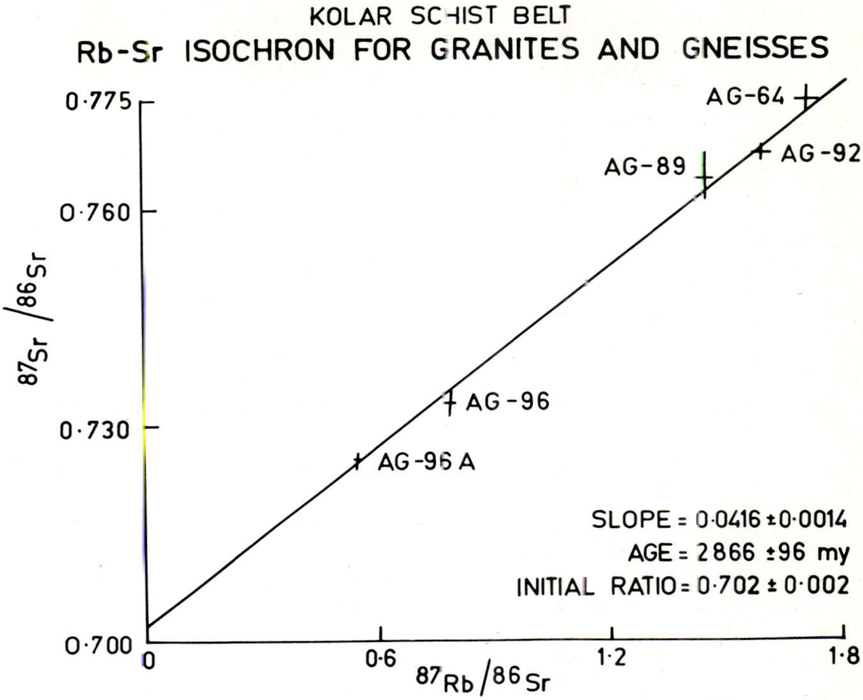

Fig. 2. Rb-Sr isochron for granites and gneisses of the Kolar schist belt.

The position of the Champion gneiss itself in the Precambrian geology of South India is controversial. Rama Rao (1940, 1962) considered it to represent the earliest granitic rock of the Peninsular shield; Pichamuthu (1971), however, equated it with the Peninsular gneiss. Different views have been expressed with respect to the granites bordering the Kolar belt: they were considered as separate granitic bodies by earlier workers but later equated to the Peninsular gneiss (Rama Rao, 1962). Recent workers (Ramakrishna et al., 1976) suggest that these granites are the product of granitisation of pre-existing rocks.

From the data presented in this paper it can be seen that the granites within and bordering the Kolar schist belt are slightly older than the Peninsular gneiss which has an age of c. 2600 my; this confirms the polycrogenic character of the Peninsular gneiss. The results also point to the need for a systematic study of more samples which is likely to bring out a possible chronological zonation within the Peninsular gneiss in relation to the schistose formations.

ACKNOWLEDGEMENTS

The authors thank M/s M.A. Ahmad and R.N. Prasad for their help in the mass spectrometric analysis and Dr D. Mukerjee and Mr K. Suryanarayana for petrological studies. They are indebted to M/s T.M. Mahadevan, C.S. Rao, Yudhishter Lal and Dr S. Viswanathan for discussions and useful suggestions and to Dr G.R. Udas, Director A.M.D., for his constant encouragement and guidance throughout the work.

REFERENCES

Compston, W., Lovering, J.F. and Vernon, M.J., 1965. The rubidium-strontium age of the Bishopville norite and its component enstatite and feldspar. Geochim. Cosmoch. Acta, 29: 1085-1099.

Crawford, A.R., 1969. Reconnaissance Rb-Sr dating of the Precambrian rocks of southern Peninsular India. J. geol. Soc. India, 10: 117-166.

Pichamuthu, C.S., 1967. India. In: The Precambrian, K. Rankama (Editor), Vol. 3, Interscience, New York.

Pichamuthu, C.S., 1971. Precambrian geochronology of Peninsular India. J. geol. Soc. India, 12: 262-273.

Pichamuthu, C.S., 1976. Some problems pertaining to the Peninsular gneissic complex. J. geol. Soc. India, 17.

Radhakrishna, B.P., 1974. Peninsular gneissic complex - a suggested model for its evolution. J. geol. Soc. India, 15: 439-456.

Radhakrishna, B.P. and Vasudeva, V.N., 1977. The early Precambrian of the southern Indian Shield. J. geol. Soc. India, 18, No. 10.

Ramakrishna, M., Viswanatha, M.N. and Swaminath, J., 1976. Basement-cover relationship of Peninsular gneiss with high-grade schists and greenstone belts of southern Karnataka. J. geol. Soc. India, 17, No. 1.

Ramamurthy, V. and Sadashivaiah, M.S., 1967. Rubidium-strontium ages of charnockite and other rocks from the Precambrian complex, Mysore, India. In: Geochronology of Precambrian stratified rocks (Abstr.), Conf. at Edmonton, p. 73.

Rama Rao, N., 1935. The distinguishing characteristics of the Champion gneiss and the Peninsular gneiss of Mysore State and their probable mode of origin. Rec. Mysore Geol. Dept., 33: 80.

Rama Rao, B., 1940. The Archaean complex of Mysore. Bull. Mysore Geol. Dept., 17.

Rama Rao, B., 1962. A Handbook of the Geology of Mysore State, Southern India. Bangalore Printing and Publishing Co., Bangalore.

Steiger, R.H. and Jager, E., 1977. Subcommission on geochronology: convention on the use of decay constants in geo- and cosmochronology. Earth Planet. Sci. Lett., 36: 359-362.

Venkatasubramanian, V.S., Iyer, S.S. and Pal, S., 1971. Studies on the Rb-Sr geochronology of the Precambrian formations of Mysore State, India. Amer. J. Sci., 270: 43-53.

Williamson, J.H., 1968. Least squares fitting of a straight line. Can. J. Phys., 46: 1845-1847.

Ziauddin, A. and Narayanaswami, S., 1974. Gold resources of India. Bull. geol. Surv. India, 38, No. 1.

GREENSTONE SCHIST BELTS

THE ARCHAEAN GREENSTONE BELTS OF KARELIA (EASTERN FINLAND) AND THEIR KOMATIITIC AND THOLEIITIC SERIES

by

S. Blais, B. Auvray, R. Capdevila, B.M. Jahn, J. Hameurt
Centre Armoricain d'Etude Structurale des Socles,
Institut de Geologie, 35042 - RENNES Cedex (France)

and

J.M. Bertrand
Centre Geologique et Geophysique, U.S.T.L.,
Place Eugene Bataillon, 34060 - MONTPELLIER Cedex (France)

Abstract

The geological and geochemical features of three greenstone belts of Eastern Finland (Suomussalmi, Kuhmo and Tipasjarvi) have been studied. We have analysed about 70 metavolcanic rocks from the low greenstone sequence for their major and trace element compositions. The field relationships between various volcanic rocks are rather obscure; but the chemical data allow us to distinguish two general magmatic series, namely, the komatiitic and the tholeiitic series. We have concluded from this preliminary geochemical study that most volcanic rocks in each may have been derived from fractional crystallization of some parental magma. The REE data, mainly presented for the rocks from the Tipasjarvi belt, provide a strong evidence for a "depleted" mantle source, a feature very similar to that of Abitibi, Canada. The REE data also suggest that not all rocks are formed by fractional crystallization; a mechanism of different degree of partial melting is called to account for some rock types.

Although the Baltic Shield is one of the first regions in which the Archaean rocks were described (Sederholm, 1897, 1932), it is only recently that the existence of greenstone belts has been clearly demonstrated (Blais et al., 1976, 1977 and in print a and b ; Bowes, 1976 ; Gaal et al., 1976 and in print; Lobach-Zhuckenko et al., 1976 ; Mutanen, 1976; Suslova, 1976).

In this contribution, the principal characteristics of the belts, which we have studied in Finland, are briefly described and we expose the present state of our petrological and geochemical research in this context.

GEOLOGICAL SETTING

The Archaean rocks crop out widely in northern and central Finland (Simonen, 1971). In this latter region, the Archaean crust comprises a gneissic basement, largely migmatised, and supacrustal formations which constitute the greenstone belts. The gneissic basement and the cover of metasediments and greenstones have undergone a common tectonometamorphic history during the time range 2.6 to 2.7 b.y. (Kuvo and Tilton, 1966 ; Blais et al., in press a).

The greenstone belts for which we presently have the most information are those of Suomussalmi, Kuhmo and Tipasjarvi, located in eastern Finland. The Suomussalmi and Kuhmo belts lie along the same approximately north-south line, extending lengthwise over 200 km, and never wider than 20 km. They were mapped by Wilkman (1924), Matisto (1958), Vartiainen (1970) and Hypponen (1973; 1976) (Fig. 1).

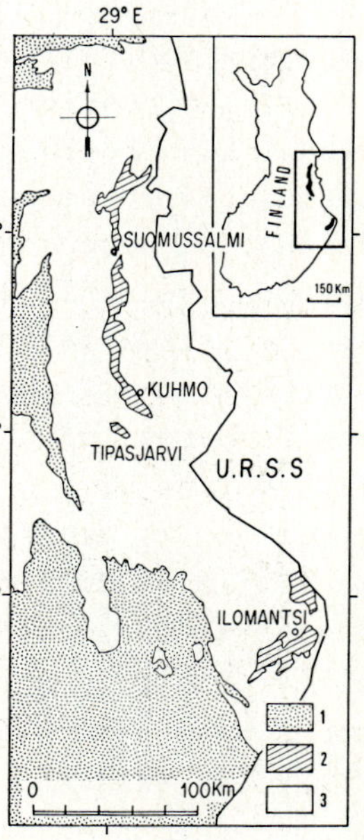

Fig. 1. Geological sketch-map of eastern Finland. 1= Karelian formations; 2= Greenstone belts; 3= High-grade metamorphic terrains (basement).

The principal characteristics of these belts are as follows:

1. Lithostratigraphy

The most complete stratigraphic column which we have found to date corresponds to the section in the northern part of the Suomussalmi belt. Its thickness is estimated at between 3000 m and 5000 m. We have distinguished three lithostratigraphic units, passing from bottom to top :

(a) An early magmatic cycle, composed initially of ultrabasic lava flows and small-scale intrusions, massive basaltic lava flows and pillow lavas flows, volcanic breccias and basic sheets.

(b) An essentially metasedimentary formation represented by pelitic mica schists, graphitic schists and greywackes, sometimes associated with quartzites, conglomerates, and volcanic rocks (mostly acidic tuffs and some minor basic lavas). The clasts of the metasediments are of mixed origin; some are derived from the greenstone belt volcanics while others show an adjacent basement provenance.

(c) A second volcanic cycle, composed of feldspathic tuffaceous rocks of intermediate composition, associated with subordinate lavas.

The members of this succession have a tectonic contact with the gneissic basement. This latter is principally composed of fine- to medium-grained grey gneisses, augen orthogneisses and migmatites, mainly heterogeneous diatexites.

2. Volcanism

The volcanism is, to a first approximation, at least bimodal in nature.

(a) The Lower volcanic sequence comprises :

i) Peridotitic komatiites (MgO $>$ 30%) which are olivine-rich cumulates; picritic komatiites (30% $>$ MgO $>$ 20%) in which quench and spinifex textures are sometimes preserved; picritic basalts or gabbros (20% $<$ MgO $<$ 12%) and basaltic komatiites (MgO $<$ 12%).

ii) Tholeiitic basalt sensu-stricto, depleted in potassium, sometimes accompanied by clino-pyroxene-bearing cumulates.

(b) The Upper volcanic sequence comprises calc-alkaline felsitic rocks, all highly sodic and of andesite to rhyolite composition.

The two sequences, tholeiitic sensu-lato and calc-alkaline, are well distinguished in the classical diagrams of Figs. 2 and 3. Their respective emplacements are clearly separated in time by the deposition of sediments.

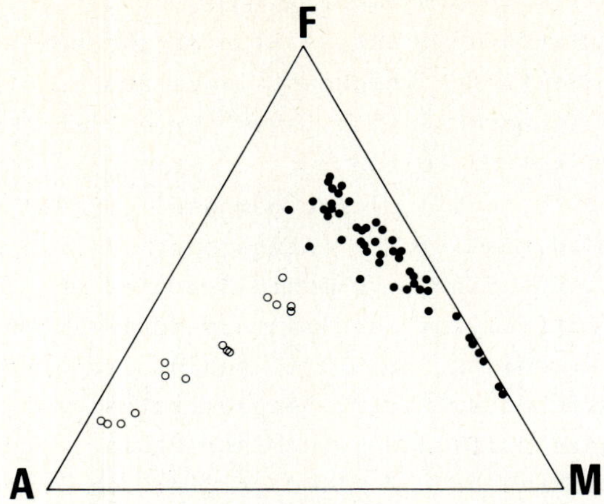

Fig. 2. Location of the two volcanic sequences in a ternary AFM diagram; filled circles = tholeiitic (s.l.) sequence; open circles = calc-alkaline sequence.

3. Structural evolution

The gneissic basement and greenstone belts of eastern Finland underwent a common structural evolution, the sequence of which in chronological order is as follows:

Phase 1 : Development of isoclinal folds deforming the volcano-sedimentary bedding of the belts and the first metamorphic banding of the basement.

Phase 2 : This is the most evident deformation in both the belts and the basement and is responsible for the general geographical distribution of the belts. Folds with subvertical axial planes and steeply plunging axes, often carry an axial planar foliation and a well-developed mineral lineation. Minor shear-zones frequently develop in the limbs of asymmetrical folds.

Phase 3 : This is a complex event, non-penetrative on the regional scale. It principally comprises early eastward thrusts, well seen in the gneissic basement, followed by major faults whose horizontal displacement components are generally sinistral. The phase 3 structures affect the Proterozoic Karelian schists and therefore are post-Archaean.

Phases 1 and 2 correspond to the end of Archaean history of the gneissic basement and greenstone belts. The tectonometamorphic banding of the basement, deformed during phase 1, corresponds to a structural evolution which predates the formation of the belts.

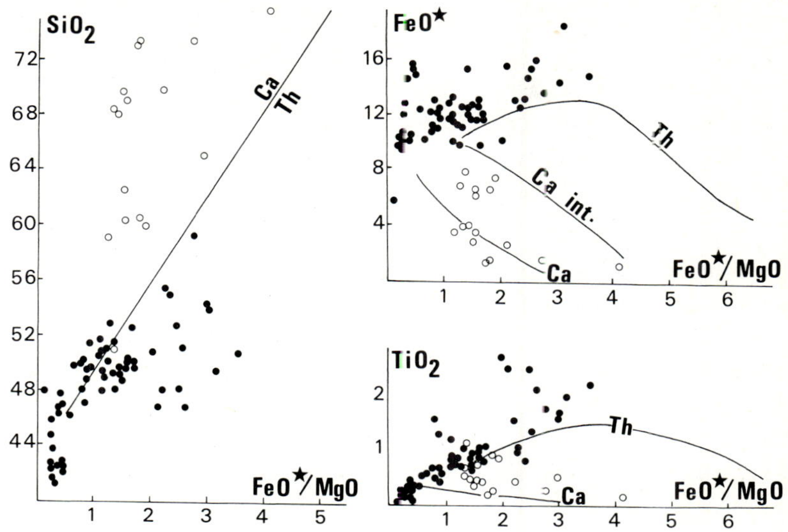

Fig. 3. Location of the two volcanic sequences in the diagrams of Miyashiro (1975); same symbols as in Fig. 2.

4. Metamorphism

All the members of the belts are affected by a regional metamorphism of weak to intermediate intensity. The most frequent paragenesis are :

(a) Quartz + biotite + muscovite + albite or oligoclase ± chlorite ± garnet in the metapelites.

(b) Chlorite + epidote + actinolite + plagioclase ± biotite ± hornblende ± garnet ± diopside in the metabasites.

(c) Antigorite ± tremolite ± chlorite ± talc ± carbonates in the meta-ultrabasites.

The rocks of the belts are all metamorphosed to at least biotite grade. According to location, they correspond to the upper greenschist facies or lower garnet amphibolite facies. This metamorphism is apparently of intermediate pressure. The analysis of deformation/blastesis relations shows that the metamorphism commences with phase 1 and reaches a maximum during phase 2.

5. Important acid and intermediate magmatic emplacement is

related to orogenic evolution of the greenstone belts. It consists of :

(a) early granodiorites, emplaced post-phase 1 and pre-phase 2. These are principally phenocrystic granodiorites, with biotite or biotite and amphibole, relatively enriched in CaO and Na_2O and poor in K_2O.

(b) late events, post-phase 2 and pre-phase 3, amongst which can be distinguished :

i) medium-grained, biotite and amphibole granodiorites and tonalites, rich in CaO and MgO and poor in SiO_2 and K_2O;

ii) fine- or medium-grained, biotite muscovite leucogranites and leucogranodiorites of a pink colour, rich in SiO_2 and poor in CaO and relatively poor in K_2O.

iii) large bodies of biotite muscovite pegmatites.

The different plutonic bodies appear either in the greenstone belt or at the greenstone belt/basement boundary or within the basement, close to this last. It is of note that all these plutonic rocks are allochtonous and are poor in potassium.

6. Conclusions

The geological characteristics of the greenstone belts of Eastern Finland show that they correspond to the classical model of Archaean belts. From a geodynamic point of view two fundamental stages in their evolution can be distinguished :

i) an early stage of opening of proto-oceanic rifts, during which the komatiitic and tholeiitic rocks of the Lower volcanic sequence were erupted and the sediments were deposited.

ii) a later stage of closing, during which the calc-alkaline rocks were erupted, followed by the orogeny.

Before deformation, the belts studied here may therefore have corresponded to ensimatic island-arcs, adjacent to continental areas.

DATA ON KOMATIITIC AND THOLEIITIC SERIES OF EASTERN FINLAND

The following section presents data on the komatiitic and tholeiitic components of the Lower volcanic sequence of the East Finland greenstone belts :

1. Occurrence

The following lithological units have been distinguished :

(a) Slices of schistose ultrabasites, highly serpentinized ("soapstones").

(b) Massive ultrabasites, some with magmatic layering.

(c) Massive, generally gabbroic, sometimes ultrabasic rocks with relict igneous (gabbroic), intersertal or quench (spinifex) textures, most commonly as sills, more rarely as originally discordant veins or as lava flows with occasionally scoriaceous surfaces.

(d) Fine-grained amphibolites without relict textures, corresponding to lava flows or tuff horizons.

(e) Metabasalts sensu-lato in which the most characteristic texture is defined by radiating tufts of fine amphibole fibres, with local relicts of vacuole structures. These represent pillow lavas whose original form is often well preserved.

The isolation of outcrops and absence of normal contacts renders the interpretation of the relationships between these various rock types difficult.

2. Petrology and geochemistry

As indicated above, the basic and ultrabasic rocks of the Lower volcanic sequence have been metamorphosed in the upper greenschist facies to lower garnet-amphibolite facies. Elsewhere, certain ultrabasic rocks have undergone metasomatic modification, some "soapstones" containing up to 20% volatiles.

Under these conditions, most of the primary minerals and glasses are no longer directly visible and this study of the magmatic evolution is rendered that much more difficult. Indeed, we are restricted to the study of a few relict minerals and the geochemistry of immobile major and trace elements of the rocks.

(a) Relict minerals : only relict minerals in rocks of cumulate texture have been observed.

The ultrabasic cumulates of the komatiitic series frequently contain residual, partially serpentinized, olivines and occasionally, clinopyroxenes and chromite. These minerals are at present the object of study.

The picritic cumulates of the tholeiitic series frequently contain clinopyroxene, the microprobe analysis of which identifies them as salite. In certain olivine-bearing cumulates of the same series, microprobe analysis shows a relative enrichment in iron (Fo_{84}).

(b) Mineral assemblages : according to their composition and degree of metamorphism, the various rock types have the following assemblages:

i) Soapstones : antigorite + talc + carbonates (breunnerite, stichtite) + magnetite with, in minor amounts, Mg - chlorite, penninite, tremolite and graphite.

ii) Massive metaperidotites : residual olivine and clinopyroxene in variable proportions + antigorite + tremolite + sulphides, and, in minor amounts, Mg - chlorite, penninite, magnesite and ± chromite ± opaques ± chrysotile in cross-cutting veinlets.

These rocks often contain olivine cumulate texture and have the normative compositions of harzburgites or lherzolites of dunite affinity.

iii) Metapicrites : tremolite + Mg-chlorite ± talc ± opaques ± carbonates ± penninite. The observed spinifex textures in these rocks demonstrate the replacement of olivine by fibrous Mg-chlorite; the spaces between the network formed by the Mg-chlorite is filled by acicular tremolite, chlorites and carbonates.

iv) Metagabbros and metadolerites: the most common assemblages are : a) actinolite + albite or oligoclase + epidote ± chlorite ± biotite ± quartz.

b) green hornblende + oligoclase or andesine ± chlorite ± epidote ± biotite ± actinolite ± garnet. In some samples, actinolite and hornblende have been observed together. Exceptionally, metamorphic associations with clinopyroxene or scapolites occur.

Certain metagabbros contain residual clinopyroxene in amphibole cores. Their chemical composition shows that, in these instances, the rocks are probably former tholeiitic clinopyroxene cumulates.

v) Fine-grained amphibolites : these present the same assemblages as the metagabbros but without relict textures. Furthermore, their plagioclase proportions are highly variable from one rock to the next; compositionally, they correspond to picritic basalts, olivine tholeiites (the most abundant) and mildly saturated tholeiites.

vi) The pillow lavas : these generally comprise fine-grained tremolite, epidote and chlorite, lack any plagioclase and have a picritic basalt composition.

(c) Major element geochemistry : the major element composition has been measured for seventy samples, representative of the observed rock types. The average compositions are given in Table 1.

i) Classification adopted : the content of MgO, being the most variable of the oxides (4% < MgO < 46%), has been chosen as the basis for initial subdivision. The peridotites and "soapstones" have MgO contents falling between 33% and 46%. The massive peridotites divide into two groups whose MgO contents vary about

TABLE 1: Average compositions (anhydrous) of the main rock types in the Suomussalmi, the Kuhmo and the Tipasjarvi greenstone belts

	a	b	c	d	e	f	g	h	i	j
No. of analyses	3	4	9	1	5	6	19	2	2	11
SiO_2	44.12	43.25	43.74	47.84	46.85	50.21	50.54	51.15	48.00	50.90
Al_2O_3	2.27	3.93	2.66	6.86	8.08	11.77	14.80	6.29	10.84	14.04
FeO	8.86	14.88	10.56	10.73	10.75	10.74	11.87	11.90	12.59	14.94
MnO	0.14	0.25	0.09	0.20	0.17	0.22	0.20	0.27	0.21	0.24
MgO	44.27	34.23	40.89	26.67	26.01	13.91	8.06	11.84	14.77	5.63
CaO	0.13	3.08	1.90	7.21	7.45	11.01	11.21	15.28	10.11	8.90
Na_2O	-	-	-	0.01	0.18	1.23	2.14	1.65	1.00	2.81
K_2O	-	0.06	-	0.10	0.03	0.33	0.23	0.32	0.88	0.37
TiO_2	0.18	0.32	0.16	0.29	0.40	0.54	0.84	1.22	1.44	1.98
P_2O_5	-	-	-	0.11	0.07	0.07	0.09	0.10	0.14	0.19

a: peridotites (cumulate) with MgO > 40%; b: peridotites (cumulate) with MgO = 30-40%; c: soapstones; d: picritic komatiite with spinifex texture (S833); e: picritic komatiites (MgO = 20-30%); f: picritic basaltic komatiites (MgO = 12-20%); g: basaltic komatiites (MgO < 12%); h: tholeiites with clinopyroxene cumulates; i: tholeiites with possible olivine cumulates; j: tholeiites. Total iron oxides as FeO.

Analyst: F. Vidal, Centre Armoricain d'Etude Structurale des Socles, Univ. de Rennes.

44% and 34%, corresponding to two types of cumulates of different olivine proportions. The MgO content of the picrites falls between 21% and 28%, with an average close to 26%. We have grouped the basalts and picritic gabbros whose MgO content is between 12% and 16%. Some of these rocks are clinopyroxene cumulates, others are gabbros probably enriched in olivine. Finally, we have considered a very extensive group of rocks, whose MgO content lies between 4% and 12%, to be basalts.

ii) Magmatic series : the basaltic rocks have compositions of oceanic affinities. These are ocean-floor basalts and low-K tholeiites; however, it is not possible at present to discriminate between those rocks of ridge and those of island-arc types. As is often the case in the Archaean belts (Arndt et al., 1977) and recent oceanic environments (Clarke, 1970 ; Dietrich et al., 1977), the rocks studied correspond to two series of distinct composition and evolution characterized in particular by different TiO_2 concentration. Thus, the components of the Lower volcanic sequence of the belts in Karelia can be subdivided into a komatiitic series and a tholeiitic series.

The komatiitic series includes a fairly continuous series of rocks passing from peridotites to basalts, characterized by high CaO/Al_2O_3 ratio at constant MgO and low alkalis and titanium contents.

The tholeiitic series is, albeit, exclusively composed of basalts with associated rare clinopyroxene and perhaps olivine cumulates. Compared with the previous series, the tholeiitic series, though relatively poor in potassium, is richer in alkalis, relatively depleted in CaO and MgO and is especially characterized by a great iron and titanium abundance.

Fig. 4 shows the discrimination between these two series on a TiO_2 versus Al_2O_3 diagram. Except for minor details, these series are very similar to those of South Africa (Viljoen and Viljoen, 1969 a,b), Ontario (Arndt et al., 1977) and Western Australia (Naldrett and Turner, 1977) which are well known.

iii) Trend of crystallization : the representative points of the rock compositions of the two series plotted on oxide versus oxide diagrams show trends attributable to magma fractionation. Taking account of these trends and of the measured or estimated composition of the residual minerals, we can make the following observations.

<u>Tholeiitic series</u> (Fig. 5) : the principal diagrams in terms

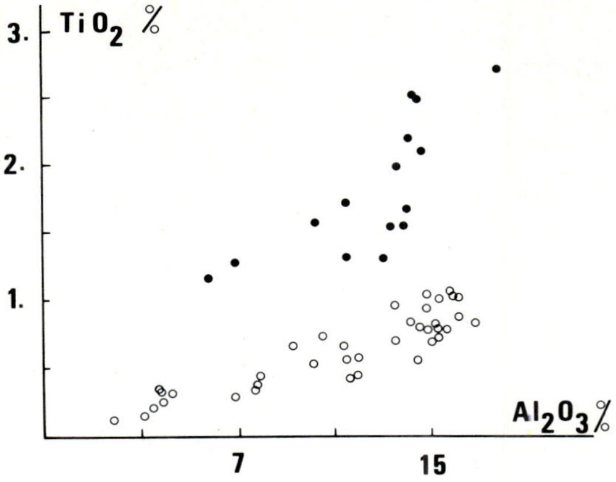

Fig. 4. TiO_2 vs Al_2O_3 diagram showing the opposition between tholeiites (s.s.) and komatiites, filled circles = tholeiites, open circles = komatiites.

of MgO and SiO_2 show that, except for the cumulates, rocks of the tholeiitic series present a unique linear differentiation trend. This trend is characterized by the regular decrease of Al_2O_3, FeO, MgO and CaO and the increase of SiO_2 and Na_2O during the crystallization.

If we consider that the rocks richest in MgO and poorest in SiO_2 are close to the primary liquid composition, this magmatic evolution is therefore mainly controlled by plagioclase and clinopyroxene fractionation. However, the primary liquid being relatively depleted in SiO_2 compared to the plagioclase-clinopyroxene mixtures, some small quantities of olivine and opaque minerals must also have played a role, at least in the early stages of the differentiation. The role of an opaque mineral rich in iron is confirmed by the fact that all of the tholeiitic series rocks are richer in total iron than mixtures plagioclase + clinopyroxene + olivine.

The observed cumulates belong to two types. Some are cumulates with evident clinopyroxene, their chemical composition confirming the textural and mineralogical observations; others are probably early cumulates, lightly enriched in olivine and clinopyroxene, as deduced from their chemical composition, but without microscopic evidence.

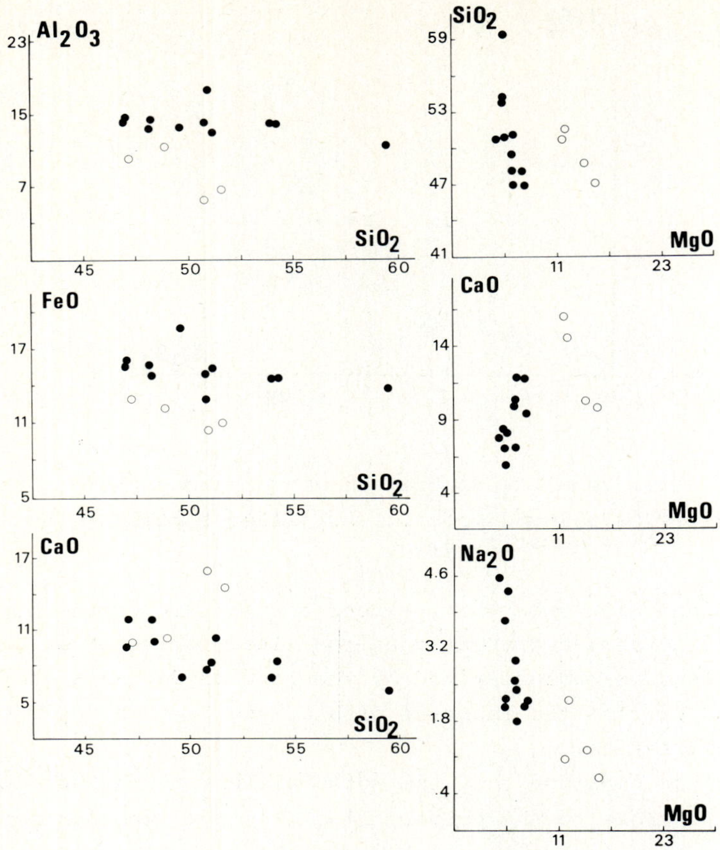

Fig. 5. Different oxides vs MgO and SiO_2 diagrams in the tholeiitic series. Filled circles = tholeiites without cumulate textures; open circles = cumulates.

Komatiitic series (Fig. 6) : the crystallization trend of the komatiitic series is much more complex. Here, the diagrams include all the rocks of Suomussalmi, Kuhmo and Tipasjarvi belts whereas the trends specific to each belt may present some slight differences. For the moment, our observations only constitute a first approximation.

The diagrams, oxide versus MgO, show continuous and regular enrichment in SiO_2, Al_2O_3 and CaO for MgO values between 45 and 15%. This enrichment suggests that the fractionation of this part

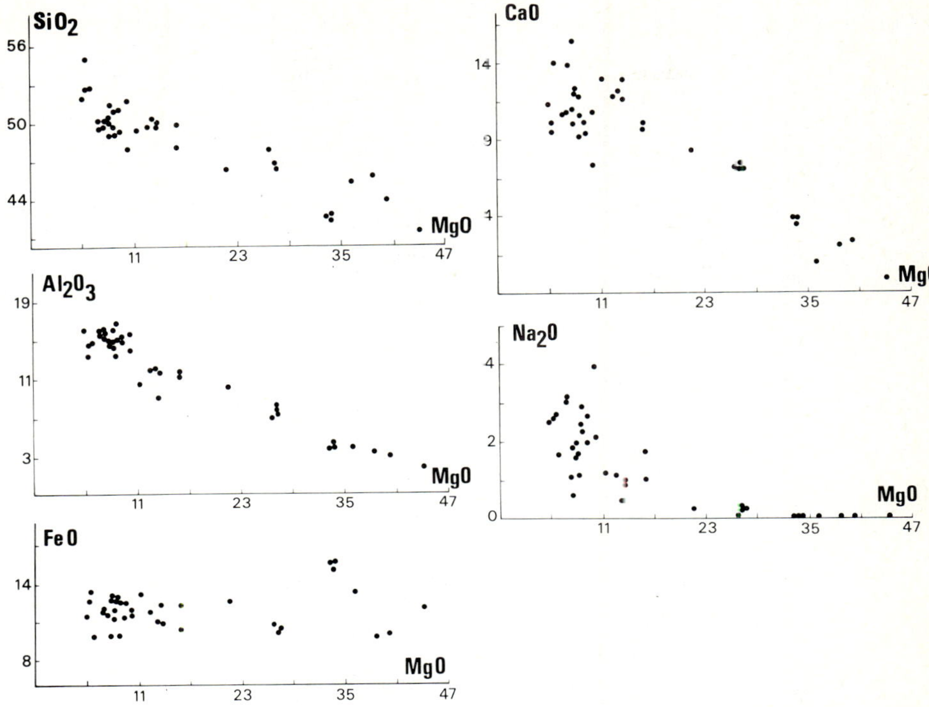

Fig. 6. Different oxides vs MgO in the komatiitic series. Cumulates are not distinguished in these diagrams.

of the suite was essentially controlled by olivine, starting from a primary liquid situated at about 26-27% MgO, a composition that corresponds to the micro-spinifex rocks. The FeO content being quite stable during the early crystallization, described above, it is possible that a small quantity of opaque iron ores crystallized at the same time as the olivine.

Between 15% and 10% of MgO, the contents of SiO_2, Al_2O_3, FeO and Na_2O are generally constant, while the CaO content increases somewhat, corresponding to a concomitant fractionation of olivine and clinopyroxene.

From about 11% MgO the general trend splits, principally into three micro-trends, mainly corresponding to liquids depleted in clinopyroxene and plagioclase and to liquids lightly enriched in clinopyroxene or clinopyroxene and plagioclase.

As may be seen, the major element geochemistry in the komatiitic series of eastern Finland is compatible with simple mechanisms of fractional crystallization. The trends agree with the experimental data of Arndt (1976).

(d) Geochemistry of some transition elements : only fragmentary data for trace elements is available (Ni, Co, V, Cr, Zr, Y), the only systematic study being that of the Tipasjarvi belt.

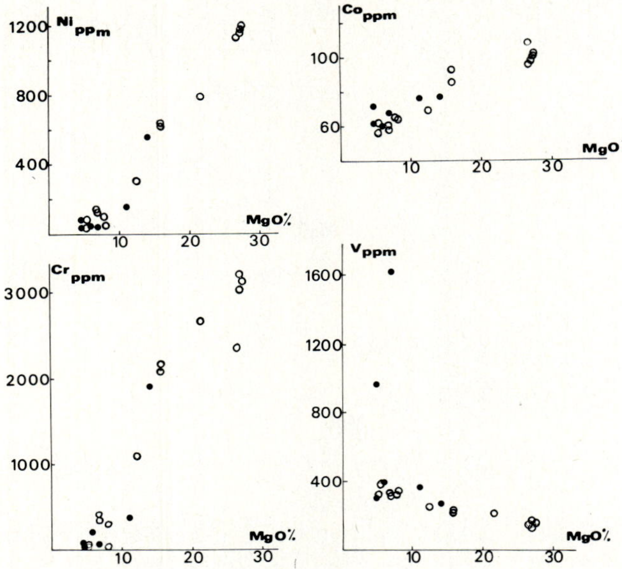

Fig. 7. Transition-elements (Ni, Co, V, Cr) vs MgO diagrams in the volcanites of Tipasjarvi greenstone belt. Filled circles = tholeiites; open circles = komatiites.

In Fig. 7, the Ni, Co, V and Cr values have been plotted vs the MgO content, this latter being considered as an index of degree of differentiation. Concerning the komatiitic rocks, Ni, Co and Cr show a good positive correlation with MgO, and this seems to indicate that they are mainly concentrated (especially Ni and Cr) during the first stages of crystallization (incorporation in olivine and perhaps, also, in opaque phases, oxides or sulphides). On the other hand, V shows a strong negative correlation; it seems to be concentrated in the residual liquid with advancing fractional crystallization. From this point of view V appears to be a more incompatible element than Ni, Co or Cr.

Correlations are less clear cut for the rocks belonging to the tholeiitic series, and this can be explained by the concomitant crystallization of several phases (clinopyroxene, plagioclase, olivine and opaques).

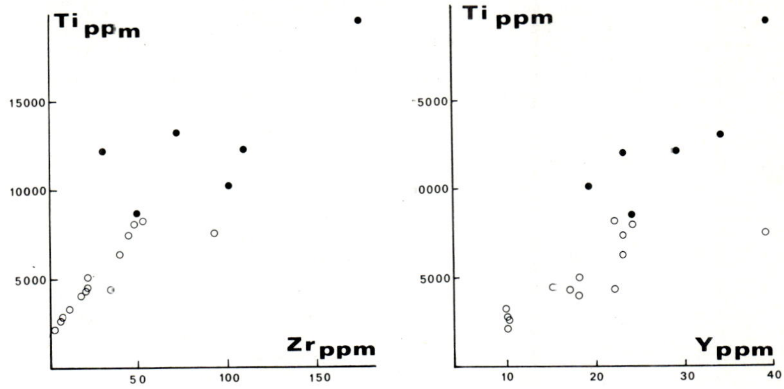

Fig. 8. Ti vs Zr and Ti vs Y diagrams in the volcanites of Tipasjarvi greenstone belt. Same symbols as in Fig. 7.

Fig. 8 demonstrates the good correlation, for the komatiitic rocks, both between Ti vs Zr and Ti vs Y ; this is a general feature of Archaean komatiites (Nesbitt and Sun, 1976). Another common characteristic of the Archaean rocks is the scattering of the plots of tholeiitic rocks in such diagrams and this is well observed for Tipasjarvi. Nesbitt and Sun (op. cit.) explained this scattering by the influence, in the tholeiitic rocks, of both the clinopyroxene and the opaque minerals; it is likely that the same interpretation applies equally to the case studied here. Lacking accurate data on the content of trace elements of minerals involved in the evolution of komatiitic and tholeiitic series, it is, for the moment, impossible to identify more accurately the mechanisms of behaviour of trace elements during fractional crystallization of the two series. However, in respect of their amount and trends of evolution, trace elements in the Tipasjarvi greenstone belt show features very similar to those described in volcanic rocks of other Archaean greenstone belts (for example, see Nesbitt and Sun, op. cit.; Arndt et al., 1977; Naldrett and Turner, 1977).

(e) Rare-earth element characteristics : the REE abundances in 35 basaltic rocks from the three belts (Tipasjarvi, Kuhmo and

Suomussalmi) were determined by the isotopic dilution method at Universite de Rennes. Because the rocks from the Tipasjarvi belt possess most of the features characterizing the evolutionary processes and the source natures, only the results obtained for this belt are discussed. A more detailed account for the REE geochemistry on all three belts will be reported elsewhere (Jahn et al., in prep.). In the following, the REE concentrations have been normalized against the chondritic values determined by Masuda et al. (1973), but further divided by a factor of 1.2 and the results are plotted in the conventional REE variation diagrams (Fig. 9).

In the komatiitic series of the Tipasjarvi belt, three types of REE patterns are recognized :

i) REE patterns with strong LREE depletion, $(La/Sm)_N$ = 0.27 - 0.68, but essentially flat HREE, $(Ga/Yb)_N \sim 1.0$; HREE are 4 to 12 times chondritic abundances. This type of pattern has been found for some Archaean volcanic rocks from the Munro Township, Ontario (Arth et al., 1977) and from Noranda, Ontario (Jahn, unpublished).

ii) REE patterns with slight to moderate LREE depletion but with sloping HREE, $(Ga/Yb)_N$ = 1.2 to 2.0. These two types may be related to each other with a similar mantle source characteristic, i.e. the source has previously undergone a severe LREE depletion. However, type (i) rocks, with flat HREE, seem to be derived by partial melting of the source in which no garnet remained in the residue. In contrast, type (ii) rocks, with sloping HREE and smaller degree of LREE depletion, require that they be derived from a similar source but with garnet in the residue. For type (i) rocks, the four patterns may be derived by various degrees of partial melting. However a crystal fractionation model can equally be applied. Using proper distribution coefficients for REE in basaltic melt, it can be estimated that 30% of olivine removal is required to produce rock S 831 from a melt of composition equivalent to rock S 847. In turn, 60% of fractionation of olivine and pyroxene (1:1 ratio) is required to produce rocks S 818 and S 828 from S 831. Plagioclase separation, if any, is rather insensitive in increasing the HREE abundances. Moreover, the role of plagioclase is probably insignificant because both S 818 and S 828 have smaller negative Eu anomalies than both S 831 and S 847. For type (ii) rocks, because of their similar MgO contents (13.5 and 12.5%) but very different REE abundances, a model of different degree of partial melting can better explain the results.

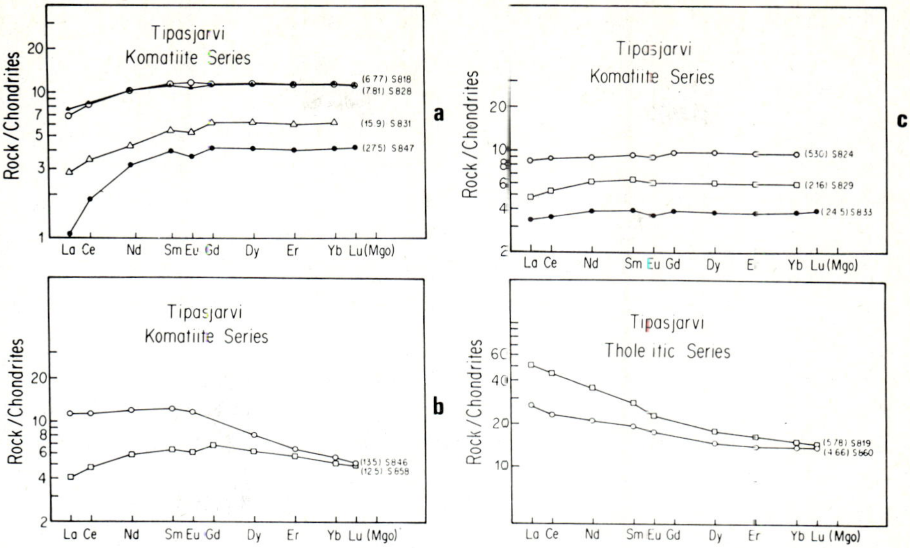

Fig. 9. Chondrite normalized rare earth distribution patterns for komatiitic and tholeiitic rocks from the Tipasjarvi greenstone belts.

(iii) REE patterns of about 4-10 x chondrites with very slight LREE depletion, $(La/Sm)_N$ = 0.75 to 0.90, and essentially flat HREE $(Ga/Yb)_N \sim 1.0$; included in this group is the rock S 833 in which a microspinifex texture is found. In a crystal fractionation model, to produce rock S 824 from S 829, 40% of fractionation of olivine + clinopyroxene (1:1 ratio) is required, and from S 833 to S 829, 40% of olivine fractionation. As in the case of types (i) and (ii), type (iii) rocks could also be derived from different degrees of partial melting from a mantle source with slight LREE depletion; further, garnet has not remained in the residue.

In the tholeiitic series, the REE patterns are characterized by the enriched LREE and sloping HREE $(La)_N$ = 26-52 x, $(Lu)_N$ = 14-15x. These are comparable to the results of Abitibi tholeiites (Arth et al., 1977). If the two rocks are related by crystal fractionation, clinopyroxene probably played the dominant role. Olivine fractionation is not important because the MgO contents of the liquids are low (MgO = 4.7 and 5.8%); plagioclase separation is not possible because it would cause the $(La/Lu)_N$ ratio to decrease from S 860 to S 819, whereas the opposite is observed. The production of tholeiitic melt could be through the partial melting of a mantle source of relatively undepleted nature.

Arth et al. (1977) formulated a unified model in which the

tholeiitic and the komatiitic series of Munro Township are thought to be genetically related simply because of their intimate spatial relationship. Almost exactly the same conclusion has been obtained from the present work. The greenstone belt of Tipasjarvi has a long dimension of about 10 km and all samples were collected within a distance of about 5 km. Most of the tholeiitic rocks from this and the Kuhmo belt have $(La/Sm)_N$ and $(Ga/Yb)_N$ ratios greater than 1.0. Early extraction of tholeiitic liquids from a mantle source characterized by a flat chondritic REE pattern would leave the residue strongly depleted in LREE and $(Gd/Yb)_N$ slightly less than 1.0. This residual mantle may serve as the source for some komatiitic rocks of types (i) and (ii) described earlier. Subsequent melting of this residue, involving garnet or not in the second generation residue, produces the variance in the REE patterns of types (i) and (ii).

It is important to note that the highly LREE-depleted nature of the mantle source is found for only the second time in Archaean greenstone belts, the first case being the Abitibi (Arth et al., 1977; Sun and Nesbitt, 1977 ; Jahn, unpublished). Whether this type of depletion is world-wide remains to be explored.

However, the significance of this depletion must be emphasized. If the depletion is world-wide and large scale, this phenomenon may have to be related with the known and tremendous production of granitic liquid from the mantle sources. The granitic basement rocks surrounding the greenstone belts are emplaced pene-contemporaneously (Vidal and Blais, 1977). If the belts and granitic basement are genetically related, the depletion nature is then due to the earlier separation of granitic liquids, rather than from the extraction of tholeiitic liquids as mentioned before.

The depletion nature of the mantle source has another aspect of importance. It clearly indicates that the mantle heterogeneity has existed ever since at least 2.7 b.y. ago. The available Sr isotopic data show that the initial Sr^{87}/Sr^{86} ratios in the 2.7 b.y. terrains are variable. This suggests that the mantle heterogeneity was already created at least 3.5 b.y. ago (see Jahn and Nyquist, 1976).

CONCLUSIONS

The study of Archaean greenstone belts in eastern Finland, initiated three years ago, is still very fragmentary and the comparison between the interpretations based on the major element geochemistry and the rare earth element geochemistry may occasionally be contradictory, as in the case of the tholeiitic series.

However, even at this early stage, it is possible to present certain conclusions : the Archaean greenstone belts reported as such in the Baltic Shield in 1976 are the first so identified in Europe; they are very similar in many respects to the other Archaean greenstone belts described in Africa, North America, Australia and India (Windley, 1973; Hunter, 1974; Anhaeusser, 1975; Glikson, 1976 and others). The formation of the Finnish greenstone belts appears to be compatible with a model combining the proto-oceanic rift model of Windley (1973) and the ensimatic island-arc model of Anhaeusser (1973). However they show features which distinguish them from more classical greenstone belts : the regional metamorphism is of higher grade than in the other belts, and the erosion seems to have been deeper than in other greenstone belts because the lower sequence constitutes the essential part of the outcrops. In consequence, the Finnish greenstone belts may provide a better chance for the study of deeper levels of the Archaean greenstone belts.

Another important characteristic is that these belts are the super-structure of vast orogens, at least on the scale of the Baltic Shield. The fact that the gneissic basement and the belts themselves all formed within about 100 m. y. of each other is evidence for the extreme rapidity of geodynamic processes. Finally, it is clear that in Finland the mantle showed an important heterogeneity, at least 2.7 b. y. ago or earlier.

REFERENCES

Anhaeusser, C.R., 1973. The evolution of the early Precambrian crust of Southern Africa. Phil. Trans. R. Soc. London A, 273: 359-388.

Anhaeusser, C.R., 1975. Precambrian tectonic environments. An. Rev. Earth Planet. Sci., 3: 31-53.

Arndt, N.T., 1976. Melting relations of metamorphic lavas (komatiites) at 1 atm and high Pressure. Carng. Inst. Yb, 75: 555-562.

Arndt, N.T., Naldrett, A.J. et Pyke, D.R., 1977. Komatiitic and Iron-rich tholeiitic lavas of Munro Township, Northeast Ontario. J. Petrol, 18: 319-369.

Arth, J.G., Arndt, N.T. and Naldrett, A.J., 1977. Genesis of Archaean komatiites from Munro Township, Ontario : trace element evidence. Geology, 5: 590-594.

Blais, S., Noiret, G., Auvray, B., Capdevila, R. et Hameurt, J., 1976. Un nouvel exemple de ceinture archeenne de roches vertes : la ceinture ophiolitique de Suomussalmi (Finlande). Reun. An. Sci. Terre, Paris, 56.

Blais, S., Auvray, B., Capdevila, R., Jahn, B.M., Hameurt, J. 1977. Geochimie des series volcaniques des ceintures archeennes de roches vertes de Finlande orientale. Reun. Ann. Sci. Terre, Rennes, 82.

Blais, S., Auvray, B., Bertrand, J.M., Capdevila, R., Hameurt, J. and Vidal, Ph., in press a. Les grands traits geologiques de la ceinture archeenne de roches vertes de Suomussalmi-Kuhmo (Finlande orientale). Bull. Soc. geol. Fr.

Blais, S., Auvray, B., Capdevila, R. and Hameurt, J. in press b. Les series komatiitiques et tholeiitiques des ceintures archeennes de roches vertes de Finlande orientale. Bull. Soc. geol. Fr.

Bowes, D.R., 1976. Archaean crustal history in the Baltic shield. in : B.F. Windley (Editor), The Early History of the Earth. Wiley, London, pp. 481-488.

Clarke, D.B., 1970. Tertiary basalts of Baffin Bay : possible primary magma from the mantle. Contr. Mineral. Petrol., 25: 203-224.

Dietrich, V., Emmermann, R., Keller, J. and Puchelt, H., 1977. Tholeiitic basalts from the Tyrrhenian sea floor. Earth Planet. Sci. Lett., 36: 285-296.

Gaal, G., Mikkola, A. and Soderholm, B., 1976. Development of the Archaean crust in Finland. 25th Intern. Geol. Congress, Sidney, 1, p. 8.

Gaal, G., Mikkola, A. and Soderholm, B., in press. Evolution of the Archaean crust in Finland. Precambrian Res.

Glikson, A.Y., 1976. Stratigraphy and evolution of primary and secondary greenstones : significance of data from shields of the Southern hemisphere. In: B.F. Windley (Editor), The Early History of the Earth. Wiley, London. pp. 257-277.

Hunter, D.R., 1974. Crustal development in the Kaapvaal Craton, I, The Archean. Precambrian Res., 1: 259-294.

Hypponen, V., 1973. Suomen geologinen Kartta. Lehti 4412, Hiisijarvi.

Hypponen, V., 1976. Suomen geologinen Kartta. Lehti 4411, Ontojoki.

Jahn, B.M. and Nyquist, L.E., 1976. Crustal evolution in the Early Earth-Moon System : Constraints from Rb-Sr studies. In: B.F. Windley (Editor), The Early History of the Earth. Wiley, London. pp. 55-76.

Kouvo, O. et Tilton, F.R., 1966. Mineral ages from the Finnish Precambrian. J. Geol., 74: 421-442.

Lobach-Zhuckenko, S.B., Krylov, I.N., Baykova, V.S., Chekulaev, V.P., Arestova, N.A. et Shuleshko, I.K., 1976. The geological evolution of the Karelian greenstone terrain of the Baltic shield. 25th Intern. geol. Congress, Sidney, 1: 16-17.

Masuda, A. and Nakamura, N., 1973. Fine structures of mutually normalized rare-earth patterns of chondrites. Geochim. Cosmochim. Acta, 37: 239-248.

Matisto, A., 1958. Suomen geologinen yleiskartta, Lehti D 5., Suomussalmi. Kivilajikartan Selitys.

Mutanen, T., 1976. Komatiites and komatiite provinces in Finland. Geologi, 28, 4-5: 49-56.

Naldrett, A.J. and Turner, A.R., 1977. The geology and the petrogenesis of a greenstone belt and related nickel sulfide mineralization at Yakabindie, Western Australia. Precambrian Res., 5: 43-103.

Nesbitt, R.W. and Sun, S.S., 1976. Geochemistry of Archean spinifex - textured peridotites and magnesian and low-magnesian tholeiites. Earth Planet. Sci. Lett., 31: 433-453.

Sederholm, J.J., 1897. Uber eine archaische sediment formation in Sudwestlichen Finnland und ihre Bedeutung fur die Erklarung der Entstehungsweise des Grundgebierges. Bull. Comm. geol. Finl., 6.

Sederholm, J.J., 1932. On the geology of Fennoscandia. Bull. Comm. geol. Finl., 98.

Simonen, A., 1971. Das finnische Grundegebirge. Geol. Rundschau, 60, 4: 1406-1421.

Sun, S.S. and Nesbitt, R.W., in press. Petrogenesis of Archean ultrabasic and basic volcanics. Evidence from rare earth elements. Contr. Miner. Petrol.

Suslova, S.N., 1976. Komatiites des series volcaniques metamorphiques du Precambrien inferieur de la Peninsule de Kola (en russe). Dokl. Acad. Nauk. SSSR, 228, 3: 697-700.

Vartianen, H., 1970. Schist belt of Tipasjarvi in the Parish of Sotkamo, Finland. Bull. geol. Soc. Finl., 42: 13-22.

Vidal, Ph. and Blais, S., 1977. Donnees geochronologiques sur la ceinture de roches vertes de Suomussalmi (Finlande). Reun. Ann. Sci. Terre, Rennes, 459.

Viljoen, M.J. and Viljoen, R.P., 1969a. The geology and geochemistry of the lower ultramafic unit of Onverwacht Group and a proposed new class of igneous rock. Spec. Publ. geol. Soc. Afr., 2: 55-85.

Viljoen, M.J. and Viljoen, R.P., 1969b. Evidence for the existence of a mobile extrusive peridotitic magma from the Komati formation of the Onverwacht Group. Spec. Publ. geol. Soc. Afr., 2: 87-112.

Wilkman, W.W., 1924. Suomen geologinen yleiskartta, Lehti D 4., Nurmes Vuorilajirkartan Selitys. Geologinen Kommission.

Windley, B.F., 1973. Crustal development in the Precambrian. Phil. Trans. R. Soc. Lond., A273: 321-341.

GEOLOGY AND GEOCHEMISTRY OF THE HOLENARASIPUR SCHIST BELT AND ITS PLACE IN THE EVOLUTIONARY HISTORY OF THE INDIAN PENINSULA

S.M. Naqvi
National Geophysical Research Institute, Hyderabad, India

S. Viswanathan
Geological Survey of India, Hyderabad, India

and M.N. Viswanatha
Atomic Minerals Division, Hyderabad, India

Abstract

The Holenarasipur schist belt is one of the most critical, complicated and oldest (3.2-3.5 by) supracrustal belts in the Dharwar craton of India. Two lithostratigraphic groups, the Sargur and Dharwar, are separated by an angular unconformity. The Sargur group starts with a basic-ultrabasic base which is overlain by metasediments of various compositions, whereas the Dharwar group develops upward from a basal conglomerate into a sequence of amphibolites, quartzites and banded magnetite quartzites. The Dharwar belt as a whole is metamorphosed to an amphibolite facies grade, but the Sargur group shows a higher grade (kyanite zone). Recumbent, isoclinal, doubly plunging folds are common, but a regional northward plunge is dominant. Unlike the Dharwars there is no conglomerate at the base or anywhere in the succession of the Sargur group. The meta-ultramafics/mafics of the Sargur group are in places interbedded with fuchsite quartzite and show deformed pillow lavas and microspinifex textures. Their composition is similar to peridotitic, pyroxenitic and basaltic komatiites. The anorthosites in the ultramafites are highly calcic (An 97%) and resemble lunar anorthosites in many respects. Metapelites are enriched in Al, Mg, Fe, Ti, Ca, Ni, Cr and depleted in Si, K, Rb and Sr. The quartzites of the Sargur group are mature sandstones. The basic rocks (amphibolites) of the Dharwar group are tholeiitic. From our geochemical data and from a comparison with those of younger formations, we suggest that the Sargur group of the Holenarasipur schist belt consists of a volcano-sedimentary sequence which was probably not repeated in subsequent geological periods. It is difficult to consider this volcano-sedimentary complex as a platformal or geosynclinal sequence; instead, the belt probably formed in a "nuclear stage" of the Dharwar craton. Platforms and geosynclines appear to be phenomean common in the Proterozoic of the Indian Shield, when the sialic crust had matured and stabilised. The vast post-Sargur tonalitic activity appears to have provided sufficient moderate stability to generate thin and small platformal sediments at the top of the "nuclear stage" formations. These tonalites together with later granites and pegmatites appear to have added Si, K, Rb and Sr to other rocks.

INTRODUCTION

Among the lower members of the greenschists of the Dharwar craton the Holenarasipur schist belt is one of the best exposed, most critical and relatively well studied. For various geological reasons it is considered to represent one of the oldest (3.2-3.5 by?) greenschists of the craton. Though radioactive ages are still not available, it is inferred that the schist belt is older than 3.2 by from the 3.2 by age of the tonalitic pebbles in the Kaldurga conglomerate (Venkatasubramanian and Narayanaswamy, 1974) in the Chitradurga group of the Bababudan schist belt, and from the intrusive relationship of similar tonalites with the Holenarasipur schist belt (Naqvi et al., 1978a). The schist belt consists of two sequences separated by a break: the lower sequence is considered to be a member of the Sargur group, the upper is thought to belong to the Dharwar group (Viswanatha and Ramakrishnan, 1975). Both groups are highly deformed, metamorphosed to amphibolite facies and are made up of ultramafics-mafics, anorthosites, metasediments and gneisses. Though the area concerned has been studied for the past 80 years by different workers (Rama Rao, 1962), many problems are unsolved and it still remains a very critical and controversial area in the Dharwar craton. Because it contains a varied rock assemblage of great antiquity and geochemical interest and because quite considerable geological data are available, the area was taken up by the working groups of the GSI, NGRI and AMD for detailed geological and geochemical studies in order to find out (a) the nature, composition and origin of the constituent rock types, (b) their similarities and dissimilarities with those of younger formations, (c) the relationship with the gneisses, and (d) to elucidate the nature and composition of the Archaean crust of the Dharwar craton. 77 samples were chosen for petrological and geochemical studies.

GEOLOGY OF THE HOLENARASIPUR SCHIST BELT

1. Stratigraphy and structure

The cusp-shaped schist belt is surrounded by tonalitic and trondhjemitic gneisses which contain ultramafic and mafic enclaves of various shapes and sizes. The stratigraphic succession envisaged by Viswanatha and Ramakrishnan (1975) is given in Table 1. The lower sequence (viz. the Sargur group of rocks) consists mainly of tremolite-actinolite-talc schists, chlorite schists, chloritoid-magnetite schists, serpentinites, dunites, peridotites, anorthosites, staurolite-kyanite-garnet schists, mica-quartz schist, quartzites (?), garnetiferous and non-garnetiferous amphibolites and garnet-grunerite-magnetite-hornblende-quartzites.

TABLE 1. Stratigraphic column for the Holenarasipur schist belt

	LITHOLOGY		GENETIC DESCRIPTION
INTRUSIVES		Dolerite and gabbro amphibolites Pegmatites and schorl veins	Acid and basic intrusives (Post to late tectonic igneous activity)
DHARWAR SUPERGROUP	Mallappan-Betta Formation	Meta-ultramafics, ironstones, chlorite-biotite-garnet schist with Dodgudda and Sannenahalli conglomerate	Ultramafic intrusives Exhalative sediments Ultramafic activity
	Satnahhalli Formation	Metavolcanic rocks with associated quartzites, pelitic schists and ultramafics, Kunkuma Hosur conglomerates	Contemporaneous volcanism and sedimentation Oligomict quartz-pebble conglomerate (basal)
		UNCONFORMITY	
	Peninsular Gneissic Complex	Migmatite and granite	Pantectogenesis
SARGUR SUPERGROUP TIRUMAPUR GROUP	Yenneholerangan-betta ultramafic	Serpentinite, metaperidotite, pyroxenite and dunite	Metamorphosed ultrabasics
	Biranahalli iron Formation	Garnet-grunerite quartzite and hornblende quartz garnet rock	Metasediments
	Karali amphibolite	Para-amphibolite and calc silicate rocks, amphibolites, garnet amphibolites and meta-anorthosites	Ortho- and para-amphibolites
	Tivadahalli Formation	Tattekere conglomerate, kyanite-staurolite-garnet-biotite-muscovite schist	Oligomict conglomerate with associated psammites, pelites and psammo-pelites
		Base possibly gneissic complex (migmatite and granitoids)	

The upper sequence, namely the Dharwar group (Bababudan group), consists of amphibolite and interbedded ripple-marked quartzites, sericitic phyllites and banded magnetite quartzites. These two groups are separated by a basal conglomerate (?) made up of quartzite pebbles and a fuchsite quartzite matrix (Viswanatha and Ramakrishnan, 1975). The stratigraphy of the schist belt is not agreed between the different workers of the working group. Viswanatha and Ramakrishnan (1975) consider that the metasediments are the lowermost members of the Sargur group and that the mafic-ultramafic units intrude into them. The ultramafic units in most places are directly in contact with the gneissic tonalites. The exhibit deformed pillow structures, are found interbedded with fuchsite quartzite, and show micro-spinifex textures. There are pillow laves in the same horizon in the Nuggihalli schist belt, another member of the Sargur group about 50 miles NE of this belt (Srinivas and Sreenivas, 1972), and in the

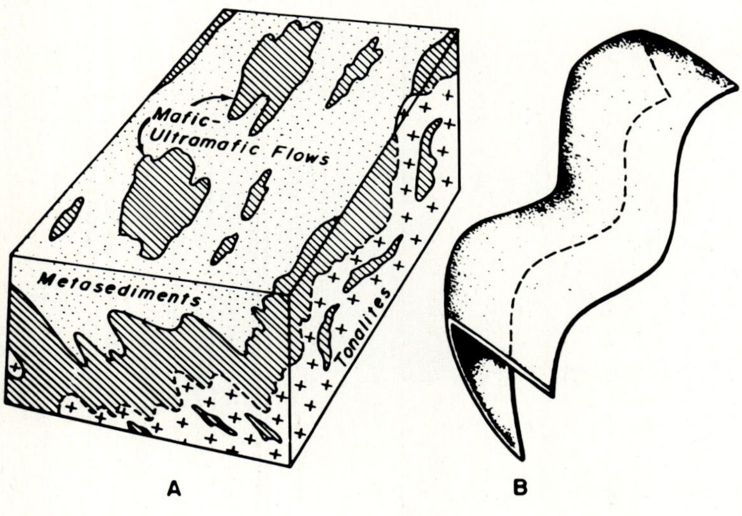

Fig. 1A. Schematic three-dimensional model of Yenneholeranganabetta Hill. The model shows how the lowermost ultramafic member is misinterpreted as an intrusion in the anticlinal cores.

1B. Type of folding predominent in the area.

same horizon of the Sargur group there are spinifex textures at Kummanghata in the Ghatti Hosahalli band near Chitradurga (Viswanatha, 1977; Viswanatha et al., 1977). Therefore the ultramafics (Sargur group) of the Holenarasipur schist belt are most probably subaqueous flows and not intrusive (Naqvi, 1978; Naqvi et al., 1978b). The area exhibits isoclinal recumbent doubly plunging folds (Fig. 1) in the cores of which mafic and ultramafic flows are exposed, these flows thus being the lowermost members of the Sargur group of the Holenarasipur schist belt. The fact that these mafic-ultramafic units are exposed in anticlinal cores was identified by Viswanatha (1968) and later by Ramakrishnan and Viswanatha (1972). Around Yenneholeranganabetta Hill Viswanatha (1968) recognised two major anticlines with a syncline in between. He found that the western anticline was overturned to the west with serpentine in the core, while the eastern anticline had another anticlinal flexure along which serpentine was also emplaced; further east there is another overturned syncline with a core of banded magnetite. Since the ultramafic-mafic rocks are now proved to be subaqueous flows they cannot be exposed in the anticlinal cores unless they represent the lowermost or the oldest members of the schist belt (Naqvi, 1978; Naqvi et al., 1978b). Therefore the stratigraphic succession of Viswanatha and Ramakrishnan (1975, p.53) needs to be revised. The stratigraphic succession proposed by Hussain and Naqvi (Geochemistry Group, 1977) is given in Table 2.

TABLE 2. Geological succession of the Holenarasipur schist belt

11. Dykes
10. Pegmatite
9. Tonalites
8. Ironstones (banded magnetite quartzite)
7. Basic metavolcanics and interbedded quartzites
6. Basal conglomerate

- - - - - - - - - - Unconformity - - - - - - - - -

5. Pegmatite
4. Tonalites and trondhjemites
3. Ironstones (garnet-grunerite-magnetite quartzite)
2. Metasediments (kyanite-staurolite-garnet-mica quartz schist and metacherts, i.e. fuchsite-quartzite)
1. Basic-ultrabasic subaqueous flows including anorthosites with interbedded metacherts (fuchsite quartzite)

- -

Basic-ultrabasic primordial crust

The entire area including the surrounding gneisses has been subjected to a phase of deformation (2.6 by) which produced a N-S structural grain in almost every rock unit in the south and north; however, the western part of the area has escaped this trend and even the gneisses here show a swing in their trend to E-W at the margin of this part of the schist belt. The parallelism between the stress direction and differences of physical properties and response to the stress between gneisses and sediments may explain this swing.

2. Metamorphism

The structural complexities of the area are reflected in the metamorphic grade. The metamorphic mineral parageneses indicate that the Sargur group has a higher metamorphic grade than the Dharwar group (Viswanatha, 1968; Ramakrishnan and Viswanatha, 1972). The Sargur group belongs to the staurolite-kyanite zone of the amphibolite facies, whereas the Dharwar sequence does not contain staurolite and kyanite. Most assemblages are stable in the almandine zone of the amphibolite facies as well as the greenschist-amphibolite transition facies. The amphibolite facies rocks occupy the major portion of the belt, whereas the centre of the belt adjoining the Muddalgudda Hills the rocks show greenschist facies metamorphism. Therefore, Viswanatha (1968) suggested that the metamorphic isograds run parallel to the margins of the schist belt. The mineral paragenesis is a high pressure, low temperature type which probably indicates that compression has been higher at the margins. Recent studies have indicated that serpentinisation results in 60% increase in rock volume (Moeskops, 1977). A rough calculation indicates that this increase in volume will exert 5 Kb pressure at the margins of a body which has undergone serpentinisation. This should be considered to be a means of producing the high pressure, low temperature mineralogy at the margins of the schist belt.

3. Petrology and mineralogy

Since most of the rocks are schistose, modal analysis was carried out on 25 samples of granoblastic anorthosites and tonalites, etc. Petrologically, the rocks of the Sargur group can be divided into ultramafics, anorthosites, mafics, pelites, quartzites and ironstones, and those of the Dharwar group into pelites, quartzites, mafics and ironstones. The meta-ultramafics are serpentinites, dunites and actinolite-tremolite-talc schists. The relics of micro-spinifex texture were identified by Hussain and Naqvi (1978) in the serpentinites of the Sargur group from Holenarasipur. The actinolite-tremolite-talc schists occasionally exhibit relics of clino- and ortho-pyroxenes. Serpentine, magnetite, talc, chlorite, tremolite, anthophyllite, cummingtonite, calcite, graphite and chlorite are found in various assemblages in the ultramafic unit. The anorthosite, anorthositic gabbro and gabbroic anorthosite sheets in these ultramafics are made up of anorthite-bytownite, hornblende, garnet, ilmenite, magnetite and chromite. These anorthosites show a metamorphic granoblastic texture and exhibit extreme variation in grain size; crushing and recrystallisation of the larger plagioclase grains are evident in many sections. The plagioclase is highly calcic (An 97) and resembles that in the lunar anorthosites in its REE distribution and several other aspects (Geochemistry Group, 1977; Drury et al., this vol.). The amphibolites associated with the anorthosites and metamafics also contain calcic plagioclase (An 65-70) and contain Mg-rich amphiboles. The garnet in these rocks is almandine containing inclusions of plagioclase.

The metasediments (pelites, quartzites) are made up of different layers of minerals in varying proportions. The dominant assemblages contain kyanite, staurolite, amphibole, chlorite, biotite, corundum and quartz (Naqvi, this vol.). In these metasedimentary layers there are some thin-layered, medium- to fine-grained amphibolites which may be part of the sedimentary unit of the Sargur group. The quartzites form minor constituents of the metasediments and are mostly made up of quartz, fuchsite or sericite; field and laboratory evidence suggests that they are cherts or secondary quartzites (Naqvi, this vol.). The ironstones comprise garnet, quartz, grunerite and magnetite.

The amphibolites of the Dharwar group are medium to coarse grained, contain amygdules of calcite and quartz and consist of hornblende, plagioclase and epidote. Some amphibolite layers are garnetiferous and contain big crystals of garnet. The interbedded quartzites show relict sandstone textures and consist of quartz and sericite. The ironstones of the Dharwar group show the typical morphology, mineralogy and texture of banded magnetite quartzites. The gneisses show a great variation in abundance of constituent minerals; modal analysis shows they contain quartz, oligoclase, hornblende, epidote, biotite, K-feldspar, garnet (occasionally), apatite, sphene and zircon. There is another group of gneisses which has not developed a migmatitic appearance but which still exhibits a gneissic texture; they are regarded as granites by Ramakrishnan and Viswanatha (1972) and have been named by them as the Holekate granite. They show an undoubted intrusive relationship with the schist belt and are made up of quartz, highly sodic plagioclase, amphibole, biotite, minor K-feldspar and zircon.

3. Geochemistry

The averages of the different petrographic groups found in the Holenarasipur schist belt are given in Table 3. The complete analytical data are illustrated in different figures and are discussed below. The details of the analytical technique and precision of the data are given by Naqvi and Hussain (1972) and the Geochemistry Group (1977). The normative composition of the rocks shows that most of them are undersaturated (Fig. 2). Both quartz and olivine normative anorthosites are found. Anorthosites are anorthite rich whereas gabbroic anorthosites and anorthositic gabbros have relatively lower An contents. A few samples of actinolite-tremolite-schist (metapyroxenite) are quartz normative; this may be the result of addition of SiO_2 during later pegmatitic activities. Metapelites, amphibolites and chlorite schists also show both quartz and olivine normative compositions (Fig. 2). The normative Or-Ab-An ratios indicate that metapelites and gneisses have relatively higher K-feldspar abundances than the other rock types (Fig. 2). However, the petrological and modal studies of the metapelites do not indicate the presence of any K-feldspar. This suggests that the K content of these rocks is most probably present in the micas and amphiboles.

TABLE 3. Average chemical composition of rocks from the Holenarasipur schist belt

	1	2	3	4	5	6	7	8	9
N*	15	9	2	8	10	6	8	7	12
SiO_2	42.85	49.58	51.17	48.32	51.17	52.86	44.86	52.01	68.28
TiO_2	0.15	0.14	0.76	0.60	1.30	0.98	1.73	0.80	0.34
Al_2O_3	3.22	7.54	11.87	21.86	14.70	16.95	18.67	14.71	12.52
MgO	30.18	15.83	7.91	5.47	5.64	4.69	14.42	5.77	3.10
CaO	7.38	9.35	11.25	13.78	9.79	7.85	2.65	10.37	5.18
Fe_2O_3	4.43	2.58	2.38	0.74	2.26	1.33	2.78	2.06	0.98
FeO	5.61	7.32	9.18	6.28	10.47	8.12	6.91	9.98	3.26
Na_2O	1.17	1.79	3.35	2.23	2.36	3.65	1.56	3.91	4.85
K_2O	0.07	0.27	0.29	0.11	0.18	2.02	0.09	0.42	1.40
MnO	0.13	0.20	0.22	0.22	0.29	0.19	0.11	0.24	0.16
P_2O_5	0.09	0.04	0.02	0.04	0.07	0.06	0.06	0.03	0.04
H_2O	6.30	3.08	0.80	0.41	0.93	1.23	4.96	0.78	0.24
TOTAL	101.58	97.72	99.20	99.95	100.06	99.93	98.80	101.08	100.35
Co	93	68	53	31	53	44	95	52	25
Cr	4788	1964	815	273	148	232	549	160	53
Cu	23	17	42	122	109	55	40	107	13
Ni	429	249	131	227	118	119	191	150	29
V	69	122	25	281	272	192	238	245	117
Rb	10	10	10	10	10	44	11	18	54
Sr	10	52	95	159	166	162	19	166	388

* N = number of samples

1. Av. of peridotitic (serpentinite) komatiites
2. Av. of pyroxenitic komatiites (Sargur)
3. Av. of basaltic komatiites
4. Av. of anorthosites
5. Av. of amphibolites (Sargur)
6. Av. of metapelites
7. Av. of chlorite schists
8. Av. of amphibolites (Dharwar)
9. Av. of gneisses

The ratios of the different chemical constituents (Fig. 3) and the abundances of MgO (Fig. 4) indicate that the metavolcanics of this schist belt, following Arndt et al. (1977), can be grouped into peridotitic, pyroxenitic and basaltic komatiites. However, some of the amphibolites (metavolcanics) associated with these komatiites are found to be tholeiitic (Fig. 2), while the anorthosites have both komatiitic and tholeiitic affinities (Figs. 2 and 3). The metapelites show a wide variation in composition (Naqvi, this vol.). Chloritic schist is mainly made up of SiO_2, Al_2O_3 and MgO and is depleted in CaO. The composition of the ironstone of the Sargur group is entirely different, especially in Al_2O_3, MgO and TiO_2 content, from that of the banded magnetite quartzite. The amphibolites of the Dharwar group are low-K tholeiites. The Mg No (Mg/Mg+Fe atomic ratio) of

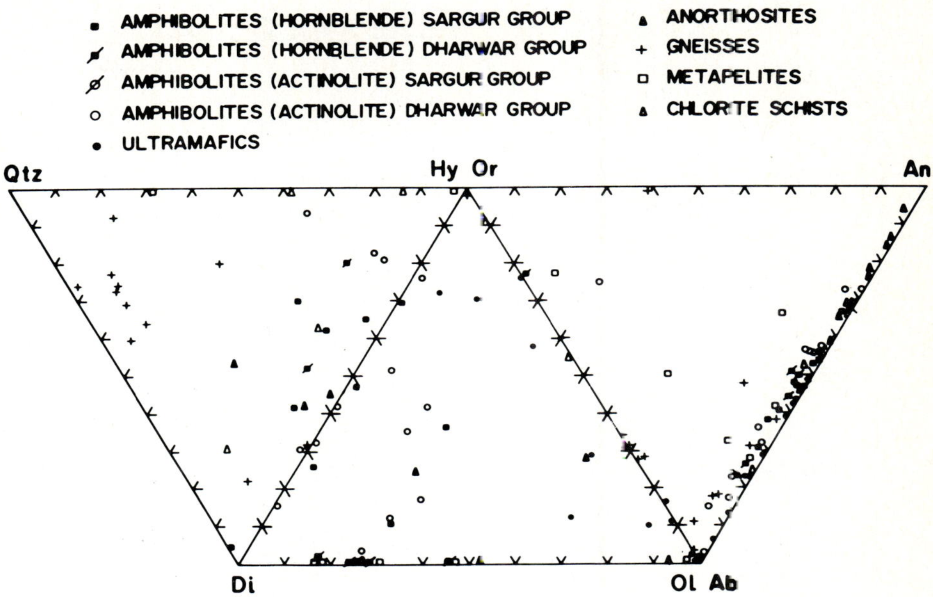

Fig. 2. Normative composition of the analysed samples

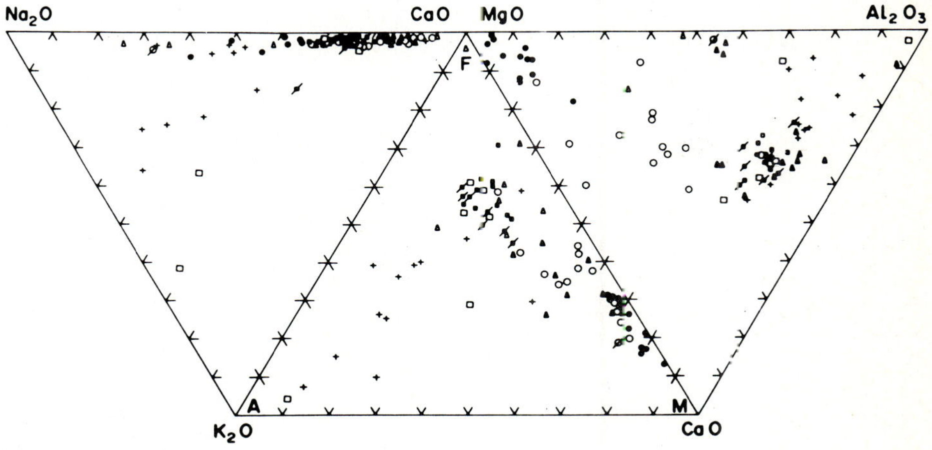

Fig. 3. Ternary oxide ratios of the analysed samples showing the peridotitic komatiitic nature of the serpentinite, pyroxenitic komatiitic composition of actinolite tremolite (amphibolite) schists and basaltic komatiitic nature of some of the hornblendic amphibolites. Dharwar amphibolites are tholeiitic. Note the position of anorthosites (same legend as in Fig 2).

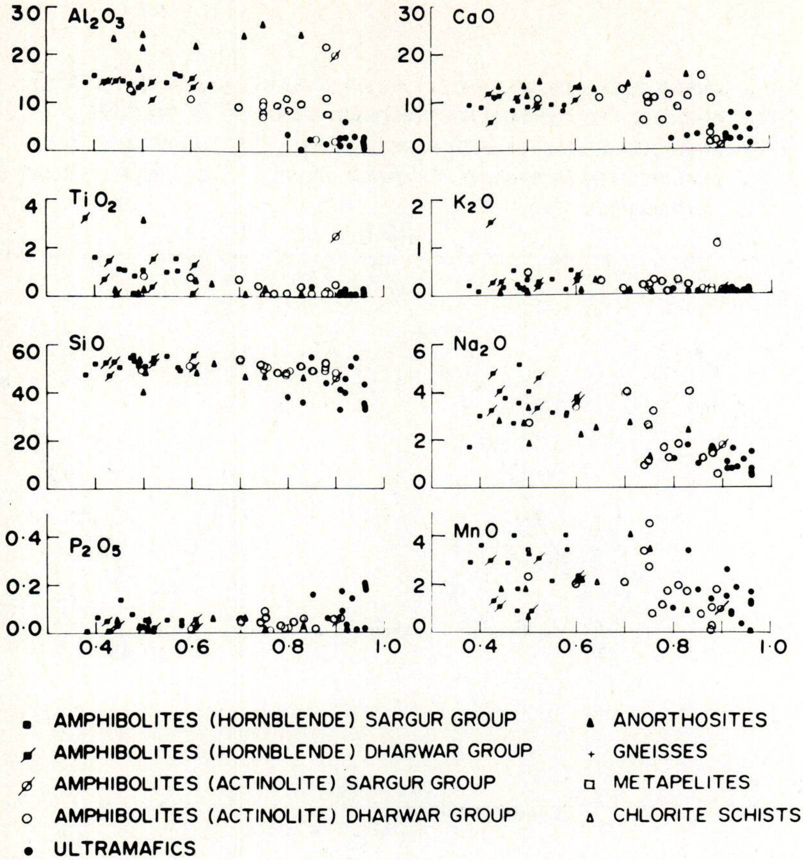

Fig. 4. Distribution of major elements of the igneous member of the Holenarasipur against their Mg No (Mg/Mg+Fe ratio).

the metavolcanics of the Holenarasipur schist belt decreases from 0.95 to 0.3. The Mg No of the peridotitic komatiites is equivalent to that generally known in cumulates, while that of the anorthosites is equivalent to that of the amphibolites (Fig. 4). The abundance of Al_2O_3, SiO_2, TiO_2 and K_2O in the metavolcanics decreases with the increase of Mg No (Fig. 5). CaO seems to have a bimodal distribution. Its maximum concentration is in anorthosite; therefore it increases up to the Mg No of anorthosite separation (?) and decreases at higher Mg No, which denotes olivine separation (Fig. 5). While MnO shows a scatter, P_2O_5 is more abundant in metaserpentinites and pyroxenites (Fig. 3). Co, Cr and Ni show an increase from 0.7 to 0.95 with Mg No, but their abundance remains almost unchanged at lower levels (Fig. 5). Copper appears to be more abundant in lower Mg No rocks. Rb and Sr remain low, especially in metaserpentinites and pyroxenites, but Sr is high in amphibolites, anorthosites (Fig. 5) and gneisses. The gneisses form a different group with respect to K/Rb, Rb/Sr, Ca/Sr and Na/Ca (Fig. 6) and these ratios show no relationship to similar ratios in the metapelites. Mg/Fe, K/Na and CaO/Al_2O_3 ratios also indicate that there is no genetic relationship between the gneisses and metasediments in the Holenarasipur schist belt (Fig. 7).

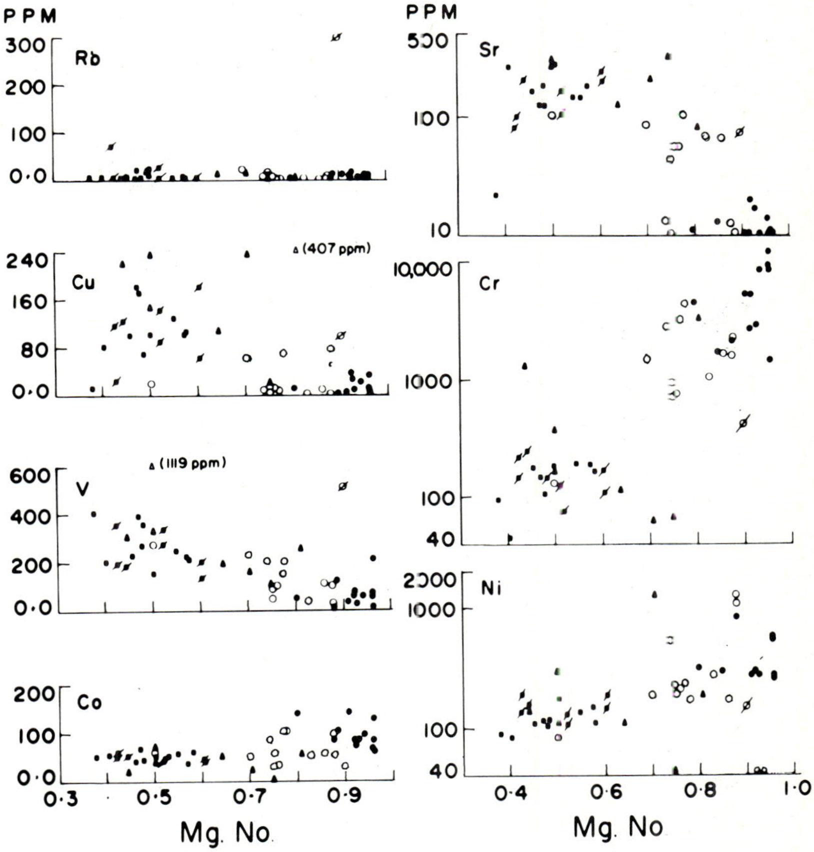

Fig. 5. Distribution of trace elements in the metavolcanics of the Holenarasipur schist belt (same legend as in Fig. 4).

GEOCHEMISTRY OF THE HOLENARASIPUR VOLCANO-SEDIMENTARY ROCKS COMPARED WITH YOUNGER FORMATIONS

The meta-ultramafics and mafics of the Sargur group are significantly different from those of the younger Bababudan and Chitradurga groups (Table 4). The

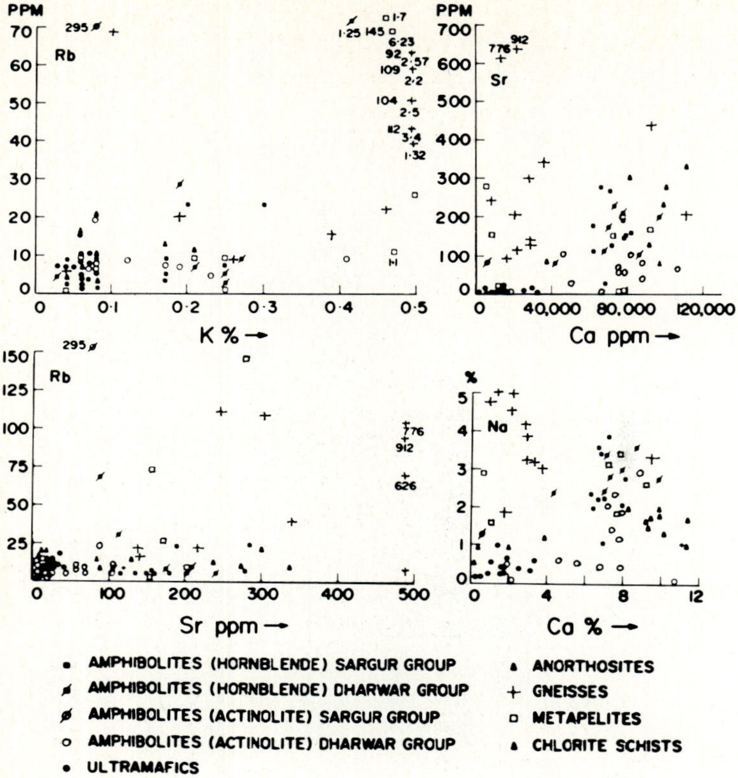

Fig. 6. K-Rb, Rb-Sr, Na-Ca and Sr-Ca plots of rocks from the Holenarasipur schist belt. Note the position of gneisses and pelites.

Bababudan metavolcanics (Bhaskar Rao and Naqvi, this vol.) vary in composition from olivine tholeiites to rhyodacites, whereas those of Chitradurga vary from olivine tholeiites to andesites (Naqvi and Hussain, 1973a,b). The volcanism of the younger belts is much more evolved and fractionated than that of the Archaean (Naqvi, 1976a,b), though the southern part of the Holenarasipur schist belt is correlated with the Sargur (Viswanatha et al.,pers. comm.). The metasediments of the lower parts of the younger belts like Chitradurga consist of current-bedded and ripple-marked quartzites and arkoses, whereas those of the upper parts are made up of greywackes (Naqvi and Hussain, 1972). These differences between the volcano-sedimentary constituents of the Holenarasipur schist belt and younger belts probably reflect different stages in the evolution of the crust.

TABLE 4. Composition of metavolcanics of other schist belts of the Dharwar craton and other shields

	1	2	3	4	5	6	7	8	9	10	11	12
SiO_2	49.04	50.17	50.62	51.78	51.40	52.70	45.11	45.26	51.99	44.72	52.73	43.70
TiO_2	1.12	0.86	1.16	1.40	0.69	0.40	0.31	0.29	0.51	0.52	0.85	0.28
Al_2O_3	14.01	14.17	15.16	14.17	14.85	13.13	5.96	5.80	15.27	3.25	9.83	6.10
MgO	8.81	6.59	6.90	5.76	6.36	7.98	31.50	27.96	9.68	25.25	10.10	27.85
CaO	9.89	11.92	8.08	8.64	12.16	10.70	5.83	7.57	11.65	6.97	9.99	6.25
Fe_2O_3	2.88	2.92	2.45	2.05	1.26	1.20	11.50	8.43	9.33	6.02	1.23	2.94
FeO	9.19	8.32	8.45	8.80	8.55	9.16	-	-	-	5.52	9.70	5.27
Na_2O	2.84	2.09	3.60	2.86	2.19	2.38	0.35	0.37	2.07	0.49	2.65	0.19
K_2O	0.36	0.18	0.31	0.62	0.51	0.16	0.03	0.02	0.19	0.05	0.46	0.02
MnO	0.34	0.42	0.37	0.38	0.53	0.13	0.19	0.19	0.16	0.19	0.22	0.20
P_2O_5	0.49	0.52	0.17	0.24	0.14	0.77	-			-	-	-

1. Av. of 8 Ol normative meta-tholeiites from Sargur schist belt (Geochemistry Group, 1977)
2. Av. of 4 Q normative meta-tholeiites from Sargur schist belt
3. Av. of 10 Ol normative meta-tholeiites from Bababudan schist belt (Bhaskar Rao and Naqvi, 1977)
4. Av. of 11 Ol normative meta-tholeiites from Bababudan schist belt (Bhaskar Rao and Naqvi, 1977)
5. Av. of 50 Chitradurga metabasalts (Naqvi and Hussain, 1973)
6. Av. of 7 ortho-amphibolites from Chitradurga schist belt (Satyanarayana, 1974)
7. Spinifex peridotitic komatiite, Yakbindie, W. Australia (Naldrett and Turner, 1977)
8. Spinifex pyroxenite komatiite, Yakbindie, W. Australia (Naldrett and Turner, 1977)
9. Basaltic komatiites, Yakbindie, W. Australia (Naldrett and Turner, 1977)
10. Av. of peridotitic komatiites (Viljoen and Viljoen, 1969)
11. Av. of basaltic komatiites (Viljoen and Viljoen, 1969)
12. Spinifex peridotites, W. Australia (Nesbitt, 1971)

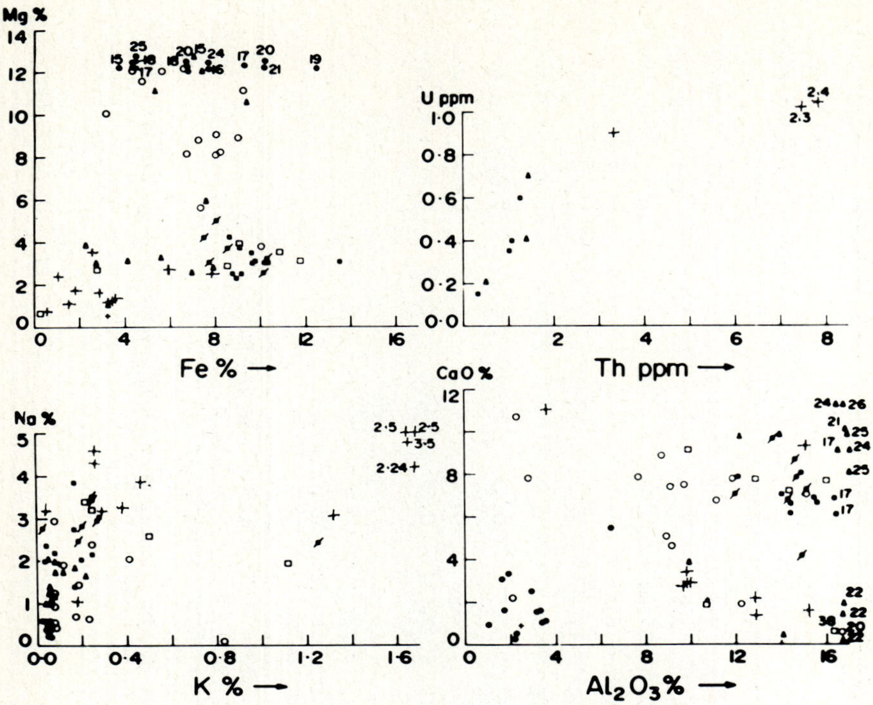

Fig. 7. Na-K, Mg-Fe, CaO-Al_2O_3 and U-Th plots of the rocks of Holenarasipur. Note the position of gneisses and metapelites (same legend as in Fig. 6).

HOLENARASIPUR SCHIST BELT AND ITS PLACE IN THE GEOLOGICAL HISTORY OF THE DHARWAR CRATON

As already discussed, the meta-ultramafics and mafics represent the lowermost members of the Sargur group. Sillimanite, corundum and other high pressure-high alumina minerals are found in the lower successions of greenstone belts in Barberton, the Aldan Shield and elsewhere (Salop, 1968; Viljoen and Viljoen, 1969; Anhaeusser, 1971, 1972; Glukhovskiy and Pavlovskiy, 1973). The composition of the subaqueous volcanics of Holenarasipur is also similar to that of komatiites (Table 4) of greenstone belts. Therefore this schist belt is perhaps not different from the lower greenstone belts of Barberton, Aldan or Australia, except probably for a slightly higher grade of metamorphism.

The bottom of the schist belt has no conglomerate. The so-called Tattakere conglomerate has been studied by many workers since 1898, most of whom found that it is nothing more than a rodded fuchsite quartzite (Rama Rao, 1924, 1962; Sreenivas and Srinivasan, 1968; Naqvi et al., 1978b). Conditions during the early history of the earth are commonly regarded as more catastrophic with rapid changes within and over the earth. These conditions should generate products of rapid weathering and deposition, like conglomerates and arkoses. Therefore, if there were a pre-existing tonalitic or sialic crust, conglomerates containing tonalitic pebbles should be deposited at the base or anywhere in the Sargur group succession, as they are in the upper greenstone belts of Chitradurga, Shimoga and elsewhere (Naqvi et al., 1978b). Alternatively, rudaceous and arenaceous sediments might occur at the base of the belt, but their absence from the Sargur Supergroup is noted by Ramakrishnan and Viswanatha (1972) who stated that "the normal association of rudaceous and arenaceous rocks, which should have underlain this (pelite) unit in the geological milieu is not found in this belt and may have to be looked for elsewhere in the vicinity" (Ramakrishnan and Viswanatha, 1972; Ramakrishnan et al., 1976). This rudaceous and arenaceous component should have been below the metapelites. Naqvi (this vol.) pointed out that there appears to be no genetic relationship between the gneisses and sediments of the Sargur group; therefore it is more logical to infer that the tonalitic gneisses and their precursor tonalites are, as a whole, a post-Sargur phenomenon and that they pre-date the tectonic event dated at 2.6 by.

GEOSYNCLINAL-PLATFORMAL OR NUCLEAR STAGE

Many workers have made assumptions about platformal and geosynclinal conditions and have applied Phanerozoic types of tectonism to the early stages of earth history, but it is difficult to fit available data on the Archaean of India into either a platformal or geosynclinal concept. The platformal volcano-sedimentary sequences have their own characteristic rock types and abundances (Garrels et al., 1971; Ronov and Migdisov, 1971; Veizer, 1973; Sibley and Wilband, 1977). The basal conglomerates and mature quartzites, so abundant in platformal sediments, are entirely missing from these belts. Platformal volcanism is well evolved, as in nepheline normative basalts (Naqvi, this vol.), but in these belts such volcanism is replaced by abortive volcanism, i.e. by highly undifferentiated melts. Similarly, the volcano-sedimentary piles of geosynclines dominated by greywackes, lutites, greywacke conglomerates and evolved volcanism of olivine tholeiite to rhyolite are absent. Therefore terms like platforms and geosynclines are not applicable to at least the lower and middle Archaean greenstone belts. On the other hand, it appears more appropriate (Glukhovskiy and Pavlovskiy, 1973) that these Archaean volcano-sedimentary belts represent a nuclear stage in the earth's development in which platforms and geosynclines did not exist. Hoffman (1973) has convincingly shown that the history of geosynclines started 2.5 by ago. Some early geosynclines are inverted and they do not contain all the features of typical geosynclines (Turner and Walker, 1973). In fact, global studies carried out by sedimentologists and sedimentary geochemists (Garrels and Mackenzie, 1971;

Ronov and Migdisov, 1971; Ronov, 1972; Veizer, 1973; Veizer and Hoefs, 1975) have shown that the early Proterozoic was dominated by geosynclines, and the middle and late Proterozoic became the era of platforms. Therefore considerable caution is urged in the interpretation of Archaean data with regard to concepts which have mainly developed on the basis of observations from Phanerozoic and middle/late Proterozoic rocks.

ACKNOWLEDGEMENTS

The joint programmes of the working group of the National Geophysical Research Institute, the Geological Survey of India and the Atomic Minerals Division, the results of which are presented in this paper, could never have succeeded without the coordination of Dr M.V.N. Murthy. Dr Hari Narain, Mr V.K.S. Vardan and Dr G.R. Udas gave full independence and encouragement to the working group and provided all facilities and funds.

Field studies and the mapping of the area were carried out by M.N. Viswanatha, M. Ramakrishnan, S.M. Naqvi and S.M. Hussain, and samples were collected by M.N. Viswanatha, M. Ramakrishnan, S. Viswanathan, S.M. Naqvi and S.M. Hussain. Petrological studies were carried out by T.V. Viswanathan and Vasudeva. Major element analysis was made by S.M. Naqvi, K. Chandrashekar, Y.J. Bhaskar Rao, S.H. Jafri, S.M. Ahmad and T. Gnaneshwar Rao, while trace element analysis was made by S. Viswanathan, A.V. Patwardhan, R. Rangaswamy, R.N. Gupta, S.K. Saxena, N. Nagrraja Rao, Y.N. Lal and B.S. Atal.

REFERENCES

Anhaeusser, C.R., 1971. Cyclic volcanicity and sedimentation in the evolutionary development of Archaean greenstone belts of shield areas. Sp. Publ. geol. Soc. Aust., 3: 57-70.

Anhaeusser, C.R., 1972. The geology of the Jamestown Hills area of the Barberton Mountainland, South Africa. Trans. geol. Soc. S.Afr., 75: 225-263.

Arndt, N.T., Naldrett, A.J. and Pyke, D.R., 1977. Komatiitic and iron-rich tholeiitic lavas of Munro Township, Northeast Ontario. J. Petrol., 18: 319-369.

Bhaskar Rao, Y.J. and Naqvi, S.M. (this vol.) Geochemistry of metavolcanics from Bababudan schist belt, a late Archaean/early Proterozoic geosynclinal pile from India.

Drury, S.A., Naqvi, S.M. and Hussain, S.M. (this vol.) REE distribution in basaltic anorthosites from the Holenarasipur greenstone belt, Karnataka, South India.

Garrels, R.M. and Mackenzie, F.T., 1971. Evolution of sedimentary rocks: a geochemical approach. Norton, New York. 397 pp.

Garrels, R.M., Mackenzie, F.T. and Siever, R., 1971. Sedimentary cycling in relation to the history of the continents and oceans. In: E.C. Robertson, J.F. Hay and L. Knapoff (Editors), The Nature of Solid Earth. pp. 93-121.

Geochemistry Group, 1977. Archaean Geochemistry. Contrib. and Prog. NGRI, Bull. geophys. Res., 15, 54 pp.

Glukhovskiy, M.Z. and Pavlovskiy, Ye.V., 1973. Early stages of the evolution of the earth. Geotectonics, 2: 61-63.

Hoffman, P., 1973. Evolution of an early Proterozoic continental margin: the Coronation geosyncline and associated aulacogens of the northwestern Canadian shield. Phil. Trans. R. Soc. Lond., A273: 547-581.

Hussain, S.M. and Naqvi, S.M., 1978. Geochemistry of Archaean basaltic and peridotitic komatiites from the Holenarasipur schist belt (pers. comm.).

Moeskops, P.G., 1977. Volume increase serpentinisation in Archaean quench texture ultramafic rocks near Kalgoorlie, Western Australia. Geol. Mag., 114: 41-46.

Naldrett, A.J. and Turner, A.R., 1977. The geology and petrogenesis of a greenstone belt and related nickel sulfide mineralisation at Yakabindie, Western Australia. Precamb. Res., 5: 43-103.

Naqvi, S.M., 1976a. Physico-chemical conditions during the Archaean as indicated by Dharwar geochemistry. In: B.F. Windley (Editor), The Early History of the Earth. Wiley, Lond., pp. 289-298.

Naqvi, S.M., 1976b. Distribution of elements in the crust and mantle during the Archaean. Abstr. 25th Int. Geol. Congr. Sydney.

Naqvi, S.M., 1978. Discussion on possible spinifex texture in a serpentinite from Karnataka (pers. comm.).

Naqvi, S.M. (this vol.) Geochemical balance between Archaean sediments and igneous activity, evidence for ANT being prominent in the Archaean basaltic crust.

Naqvi, S.M. and Hussain, S.M., 1972. Petrochemistry of early Precambrian metasediments from the central part of the Chitaldrug schist belt, Mysore, India. Chem. Geol., 10: 109-135.

Naqvi, S.M. and Hussain, S.M., 1973a. Geochemistry of Dharwar metavolcanics and composition of the primeval crust of Peninsular India. Geochim. Cosmoch. Acta, 37: 159-164.

Naqvi, S.M. and Hussain, S.M., 1973b. Relation between trace and major element composition of the Chitaldrug metabasalts, Mysore, India, and the Archaean mantle. Chem. Geol., 11: 17-30.

Naqvi, S.M., Divakara Rao, V., Rogers, J.J.W., Satyanarayana, K., Hussain, S.M. and Narayana, B.L., 1978a. The petrochemistry and geologic implications of conglomerates from Archaean geosynclinal piles of southern India. Can. J. Earth Sci. (pers. comm.).

Naqvi, S.M., Divakara Rao, V. and Hari Narain, 1978b. The primitive crust; evidence from the Indian shield. Precamb. Res. (in press).

Nesbitt, R.W., 1971. Skeletal crystal forms in the ultramafic rocks of the Yilgarn block, Western Australia: evidence for an Archaean ultramafic liquid. Sp. Publ. geol. Soc. Aust., 3: 331-347.

Rama Rao, B., 1924. Conglomerates and quartzites of the Holenarasipur schist belt. Rec. Mysore Geol. Dept., 21(2).

Rama Rao, B., 1962. A Handbook of the Geology of Mysore State, Southern India. Bangalore Printing and Publ. Co. Ltd., Bangalore, 264 pp.

Ramakrishnan, M. and Viswanatha, M.N., 1972. Holenarasipur belt. Unpubl. Rep., Geol. Surv. India.

Ramakrishnan, M., Viswanatha, M.N. and Swami Nath, J., 1976. Basement-cover relationships of Peninsular gneiss with high-grade schists and greenstone belts of southern Karnataka. J. geol. Soc. India, 17: 97-111.

Ronov, A.B., 1972. Evolution of rock composition and geochemical processes in the sedimentary shell of the earth. Sedimentology, 19: 157-172.

Ronov, A.B. and Migdisov, A.A., 1971. Geochemical history of the crystalline basement and the sedimentary cover of the Russian and North American platforms. Sedimentology, 16: 137-185.

Salop, L.I., 1968. Precambrian of U.S.S.R. Proc. 23rd Int. Geol. Congr., Sect. 4, 61-73.

Sampat Iyengar, P., 1905. Report on the survey work in the Chitaldrug district. Rec. Mysore Geol. Dept., 6(2), 57-92.

Sibley, D.F. and Wilband, J.T., 1977. Chemical balance of the earth's crust. Geochim. Cosmoch. Acta, 42: 545-554.

Sreenivas, B.L. and Srinivasan, R., 1968. Dharwar conglomerates of Mysore - a restudy. J. geol. Soc. India, 9: 197-205.

Srinivasan, R. and Sreenivas, B.L., 1972. Dharwar stratigraphy. J. geol. Soc. India, 13: 75-85.

Turner, C.C. and Walker, R.G., 1973. Sedimentology, stratigraphy, and crustal evolution of the Archaean greenstone belt near Sioux Lookout, Ontario. Can. J. Earth Sci., 10: 817-845.

Veizer, J., 1973. Sedimentation in geologic history: recycling vs. evolution or recycling with evolution. Contr. Miner. Petrol., 38: 261-278.

Veizer, J. and Hoefs, J., 1976. The nature of O^{18}/O^{16} and C^{13}/C^{12} secular trends in sedimentary carbonate rocks. Geochim. Cosmoch. Acta, 40: 1387-1395.

Venkatasubramanian, V.S. and Narayanaswamy, R., 1974. The age of some gneissic pebbles in the Kaldurga conglomerate. J. geol. Soc. India, 15: 318-319.

Viljoen, M. and Viljoen, R.P., 1969. The geology and geochemistry of the Lower Ultramafic Unit of the Onverwacht Group and a proposed new class of igneous rocks. Sp. Publ. geol. Soc. S.Afr., 2: 55-85.

Viswanatha, M.N., 1968. Unpubl. Rep., geol Surv. India.

Viswanatha, M.N., 1977. Spinifex texture in serpentinite from Karnataka. J. geol. Soc. India, 18: p. 623.

Viswanatha, M.N. and Ramakrishnan, M., 1975. The pre-Dharwar supracrustal rocks of the Sargur schist complex in southern Karnataka and their tectono-metamorphic significance. Ind. Mineral., 16: 48-65.

Viswanatha, M.N., Ramakrishnan, M. and Narayanan Kutty, T.R., 1977. Possible spinifex texture in a serpentinite from Karnataka. J. geol. Soc. India, 18: 194-197.

THE SARGUR SCHIST COMPLEX - AN ARCHAEAN HIGH-GRADE TERRAIN IN SOUTHERN INDIA

A.S. Janardhan, C. Srikantappa and H.M. Ramachandra

Geology Department, University of Mysore,
Manasa gangotri, Mysore 570 006, India

Abstract

The Sargur schist complex contains Archaean quartzo-feldspathic tonalitic gneisses of mixed origin and tight to isoclinally folded remnants of quartzite-K pelite-carbonate shelf facies sediments. Relics of layered ultramafic-gabbro-anorthosite (minor) complexes with cumulate textures occur as discontinuous linear belts, pods and patches in the gneisses and are later than the metasediments. The presence of 'included' gneiss in an ultramafic body suggests the Sargur supracrustals rested on a gneissic basement. Carbonate-bearing ultramafics (similar to 'Sagvandites') suggest a significant CO_2 pressure during metamorphism. This possibly explains the local attainment of granulite facies conditions in a predominantly upper amphibolite facies terrain. There were at least two periods of migmatisation with late (anatectic) granite sheets. Three major deformational episodes are well preserved in the quartzites.

There is much similarity between the Sargur complex and the Isua-Malene-Upernavik succession of West Greenland and Labrador. The presence of younger Dharwar greenstone belts to the north and a 'mobile belt' towards the south of the Sargur terrain is similar to relations in southern Africa. However, more coordinated absolute age/isotopic/rare earth element data are needed before a convincing model can be erected to explain the evolution of the high-grade terrain of southern India.

INTRODUCTION

Mapping of selected areas in various shields, coupled with absolute age determinations, isotopic and experimental studies mostly during the past decade, have helped in the understanding of the character and evolution of Archaean mobile belts. These studies have established the presence of greenstone belts and high-grade terrains and to some extent their interrelationships. Some workers (e.g. Burke et al., 1976; Tarney, 1976; Windley and Smith, 1976; Kröner, 1977) have evolved tectonic models for the Archaean crust based on comparison with Phanerozoic fold belts.

Work on the Precambrian of the Indian Peninsular shield and particularly of the Karnataka craton, has only been made in a broad regional manner (see Rama Rao, 1962; Pichamuthu, 1976a,b). Only recently has the study of key areas established that the younger Dharwar greenstones unconformably overlie the (older than 3 by) high-grade Sargar complex.

With a view to understanding the evolution of the high-grade terrain of southern Karnataka, the authors have mapped an area of c. 300 km^2 around Sargur (Fig. 1). Geochemical analyses have been made by wet chemical and XRF methods.

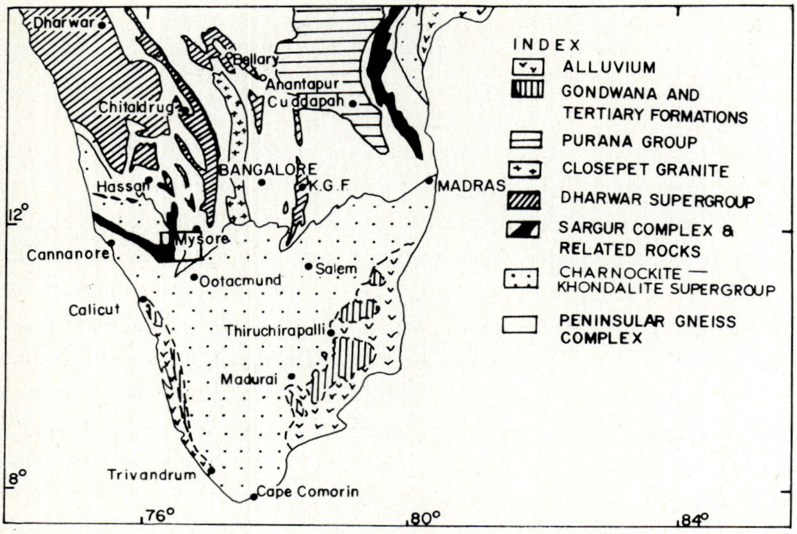

Fig. 1. Location map of the studied area on a geological map of Peninsular India (after Narayanaswamy, 1976).

LITHOSTRATIGRAPHY

The Sargur schist complex is predominantly a high-grade gneissic terrain with huge enclaves of metasediments, remnants of intrusive layered ultramafic-anorthosite-amphibolite complexes, and some late granites. Because of intense deformation, resultant interleaving (?), polyphase migmatisation and absence of any primary structures, no definite lithostratigraphic framework can be constructed. However, by correlative studies of Sargur-type rocks from different parts of Karnataka, Viswanatha and Ramakrishnan (1976) have evolved a stratigraphic sequence, and the following is a modified version from the Sargur area (in decreasing age): quartzo-feldspathic gneisses (partly volcanic), quartzites and quartz mica schists, kyanite-sillimanite schists, calcareous metasediments, banded magnetite quartzites (all will intercalated basic sills), ultramafic-anorthosite-amphibolite-metagabbro complexes, para-gneisses, hornblende-biotite gneisses, granites, late gabbros and dolerites.

STRUCTURE

The Sargur rocks follow a regional NS-NNE trend, with a regional northerly plunge (15-20°). The metasediments occur as linear enclaves with isoclinal folds outlining an original broad synformal pattern. On a minor scale early NW folds (seen clearly in the gneisses) are refolded by isoclinal NS folds and late open folds trending N 60° E. All these are superimposed by an E-W trend, particularly at the supposed "mobile belt" contact near Gundlupet about 50 km SE of the area. Banded magnetite quartzites exhibit at least three stages of deformation.

The following description of rock types of the area does not adhere strictly to the lithostratographic succession because of the constraints already explained. Metasediments, though later than some components of the gneissic complex, are dealt with first as they were deposited during a restricted time unit (more than 3 by) whereas the gneisses span a large time interval (c. 1 by).

METASEDIMENTS

The metasediments of the area conform to the quartzite-K-pelite-carbonate shelf facies of Sutton (1976). They now occur as curvilinear enclaves amidst gneisses and have been intensely deformed so that neither primary structures (except bedding) nor rigid bilateral symmetry of the folded rock types are preserved. In their field relations, rock associations, mineralogy, etc., they show a striking resemblance to the Isua supracrustals (or Akilia association, Bridgwater and Collerson, 1977) of western Greenland.

Banded magnetite quartzites are extensive. They are finely banded with alternating quartz and magnetite and occur predominantly in the southern part of the complex. They have a mineral composition of qz-mt-opx-cummingtonite/grunerite. Small bands in more migmatised areas develop ferro-hyp, gar and late hbl with almost total absence of magnetite. Fuchsite-bearing quartzites (with more than 80% qz) are intimately associated with the banded magnetite quartzites, often showing well-preserved interference fold patterns, are highly manganiferous (with gar) and are often closely associated with calcareous rocks.

Meta-pelites are represented by ky-sill-staur-qz-cor-graphite-rut$\pm$feldspar$\pm$gar schists with appreciable zircon. Ky-sill schists (c. 54% Al_2O_3) are locally mined; they are usually migmatised to para-gneisses with knots of corundum and late tectonic kyanite cross-cutting the foliation.

Calcareous varieties are present as crystalline marbles with a mineral assemblage of cal-diop-hb-phlogopite-talc$\pm$pl (An 75%) $\pm$gar, indicating an original dolomitic composition. Apart from a 4 km long band west of Bettadabidu, they usually occur as small calc-silicate (diop-pl-hb-cal or hedenbergite-pl-qz) patches amidst schists, extending for more than 20 km south of the Sargur area, as seen in the map of Jayaram (1913), or as extensive calc-gneisses.

Field relations demonstrate that the ultramafics were emplaced into the metasediments prior to the regional migmatisation episode.

The field setting, mineralogy, upper amphibolite-to-granulite grade of metamorphism and the tectonic style of the Sargur metasediments are strikingly dissimilar to the Dharwar supergroup, suggesting that they are not the deeper parts of the latter. On the other hand, the fold styles (dome-basin interference patterns) and mineralogy of the marbles and banded magnetite quartzites in particular are very similar to the Kanjamalai iron formations of the khondalite series and also to the supracrustals bordering the Sittanpundi anorthosite complex.

In a recent review, Radhakrishna and Vasudev (1977) suggest that the khondalites of the granulite terrain to the south and southeast of the area may be higher grade equivalents of the Sargur volcanics and sediments; the presence of similar layered igneous complexes in both areas strengthens this implication. Similar views are expressed by Mahabaleswar and Sadasivaiah (1976) when they state that the basic rocks and sediments of the Satnur-Halgur-Sivasamudram areas gave rise to charnockites under granulite facies conditions.

The sedimentary origin of the Sargur schists is inferred from the distribution of the major elements (Table 1) on c/al-alk and mg/c diagrams and the negative trend defined by a TiO_2/SiO_2 plot (after Tarney, 1976) supports this conclusion (figures not given).

GNEISSES

The predominant gneisses of the area, and in fact all the gneisses of the South Indian shield, have until now been discussed under the common heading 'the Peninsular gneiss', covering a time span of c. 1.5 by. Attempts at classifying these gneisses (e.g. Ramakrishnan et al., 1976; Sreenivas et al., 1976) have only been on a regional scale and not based on structure, chemistry, age or intrusive events in selected type areas. The following account of the gneisses of the Sargur area is sequentially based on structural studies related to petrography.

The earliest gneisses are quartzo-feldspathic tonalitic gneisses with a mineralogy of 50% qz, 40% pl, c. 10% bi and accessory zircon and apatite. They occur ubiquitously throughout the Sargur-type high-grade terrain. Varieties which lack zircons may represent early volcanics. But for the major part they are 'reworked' gneisses (of mixed parentage). Locally there are also augen gneisses (40% qz, 50% feldspars, 10% bi) with pink K-feldspar augen. In general, they show a highly deformed fabric in the field and in thin section and minor folds may have axial planar orientations. The field relations and mineralogy of these gneisses are similar to those in Greenland (Amitsoq-Nuk) and Labrador (Uivak I and II).

Typical migmatitic garnetiferous para-gneisses (with accessory rutile) in

TABLE 1: Chemical analyses of type metasediments of the Sargur complex

	1	2	3	4	5	6	7
SiO_2	84.18	47.96	47.46	54.40	7.69	57.00	58.17
TiO_2	0.48	0.26	0.37	3.61	0.08	2.32	0.84
Al_2O_3	10.75	3.96	12.50	29.20	0.68	27.83	17.67
Fe_2O_3	0.51	27.83	3.85	2.58	0.41	2.15	1.92
FeO	1.42	10.68	24.88	1.90	1.64	1.77	10.02
MnO	0.02	0.10	0.57	0.01	0.55	0.01	0.03
MgO	0.53	3.25	6.70	5.14	0.78	1.13	6.92
CaO	0.37	2.48	1.78	0.37	47.65	1.11	1.92
Na_2O	0.15	1.26	1.15	0.76	0.22	3.55	0.31
K_2O	-	0.52	0.52	1.74	0.39	1.36	0.81
P_2O_5	0.26	0.10	-	-	0.09	0.06	0.21
H_2O	0.42	1.50	-	-	0.74	2.25	0.67
CO_2	-	-	-	-	37.68	0.12	-
TOTAL	99.09	99.90	99.78	99.71	98.60	100.66	99.49

1 = B_2 gt. quartzite; 2 = T 27, BMQ Toravalli; 3 = SC 18 gt. cummingtonite quartzite N of Bettada bidu; 4 = I 213 kyanite schist; 5 = B crystalline limestone Bettada bidu; 6 = S 145 sill-kyan-sta-qtz-plag-schist, E of Doddabetta; 7 = SK 88 gt-sill-bi-plag-qtz-gneiss.

places with late kyanite and corundum are extensive. Sheets of pink granite with a meta-arkosic (?) appearance and a mineralogy of 50% K-feldspar perthites, 20% qz and 5% bi occur locally.

Biotite-rich pelitic gneisses are intimately associated with anthophyllite (gedritic)-hbl-rut gneisses, grade into orthopyroxene-bearing types ('acid charnockites': 30% qz, 20% opx, 25% pl, 10% antho, 5% bi) and are traversed by veins of dominantly plagioclase-rich leucosome material. The presence of orthopyroxene-bearing gneisses in the Sargur-type high-grade terrain (Janardhan and Ramachandra, 1977; Ziauddin and Yadav, 1974; Rameingar et al., 1976) indicates local attainment of granulite facies conditions. Iyer and Kutty (this volume) suggest that the greenstone-gneiss complexes of Hollenarsipur (equivalent to Sargur) and Kolar could have been the parent material for North Arcot granulites (charnockites).

The second group of gneisses are banded hornblende-biotite and biotite types derived by migmatisation of hornblende-granulites. The banding of the rock is affected by varying degrees of migmatisation, with the result that the rock is heterogeneous with reference to its mineralogy and texture. The hornblende-biotite gneisses, which resemble the 'Peninsular gneisses' around Bangalore, occur extensively to the south of the area.

The third and the youngest group consists of granites and granitic gneisses. They trend roughly N 60° E (Jayaram, 1913), occur as sheets probably related to the late N 60° E open folds, and may represent the anatectic phase of the

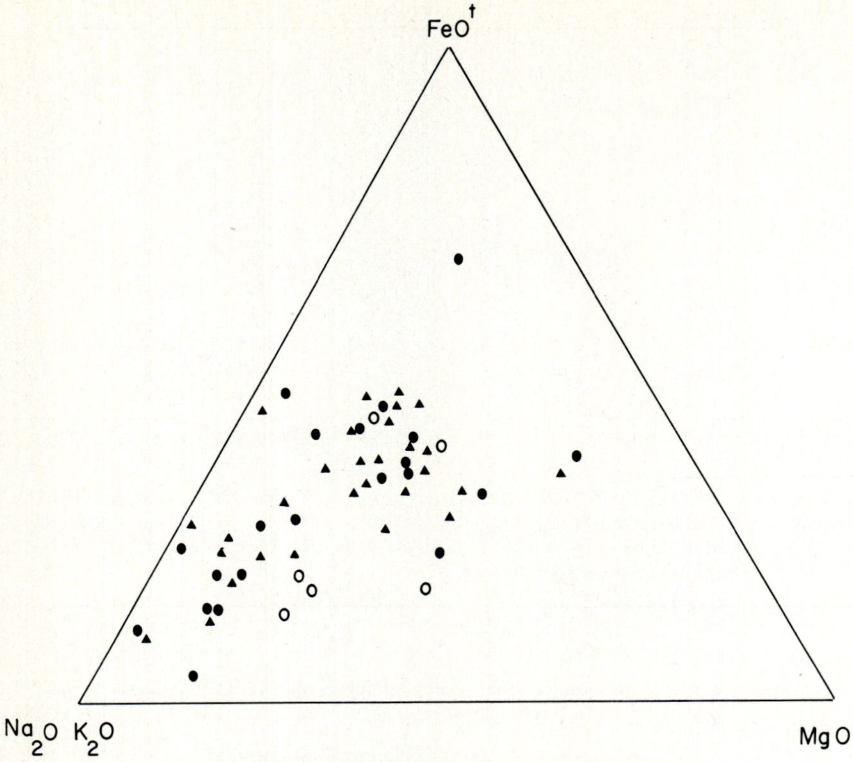

Fig. 2. Plots of Sargur gneiss along with gneiss from Tamil nadu and Peninsular gneiss showing calc-alkaline trend.

regional migmatisation episode.

Major elements (Table 2),(along with data from the Tamil nadu area, define a calc-alkaline trend in an AFM diagram (Fig. 2) and show a calc-alkaline affinity in the SiO_2/TiO_2 diagram of Tarney (1976) (figure not given). K/Rb ratios, which are 400-1000 (Fig. 3a), and K/Ba values (Fig. 3b) lie in the fields of the Ancient Gneissic Complex and Kaapvaal area in the diagrams of Lambert et al. (1976). They significantly fall away from the greenstone field and typically fall in the field of amphibolite grade gneisses.

META-BASICS

The basic rocks of the area are intimately associated with the metasediments and gneisses and can be broadly classified into three groups according to their age. They now exhibit an amphibolite-to-pyroxene granulite grade of metamorphism.

TABLE 2: Chemical analyses of gneisses

	1	2	3	4	5	6	7	8	9	10	11	12
SiO_2	72.10	70.57	71.83	49.10	72.25	70.05	78.89	71.11	71.11	61.57	74.04	53.86
TiO_2	0.18	0.22	0.39	1.10	0.28	1.14	0.15	0.40	0.55	1.35	0.26	0.19
Al_2O_3	14.20	14.50	13.22	21.10	13.14	11.41	10.09	13.84	9.08	14.02	12.83	21.41
Fe_2O_3	1.27	0.88	1.51	3.30	1.01	2.64	0.08	3.38	0.73	3.08	0.68	0.87
FeO	0.28	0.40	0.70	5.37	1.52	2.46	1.24	-	2.10	3.64	0.44	1.28
MnO	0.16	0.14	0.26	0.20	0.06	0.06	0.05	0.07	0.03	0.04	0.14	0.10
MgO	0.75	0.16	0.84	9.60	1.91	2.64	0.62	1.16	9.99	4.38	0.44	5.08
CaO	2.00	2.40	1.80	0.90	2.56	3.57	0.82	2.73	1.00	6.23	2.01	10.07
Na_2O	5.10	4.20	3.72	4.50	4.10	2.68	2.69	5.37	4.27	2.33	3.75	4.92
K_2O	1.20	4.30	3.63	3.60	0.80	1.18	3.66	1.62	0.28	1.10	3.76	0.92
P_2O_5	-	-	0.13	-	0.03	0.21	0.05	-	-	0.10	0.11	0.01
H_2O	0.94	1.82	1.51	1.16	1.57	1.27	0.56	0.92	0.68	1.68	1.20	1.01
TOTAL	98.18	99.59	99.54	99.93	99.23	99.31	98.90	100.60	99.82	99.52	99.66	99.72
Cr	-	24	16	-	51	10	10	-	57	77	-	49
Ni	-	2	24	-	29	34	28	-	6	28	-	59
Cu	-	-	14	-	26	7	19	-	-	-	-	-
Zn	-	-	52	-	49	120	17	-	-	-	-	-
Rb	-	-	83	-	18	13	42	-	-	-	-	-
Sr	-	-	239	-	118	110	39	-	-	-	-	-
Y	-	-	21	-	86	91	191	-	-	-	-	-
Ba	-	812	936	-	194	402	526	-	14	-	-	-
La	-	64	43	-	33	250	56	-	64	-	-	-
cc	-	120	70	-	61	380	96	-	145	-	-	-

1 = S332 quartzo-feldspathic gneiss, canal section, W of Motha; 2 = S 75 quartzo-feldspathic+bi+gt gneiss; 3 = LM 16 augen (pink) gneiss, Mavinahalli; 4 = S 309 kyanite gneiss, Chikkadevi betta; 5 = quartzo-feldspathic gneiss ± hb (average of 3 analyses), Sindhuvalli; 6 = bi+hb gneiss (average of 4 analyses), Doddakanya; 7 = SK 79 biotite gneiss, Doddakanya; 8 = GM 5 migmatitic gneiss, Sriramgudda, Gundlupet; 9 = K 23 (acid-looking) charnockite, SW Hullahalli; 10 = SK 187 gt+hb+bi+feldspar gneiss; 11 = SK 185 pegmatitic gneiss; 12 = SK 178 antho+plag+K-feldspar+qtz gneiss. Nos. 10-12 specimens from included gneiss in the ultramafic body, Doddakanya.

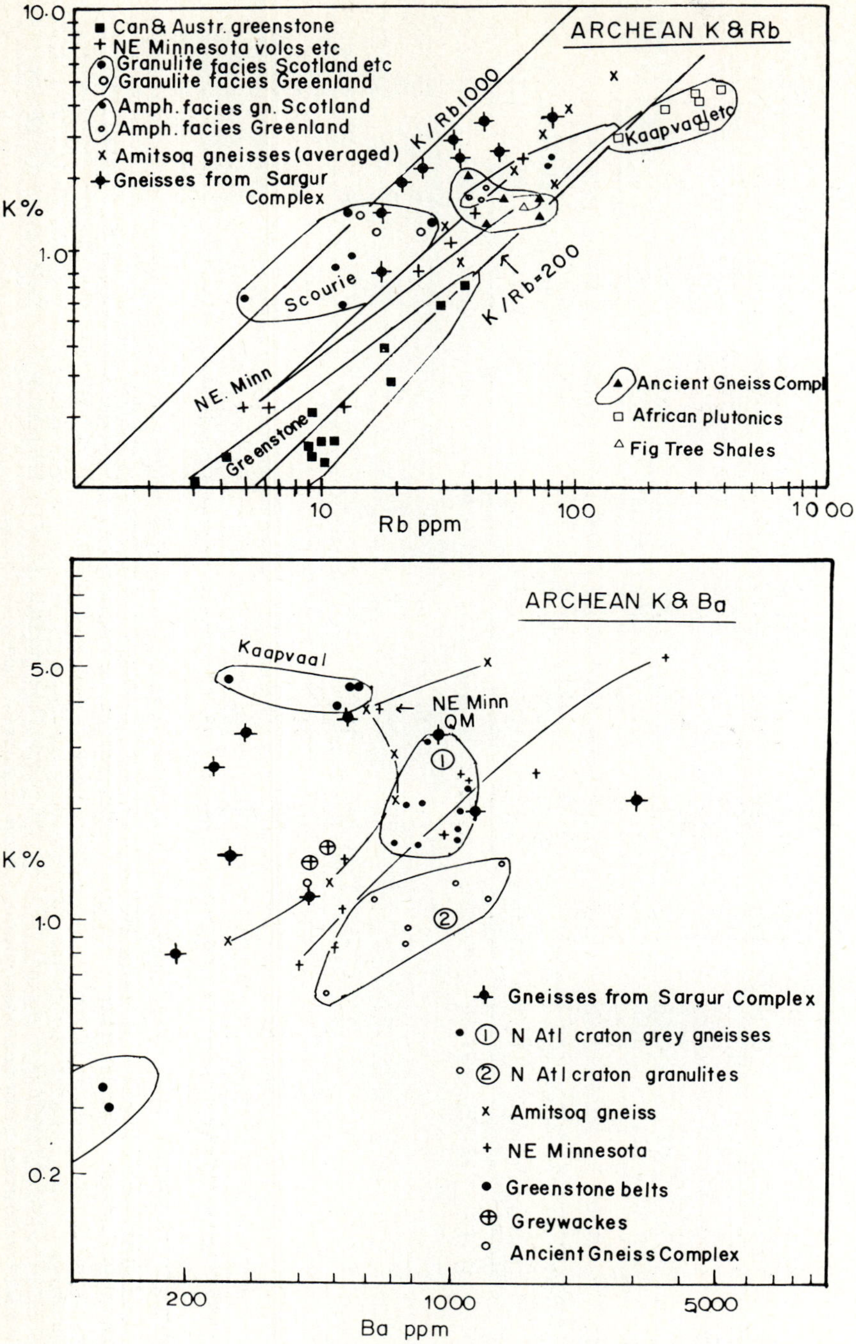

Fig. 3a and b. K vs Rb and K vs Ba plots of Sargur gneiss.

TABLE 3: Mean chemical composition of amphibolites compared with other basics

	1	2	3	4	5	6	7	8	9	10	11
SiO_2	48.98	50.96	48.94	49.07	50.59	49.07	54.00	49.90	49.20	51.20	51.60
TiO_2	1.25	0.82	0.93	1.03	0.68	0.81	1.53	1.51	1.39	0.96	0.30
Al_2O_3	15.70	14.01	11.26	14.52	12.08	14.79	12.96	17.20	15.80	15.20	15.90
Fe_2O_3	2.73	2.99	1.51	2.04	1.20	1.58	3.08	-	-	-	-
FeO	8.79	8.46	11.26	10.28	10.87	8.95	11.69	8.70	9.20	1.07	9.50
MnO	0.10	0.19	0.19	0.23	0.21	0.22	0.19	0.17	0.16	0.22	-
MgO	8.13	7.51	11.97	7.22	8.56	6.47	3.81	7.20	8.50	6.40	6.70
CaO	10.02	9.76	9.75	10.22	11.00	13.74	7.66	11.80	11.10	10.70	11.70
Na_2O	2.50	2.08	2.10	2.08	2.16	1.69	2.11	2.70	2.70	2.80	2.40
K_2O	0.50	0.45	0.24	0.39	0.74	0.38	0.40	0.16	0.26	0.20	0.44
P_2O_5	-	0.19	0.10	0.27	0.07	0.24	0.15	-	-	-	0.11
H_2O	1.86	2.01	0.88	2.18	1.37	2.22	1.75	-	-	1.20	0.45
TOTAL	100.56	99.43	99.13	99.53	99.53	100.16	99.33	99.34	98.31	89.95	99.10
Zr	169	92	-	55	45	44	50	95	100	-	70
Sr	-	171	170	134	79	121	71	130	123	102	200
Rb	-	6	3	23	26	23	109	1	-	10	5
Zn	-	93	96	93	89	83	123	-	-	57	-
Cu	-	46	46	87	71	83	51	77	87	111	-
Ba	70	70	92	116	115	86	134	14	12	-	-
Ni	56	132	420	100	147	107	53	97	123	116	30
Cr	171	212	117	337	599	505	29	297	296	400	50

1 = K20 hornblende granulite, Hullahalli; 2 = Average of 33 diop±gt-bearing amphibolites; 3 = SK 42 two pyroxene granulite, Doddakanya; 4 = Average of 14 amphibolites, Hunter belt (Collerson et al., 1976); 5 = Average of amphibolites, Sagick area (Collerson et al., 1976); 6 - Average of 6 diopside-bearing amphibolites, Hunter belt (Collerson et al., 1976); 7 = Average of 4 garnetiferous amphibolites, Hunter belt (Collerson et al., 1976); 8 = Average of 10 oceanic tholeiites (Engle et al., 1965); 9 = Average of 98 oceanic tholeiites (Cann, 1971); 10 = Average of 85 metabasalts (Glikson, 1971); 11 = Average of basalts from island arc tholeiites (Jakes and White, 1971).

The first group of highly permeated basics can be further divided into (a) hornblende-granulites (pl-cpx-brownish green hbl-gar-opx-qz with occasional antho), and (b) amphibolite (bluish-green to grass-green hbl-cpx-gar-qz-sph). The hornblende-granulites are typically associated with the high-grade migmatitic gneissic horizons and vary from remnant bands to agmatitic blocks, and can be confused with retrogressed two-pyroxene granulites. In the northern part of the area the amphibolites occur extensively with bands of crystalline limestone and quartzite.

The second group of basics are metagabbros (pl-cpx-hbl-gar-qz) which in places transect metasediments. Coarse-grained hornblendites with a little plagioclase and greenish diopside (usually altered) occur as cross-cutting dykes in the axial planes of folds. Late two-pyroxene granulites (opx-cpx-pl An 40%-brownish hbl-rut) - the basic charnockite of earlier workers - invariably exhibiting pseudotachylitic veins, traverse the quartzo-feldspathic gneisses and are folded with them. These two-pyroxene granulites show only marginal migmatisation and are connected with the ultramafic rocks both in their chemistry and field relations; they might represent the gabbroic part of the ultramafic/ anorthosite suite.

The AFM major element chemistry (Table 3) of the basic rocks shows a typical iron enrichment trend (Fig. 4), while the anorthosite-gabbro association of the area (discussed in the next section) exhibits a smooth calc-alkaline trend (Fig. 5). Similar bimodal trends are also shown by the ultramafic-gabbro anorthosite complex and the metavolcanics of the Holenarsipur schist belt (Iyer and Kutty, this volume). When plotted in the diagram of Rivalenti (1976) there is no variation in Al_2O_3-CaO-MgO levels related to FeO/FeO+MgO (Fig. 6), thus defining a tholeiitic trend. In the diagram of Kuno (1966) (Fig. 7) they scatter across the fields of tholeiitic and high alumina basalts. Geochemistry, particularly of major elements, is not an adequate criterion for deciphering an island arc provenance implied here (which to a certain extent is confirmed by K/Rb values which generally range from 400-1200 ppm, similar to the values of arc basalts, Condie, 1976), or other tectonic environments in the Archaean (Gunn, 1976). The K/Rb values define a low K-tholeiitic field (Fig. 8). The Ti/Zr and Ti-Zr-Sr/2 ratios plot in the ocean floor basalt and low potash tholeiite fields of Pearce and Cann (1973), which differ from the highly scattered and calc-alkaline nature of the Dharwar metabasalts (data from Naqvi and Hussain, 1973) (Figs. 9a and b). We present these diagrams to emphasise the different composition of the two rock suites.

ULTRAMAFICS AND THEIR ASSOCIATED ROCKS

Serpentinised and chromite-bearing ultramafics are dominant and characteristic of the Sargur high-grade terrain (Radhakrishna, 1976). They occur as linear

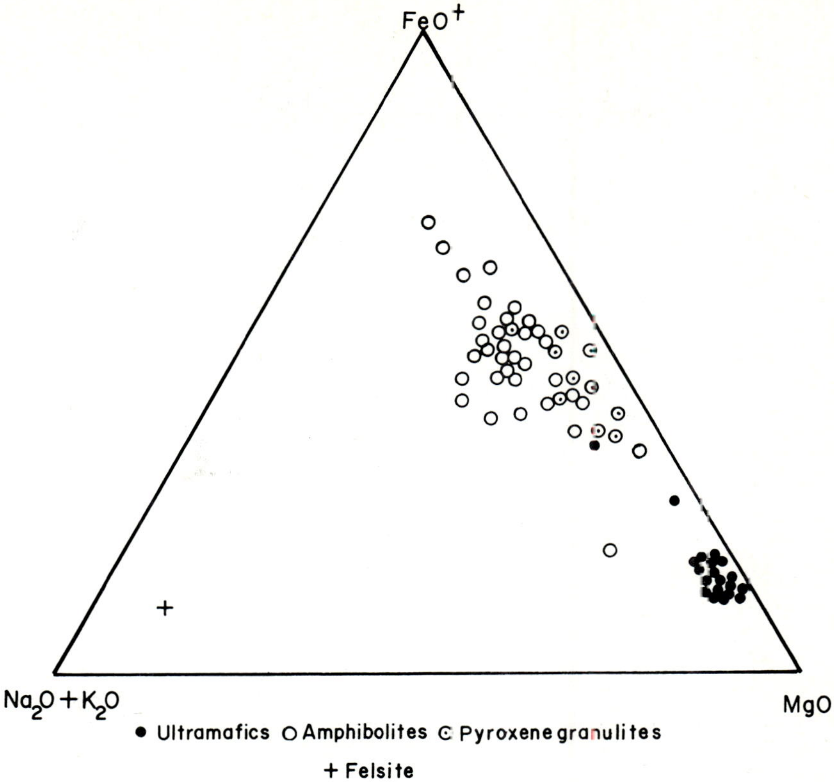

Fig. 4. Plots of ultramafics and basics of the area. Note their typical iron enrichment trend. Plot of felsite dyke cutting the ultramafic body at Doddakanya is also included.

discontinuous bodies and are well exposed in the northern parts of the area (Fig. 10) similar to those in the high-grade Holenarsipur, Nuggihalli belts (Rama Rao, 1962; Janardhan and Srikantappa, 1976; Viswanatha and Ramakrishnan, 1976). They were emplaced into the Sargur supracrustal rocks, as is demonstrated by the presence of a tectonically 'included' gneissic block within the Dodkanya ultramafic body.

The ultramafics are predominantly harzburgitic and have a crude 'zonation' with cores of dunite-harzburgite-bronzite peridotite-bronzitite and rims of titaniferous magnetite-rich gabbro (as at Nuggihalli, Pandit, 1973) and they are traversed by basic dykes (now metamorphosed to two-pyroxene granulites) as seen within the Dodkanya ultramafic body. The contacts of the ultramafic rocks are often mylonitised and in places exhibit thermal aureoles. In these respects, they resemble the Emigrant Gap ultramafic complex (James, 1971).

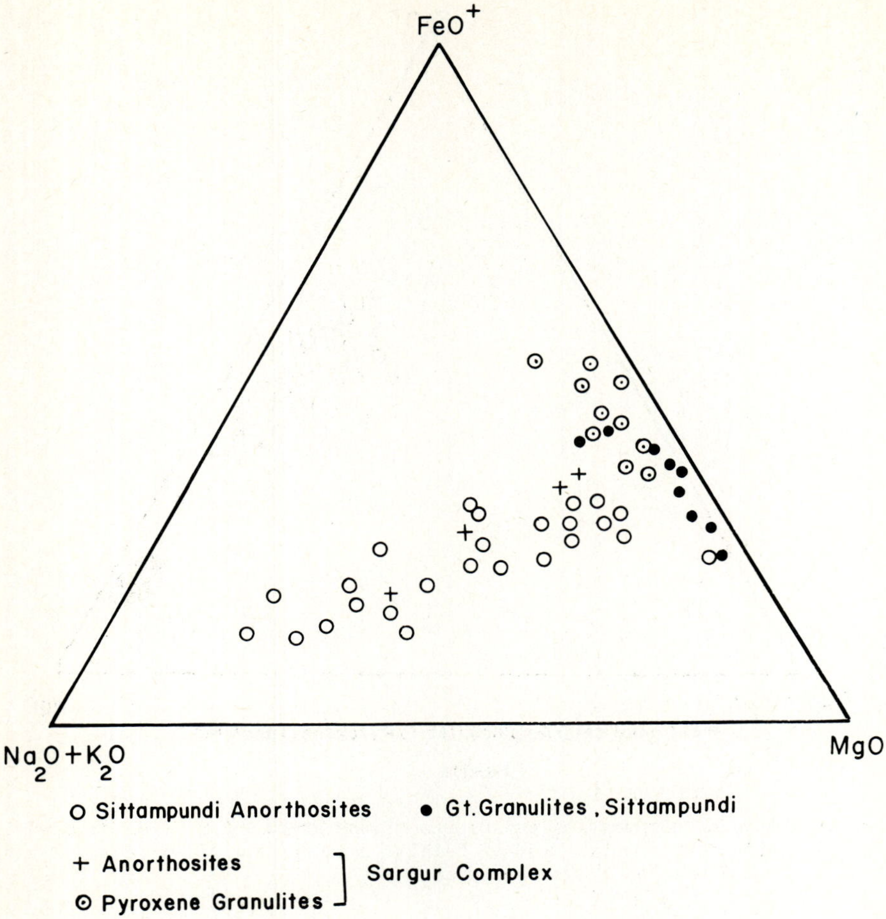

Fig. 5. Plots of anorthosites and pyroxene granulites from Sargur complex along with Sittampundi anorthosites and garnetiferous granulites indicating a smooth calc-alkaline trend.

In thin section, the ultramafics exhibit tightly packed equant olivines, often poikilitically enclosed in orthopyroxene. The common mineral assemblage is Ol (Fo 84-92%)-En with occasional cpx and late tremolite. Antigorite is the main serpentine mineral with minor brucite. Spinifex texture was reported in a serpentinite occurrence, close to the western margin of the Chitradurga schist belt by Viswanatha et al. (1977), but we have observed no such textures in the area which commonly survive relatively high-grade metamorphism and deformation (Arndt et al., 1976). The ultramafics exhibit chromite layering (layers vary in width from 0.8-1.5 cm). Chromite also occurs in pods and as veins. Inclination of the chromite layers from steep to vertical and later brecciation indicate

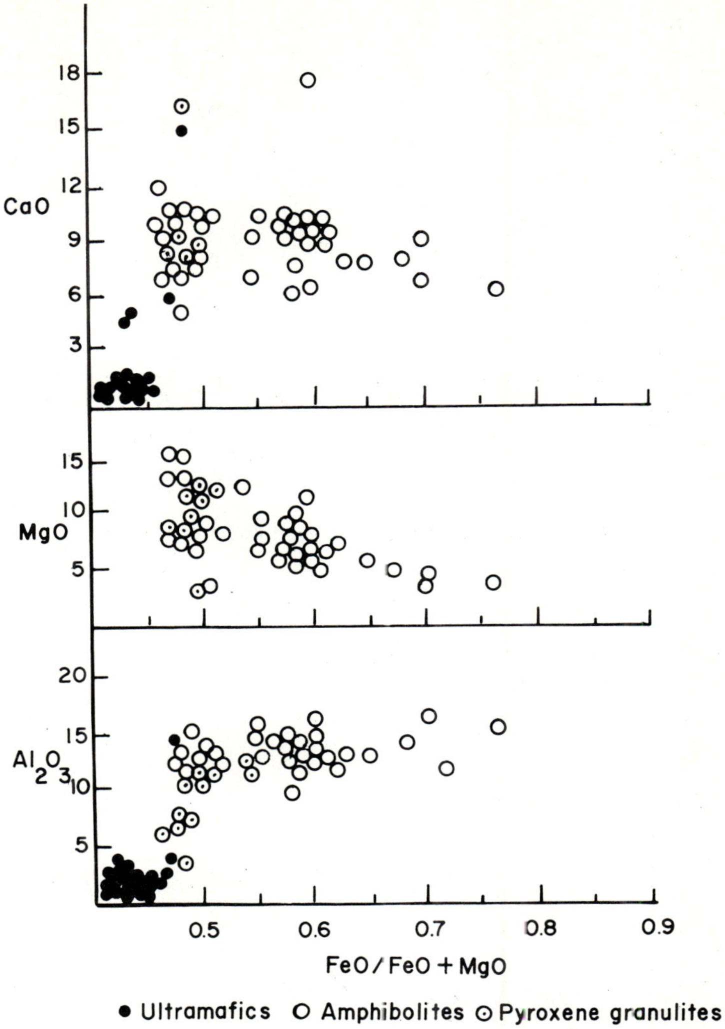

Fig. 6. FeO/FeO+MgO vs Al_2O_3, CaO, MgO (Rivalenti, 1976) diagram of ultramafics and amphibolites of the area. Note their flat trend in Al_2O_3, CaO and MgO levels indicating tholeiite nature.

post-crystallisation deformation, as suggested by Chakraborthy (1972).

In places massive greenish rocks, similar in mineralogy (opx-cr.tremolite-ol-talc-dolomite) to Norwegian Sagvandites (Schreyer et al., 1972), border the ultramafics and occur as ubiquitous relics within the gneisses (Janardhan and Srikantappa, 1977). This again demonstrates that regional migmatisation has affected and recrystallised the ultramafics and also illustrates the role of

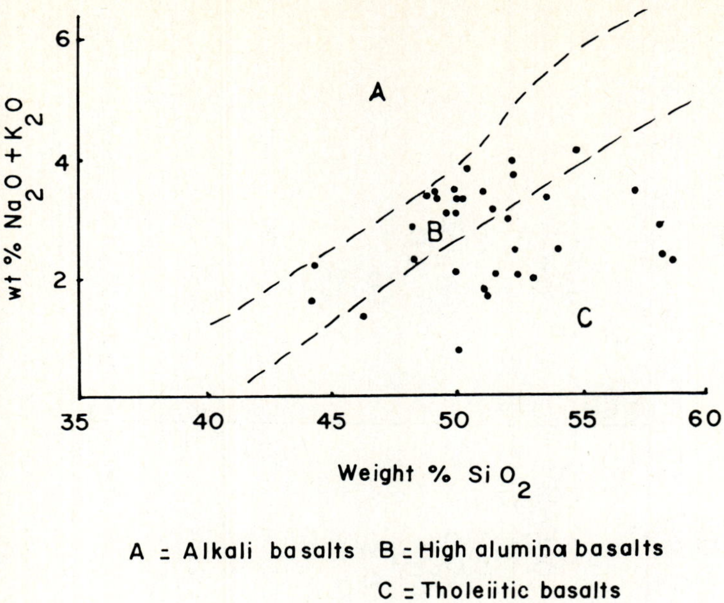

Fig. 7. SiO_2 vs Na_2O+K_2O diagram (after Kuno, 1966). Note their scatter in high alumina basalt and tholeiite field.

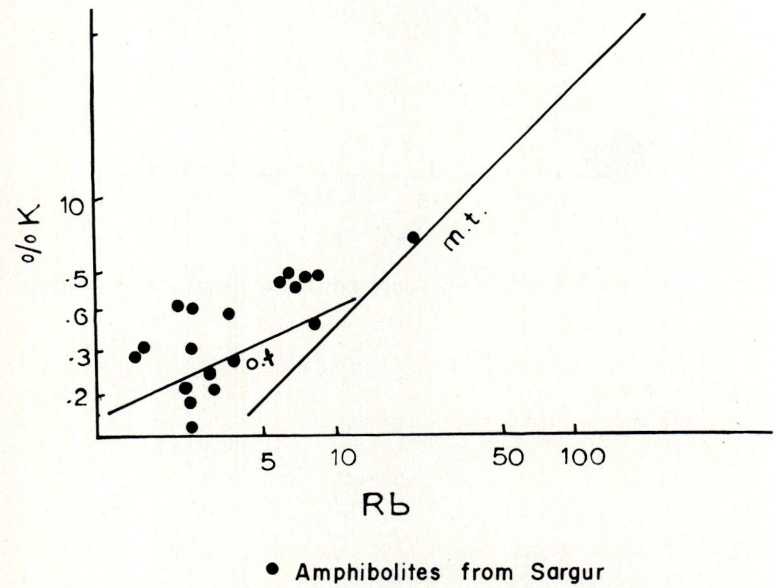

Fig. 8. K vs Rb plots for basics of the area (after Payne and Shaw, 1967) clearly showing low K-tholeiite trend.

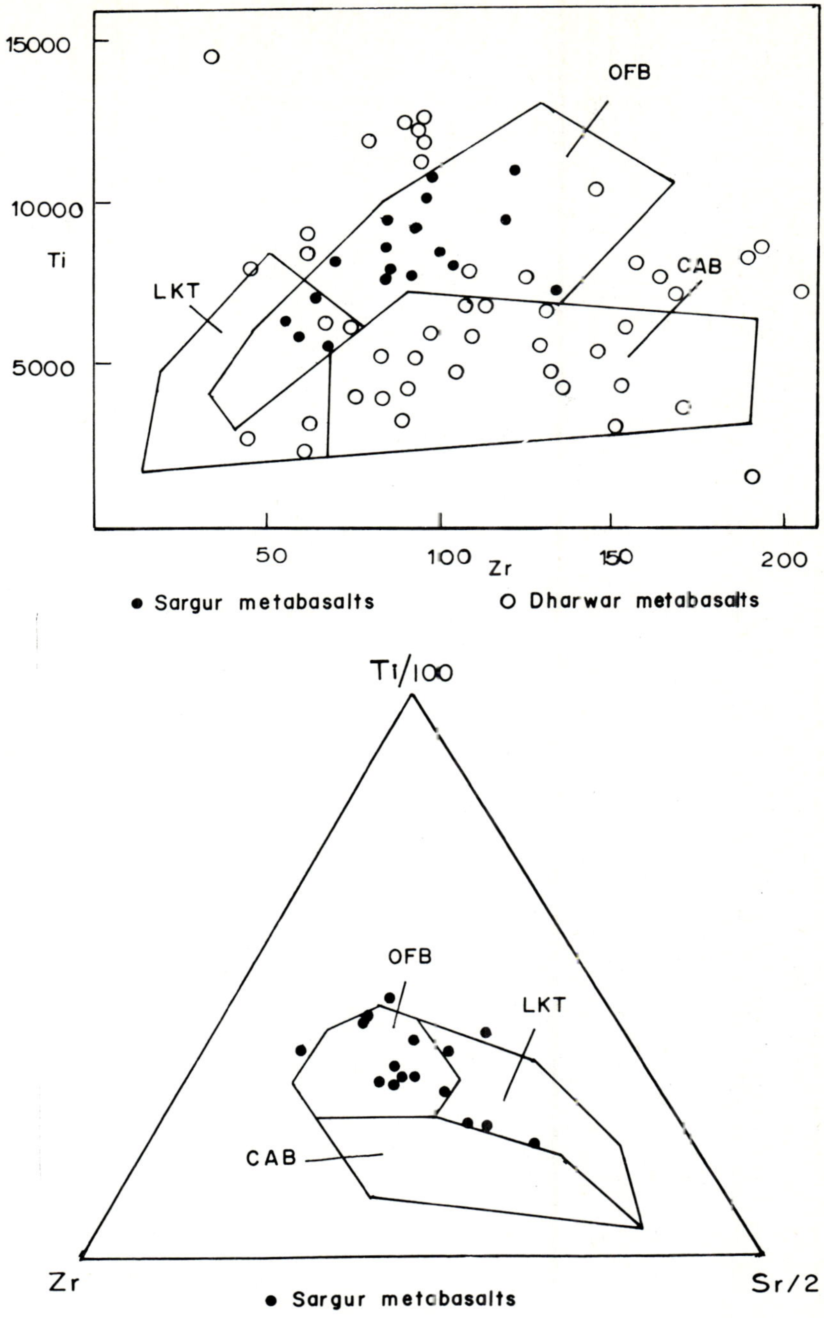

Figs. 9a and b. Ti vs Zr and Ti/100-Zr-Sr/2 plots (after Pearce and Cann, 1973) of Sargur basics and Dharwar metabasalts (data from Naqvi and Hussain, 1973).

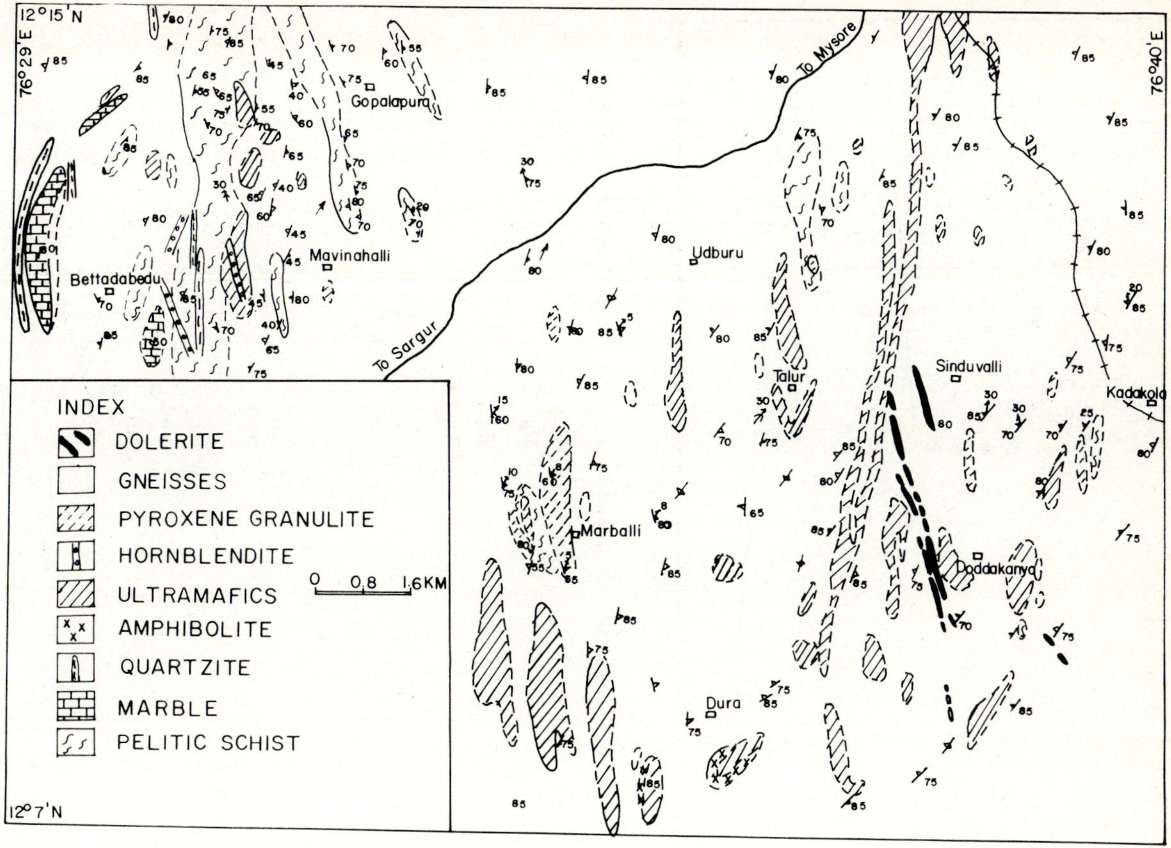

Fig. 10. Geological map of northern parts of Sargur schist complex. Note their dominant ultramafic bodies.

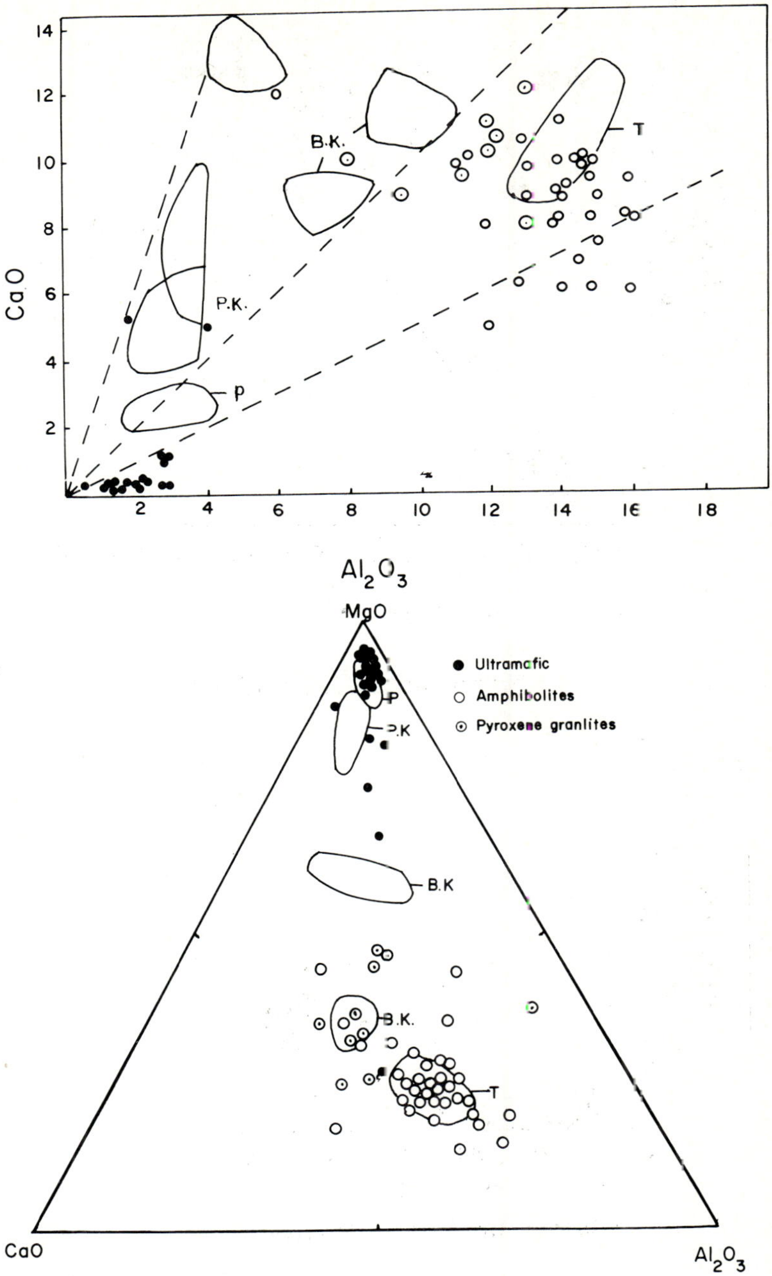

Fig. 11a,b. Al_2O_3 vs CaO and MgO–CaO–Al_2O_3 for ultramafics and basics of the Sargur complex.

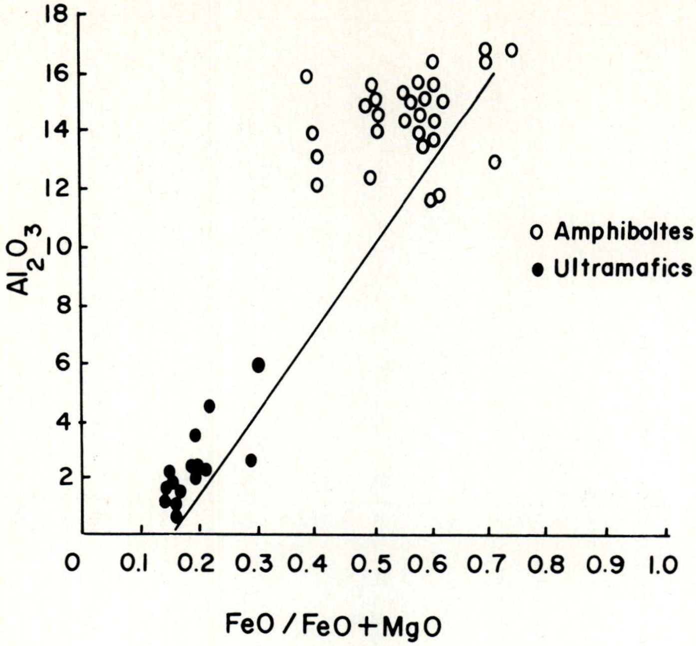

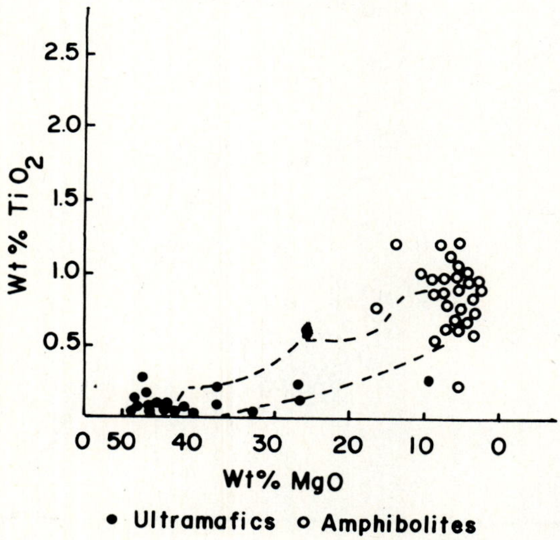

Fig. 12a,b. FeO/FeO+MgO vs Al_2O_3 and MgO vs TiO_2 diagram (Naldrett and Cabri, for Sargur ultramafics and basics exhibiting komatiitic affinities.

P CO_2 in the regional metamorphism. These assemblages, reported for the first time in this part of the Archaean of India, may have wider application as in future they may serve as marker horizons for correlative purposes (as suggested by Moore, 1977).

After the discovery of komatiites in the Barberton area (Viljoen and Viljoen, 1969), much attention was paid to identifying similar rocks in the Indian shield. Plots of about thirty ultramafics show an iron enrichment trend in an AFM diagram (Fig. 4). On the original diagrams of Viljoen and Viljoen (Figs. 11a,b) they clearly fall in the fields of normal peridotites and tholeiites, while some basics lie in the komatiitic field due to their highly retrogressed actinolite content. Diwakar Rao et al. (1975) reported the presence of komatiitic ultramafics from the same area on the basis of concealed averages. Nesbitt et al. (pers. comm.) categorically state that chemistry alone is not a sufficient criterion to diagnose komatiites because the same chemistry can also be shown by fractionated cumulates; the texture is the most important criterion. Rocks of komatiitic affinity rest on conglomerates above a gneissic basement in southern Africa (Bickle et al., 1975). The chemistry of the ultramafics (Table 4) of the area shown in diagrams after Naldrett and Cabri (1976) (Figs. 12a,b) is consistent with the existence of an early sialic crust below the Sargur supracrustals, which is further supported by the occurrence of 'included gneiss' within the Dodkanya ultramafic body.

The ultramafics of the area show high Ba values (average 130 ppm) and their Rb, Zr, Sr and Rb/Sr ratios are very similar to those of metaperidotites in Labrador (Collerson et al., 1976).

In recent years many gabbro-anorthosites with cumulate textures have been reported from the Sargur high-grade terrains in Karnataka and adjoining Kerala State (Viswanatha and Ramakrishnan, 1975; Nair et al., 1976; Ramakrishnan and Mallikarjuna, 1976; Vidhyadharan et al., 1977; Janardhan and Ramachandra, in press). These bodies have a mineralogy of opx-pl with An 50-80%-cpx-late hbl$\pm$gar vary from a few centimetres to kilometre-sized plutons and are spatially (closely) associated with the ultramafics; they represent the fragmented remnants of ultramafic-anorthosite-gabbro complexes.

It is significant to note here that the Sittampundi anorthosite complex (Janardhan and Leake, 1975; Windley and Selvan, 1975) was emplaced into metasediments (sill-ky-graphite pelite; ironstone; marble) similar to the Sargurs. But the Sittampundi anorthosites are chromite bearing, carry highly calcic plagioclase (An c. 92%) and are associated with insignificant nonchromite-bearing ultramafics - very different from the ultramafic-gabbro-anorthosite complexes of the Sargur terrain. Thus the Sittampundi and Sargur areas might belong to different megazones in an Archaean mobile belt (Windley, pers. comm.).

TABLE 4: Chemical analyses of ultramafic rocks from the northern part of the Sargur complex

	1	2	3	4	5	6	7	8
SiO_2	38.74	53.62	47.14	50.53	46.12	49.09	47.15	46.73
TiO_2	0.18	0.30	0.62	0.76	0.26	0.26	0.98	0.91
Al_2O_3	1.80	2.45	4.55	6.19	29.60	23.30	15.42	16.60
Fe_2O_3	4.77	1.07	0.87	1.91	0.85	1.52	2.58	0.53
FeO	3.66	8.76	12.90	8.23	1.41	3.40	7.92	8.54
MnO	0.16	0.22	0.25	0.45	0.10	0.05	0.06	0.08
MgO	39.90	28.17	23.59	16.24	3.65	6.66	12.63	12.00
CaO	0.90	2.76	6.17	12.37	13.20	8.80	7.80	9.00
Na_2O	0.22	0.46	0.51	0.44	5.04	5.06	4.20	3.94
K_2O	0.07	0.04	0.37	0.16	0.48	0.69	0.34	0.34
P_2O_5	0.02	0.22	0.05	0.08	-	-	-	-
CO_2	3.41	-	1.06	-	-	-	-	-
H_2O	5.37	1.24	1.57	2.53	0.25	0.13	0.35	0.06
TOTAL	99.20	99.31	99.65	99.89	100.96	98.96	99.43	98.73

1 = Average of 24 ultramafics (dominantly harzburgite); 2 = pyroxenite; 3 = Bronzite peridotite; 4 = Hornblendite; 5 = K 99 leuco-anorthosite; 6-8 = Gabbroic anorthosites.

This then suggests that the Sargurs could be equated with the Messina formation of southern Africa.

CONCLUSIONS

A discussion on the interrelationships between the Sargur high-grade rocks and the younger Dharwar supergroup and the charnockite-khondalite 'mobile belt' of southern India is only possible at present on an empirical basis. This is mainly because of lack of fabric studies of key areas in the transitional zones and of absolute age and isotopic data constraints. But the presence of an unconformity at Holenarsipur, between the Dharwar supergroup and the high-grade rocks, and the difference in the volume and composition of the basic rocks in the two environments is significant and negates the view that Sargur-type rocks represent the root zones of a Dharwar greenstone belt. On the other hand, the Sargur schists are comparable with the metasediments of a 'charnockite-khondalite' 'mobile belt'; they may represent once-continuous platformal associations. This study has revealed that there are marked resemblances between the Sargur and the Isua-Malene successions of West Greenland: both have platformal lithologies, lack coherent lithostratigraphy, have two types of ultramafic-gabbro-anorthosite layered intrusives and exhibit bimodal magmatic trends. On a regional scale there is much similarity with relations in southern Africa (Coward et al., 1976), where different types of greenstone belts occur with high-grade rocks in a mobile belt (Limpopo).

Following this analogy the Sargurs might represent the 'continental front' in a greenstone-high-grade-mobile belt sequence (Kröner, 1977). Whether or not the Sargurs evolved by plate movements representing the deeper levels of present-day Cordilleran belts (Windley and Smith, 1976), or by 'sagduction' as proposed by Goodwin (pers. comm.) is an open question. An unequivocal solution may only be obtained after completion of coordinated tectonic and geochemical studies.

ACKNOWLEDGEMENTS

The authors acknowledge their grateful thanks to Prof. M.N. Viswanathaiah for his encouragement and support during the course of investigation. They also wish to thank Dr B.P. Radhakrishna for encouragement and for improving the paper. The senior author wishes to place on record his deep sense of gratitude to Prof. Bernard Leake of Glasgow University, U.K., for extending his laboratory (XRF) facilities and to Dr Windley for his suggestions. Mr Srikantappa and Mr Ramachandra thank the University Grants Commission for financial aid. The assistance of Mr Ravindra Kumar (working in the nearby Gundlupet area) is gratefully acknowledged.

REFERENCES

Arndt, N.T., Naldrett, A.J. and Pyke, D.R., 1977. Komatiite and iron-rich tholeiitic lavas of Munro township, North-east Ontario. J. Petrol., 18: 319-369.

Bickle, M.J., Martin, A. and Nesbitt, E.C., 1975. Basaltic and peridotitic komatiites and stromatolites above a basal unconformity in the Belingwe greenstone belt, Rhodesia. Earth Planet. Sic. Lett., 27: 155-162.

Bridgwater, D. and Collerson, K.D., 1977. On the origin of early Archaean gneiss - a reply. Contrib. Min. Pet., 62: 179-192.

Burke, K., Dewey, J.F. and Kidd, W.S.F., 1976. Dominance of horizontal movements, arc and microcontinental collisions during the later permobile regime. In: B.F. Windley (Editor), The Early History of the Earth. Wiley, Lond., pp. 113-129.

Chakraborthy, K.L., 1972. Some primary structures in the chromitites of Orissa, India. Mineral. Deposita, 7: 280-284.

Condie, K.C., 1976. Plate tectonic and crustal evolution. Pergamon, New York, 288 pp.

Coward, M.P., James, P.R. and Wright, L., 1976. Northern margins of the Limpopo belt, S. Africa. Bull. geol. Soc. Am., 87: 601-611.

Diwakara Rao, V. and Naqvi, S.M., 1975. Geochemistry of Dharwar ultramafics and Archaean mantle. Lithos, 8: 77-81.

Gunn, B.M., 1976. A comparison of modern and Archaean oceanic crust and island arc petrochemistry. In: B.F. Windley (Editor), The Early History of the Earth. Wiley, Lond., pp. 389-403.

James, O.B., 1971. Origin and emplacement of the ultramafic rocks of the Emigrant Gap area, California. J. Petrol., 12: 523-560.

Janardhan, A.S. and Leake, B.E., 1975. The origin of the metanorthosites, gabbros and garnetiferous granulites of the Sittampundi complex, Madras, India. J. geol. Soc. India, 16: 391-408.

Janardhan, A.S. and Srikantappa, C., 1976. Geology of northern parts of the Sargur schist belt between Mavinahalli and Doddakanya, Mysore district, Karnataka. Ind. Mineral., 16: 66-75.

Janardhan, A.S. and Ramachandra, H.M., 1977. Charnockites from Hullahalli, Mysore district, Karnataka State. J. geol. Soc. India, 18: 331-337.

Janardhan, A.S. and Srikantappa, C., 1977. Carbonate-orthopyroxene-chrome tremolite rocks (Sagvandites) from Mavinahalli, Mysore district, Karnataka State. J. geol. Soc. India, 18: 617-622.

Janardhan, A.S. and Ramachandra, H.M. (in press). Anorthosites from Hullahalli, Mysore district, Karnataka State.

Jayaram, B., 1913. Report on the preliminary geological survey of the south-west corner of Mysore district. Rec. Mysore Geol. Dept., 13: 45-100.

Kröner, A., 1977. Precambrian mobile belts of southern and eastern Africa - ancient suture or sites of ensialic mobility? A case for crustal evolution towards plate tectonics. Tectonophysics, 40: 101-135.

Kuno, H., 1966. Lateral variation of basalt magma types across continental margins and island arcs. Bull. Volcan., 29: 195-222.

Mahabaleshwar and Sadasiviah, M.S., 1976. Charnockites of Sivasamudram area, Karnataka. In: Nagenna (Editor), Studies in Precambrian. Bangalore, pp. 74-89.

Moore, A.C., 1977. The petrography and possible regional significance of the Hjelmalama ultramafic body (Sagvandite), Nord More, Norway. Norsk geol. Tiddskr., 57: 55-64.

Nair, M.N., Vidhyadharan, K.T., Pawar, S.D., Sukumaran, P.V. and Murthy, Y.G.K., 1976. The structural and stratigraphic relationship of the schistose rocks and associated igneous rocks of the Tellicherry-Nanantoddy area, Cannanore district, Kerala. Ind. Mineral., 16: 89-100.

Naldrett, A.J. and Cabri, L.J., 1976. Ultramafic and related mafic rocks: their classification and genesis with special reference to the concentration of nickel sulfides and platinum group elements. Econ. Geol., 71: 1131-1158.

Naqvi, S.M. and Hussain, S.M., 1973. Relation between trace element and major element composition of the Chitadurga metabasalts, Mysore, India, and the Archaean mantle chemistry. Chem. Geol., 11: 17-30.

Pandith, A.S., 1973. Mineralogical study of titaniferous magnetite and associated sulfide minerals from Tagadur, Hassan district, Karnataka. Ind. Mineral., 14: 80-87.

Pearce, J.A. and Cann, J.R., 1973. Tectonic setting of basic volcanic rocks determined using trace element analyses. Earth Planet. Sci. Lett., 19: 290-300.

Payne, J.G. and Shaw, D.M., 1967. K-Rb relation in the Blue Mountain nepheline syenite. Earth Planet. Sci. Lett., 2: 289-291.

Pichamuthu, C.S., 1976a. Some problems pertaining to the Peninsular gneissic complex. J. geol. Soc. India, 17: 1-16.

Pichamuthu, C.S., 1976b. Are there rocks older than Dharwars in southern India? Ind. Mineral., 16: 1-12.

Radhakrishna, B.P., 1975. The two greenstone groups in the Dharwar craton. Ind. Mineral., 16: 12-15.

Radhakrishna, B.P. and Vasudev, V.N., 1977. The early Precambrian of the southern Indian shield. J. geol. Soc. India, 15: 439-456.

Rama Rao, B., 1962. Hand book of geology of Mysore State. Bangalore Publ. Co.

Ramakrishnan, M. and Mallikarjuna, C., 1976. Konkanahundi pluton. A new layered basic intrusion from Karnataka. J. geol. Soc. India, 17: 207-213.

Ramakrishnan, M., Viswanatha, M.N. and Swaminath, J., 1976. Basement-cover relationship of Peninsular gneiss with high-grade schists and greenstone belts of southern Karnataka. J. geol. Soc. India, 17: 97-111.

Ramiengar, A.S., Ramakrishnan, M. and Viswanatha, M.N., 1976. Charnockite gneiss complex relationship in southern Karnataka and its bearing on crustal development. Abstracts on the Group discussion on Granulite Facies Terrains, Madras, India.

Rivalenti, G., 1976. Geochemistry of metavolcanic amphibolites from south-west Greenland. In: B.F. Windley (Editor), The Early History of the Earth. Wiley, Lond., pp. 213-223.

Schreyer, W., Ohnmacht, W. and Mannchen, J., 1972. Carbonate-orthopyroxenites (Sagvandites) from Troms, Norway. Lithos, 5: 345-364.

Srinivasan, R., Sreenivas, B.L. and Krishna Rao, B., 1976. Evolution of Peninsular gneiss - restraints, reasons and reflections. J. geol. Soc. India, 17: 94-96.

Sutton, J., 1976. Tectonic relationship in the Archaean. In: B.F. Windley (Editor), The Early History of the Earth. Wiley, Lond., pp. 99-104.

Tarney, J., 1976. Geochemistry of Archaean high-grade gneisses with implications as to the origin and evolution of the Precambrian crust. In: B.F. Windley (Editor), The Early History of the Earth. Wiley, Lond., pp. 405-417.

Vidhyadharan, K.T., Sukumaran, R.V. and Nair, M.M., 1977. A note on the occurrence of anorthosite near Perinthatta, Talipuramba taluk, Cannanore district, Kerala. J. geol. Soc. India, 18: 519-520.

Viljoen, M.J. and Viljoen, R.P., 1969. Evidence of the existence of a mobile extrusive peridotitic magma from the Komati formation of the Onvermacht group. Trans. geol. Soc. S.Afr., Sp. Publ.,2: 87-123.

Viswanatha, M.N. and Ramakrishnan, M., 1976. The pre-Dharwar supracrustal rocks of Sargur schist complex in southern Karnataka and their tectonic-metamorphic significance. Ind. Mineral., 16: 48-65.

Viswanatha, M.N., Ramakrishnan, M. and Narayanan Kutty, T.R., 1977. Possible spinifex texture in a serpentine from Karnataka. J. geol. Soc. India, 18: 194-197.

Windley, B.F. and Selvan, T.A., 1975. Anorthosite and associated rocks of Tamilnadu, southern India. J. geol. Soc. India, 16: 209-214.

Windley, B.F. and Smith, J.V., 1976. Archaean high-grade complexes and modern continental margins. Nature, 260: 671-675.

Ziauddin, M. and Yadav, P.K., 1974. Acid charnockite (metasomatic) near Sivasamudram, Karnataka. J. geol. Soc. India, 16: 215-219.

ARCHAEAN GRANITE SERIES AND THE EARLY CRUST, KALGOORLIE SYSTEM WESTERN AUSTRALIA

A.Y. Glikson

Bureau of Mineral Resources, Geology and Geophysics
Canberra, ACT Australia

The relations between granites and greenstones in parts of the Eastern Goldfields of Western Australia are examined in the light of recent stratigraphic, geochemical and isotopic data. A general increase in the K_2O/Na_2O ratio of granites with time is indicated. Two plutonic suites are distinguished: (1) an early granodiorite-tonalite suite intruded at anticlinal positions into stratigraphically low volcanic-sedimentary sequences (early greenstones), and (2) late adamellite intrusions which in places pierce stratigraphically high sequences (late greenstones) discordantly. In the first suite low LIL (large ion lithophile) element abundances and low initial Sr^{87}/Sr^{86} ratios militate against a derivation by ensialic anatexis, and are consistent with partial melting of greenstone assemblages. The widespread occurrence of greenstone xenoliths within these plutons as well as in undated granite and gneiss terrains which surround the greenstone belts, and observed transitions between trains of such xenoliths and stratigraphically low greenstone units, confirm the view of the latter as the oldest rocks recognized in this region. The origin of gneiss enclaves within the granites is interpreted in terms of their derivation from high-grade root zones of the batholiths or formation and deformation along repeatedly dislocated marginal zones of the plutonic bodies. The parallel orientation of metamorphic isograds in greenstone belts and the margins of early granites is interpreted in terms of regional thermal effects by the latter. Because of the abundance of metastable relic plagioclase in the amphibolites the composition of feldspar can not be used as a prograde metamorphic indicator for these rocks. A comparison between plutonic, volcanic and sedimentary assemblages of Archaean and Proterozoic domains in Australia indicates a contrasting nature of their basement, interpreted as simatic and sialic, respectively. All the available evidence points to the formation of the Archaean granite-greenstone systems by progressive and diachronous cratonization of an early simatic regime.

INTRODUCTION

The relations between Archaean granites and enclosed supracrustal assemblages are the key for the interpretation of early crustal evolution. In the Eastern Goldfields of Western Australia (Fig. 1) these relations, where directly observed, are in the main intrusive (Williams, 1970, 1973; Doepel, 1973; Williams et al., 1976; Gee, 1976; Gemuts and Theron, 1976; Gower and Bunting, 1976; Glikson and Sheraton, 1972; Glikson and Lambert, 1976; Hallberg and Glikson, in press). However, where granites are faulted against, or crop out away from greenstones - their temporal significance cannot be directly determined in the field. This problem is compounded by the exceedingly poor outcrop in the Eastern Goldfields region and by uncertainties inherent in the interpretation of isotopic age data. Two contrasting schools have

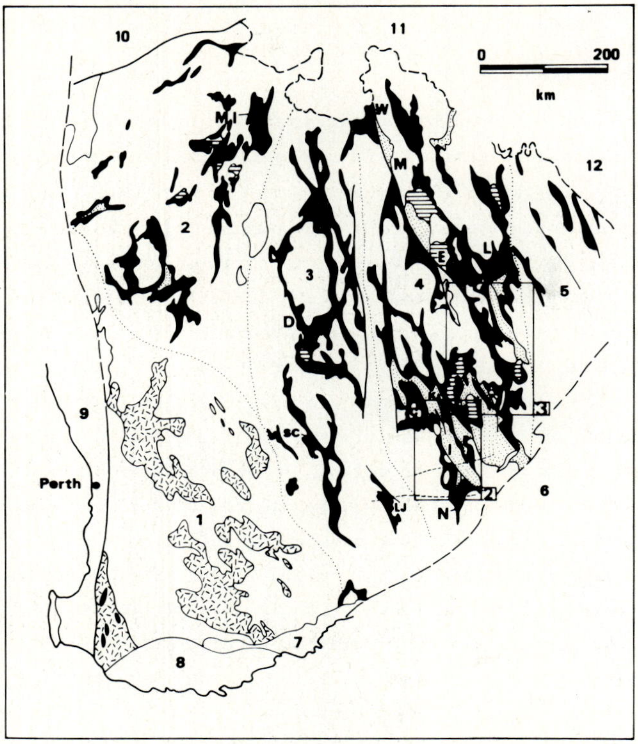

Fig. 1. distribution of granites and greenstone cycles in parts of the Eastern Goldfields region, Yilgarn Block.
1(a) - a sketch map of the Yilgarn Block (after the Tectonic Map of Australia and Gee (1976). 1 - Southwestern Province; 2 - Murchison Province; 3 - Southern Cross Subprovince of the Eastern Goldfields Province; 4 - Kalgoorlie Subprovince of the Eastern Goldfields Province; 5 - Laverton Subprovince of the Eastern Goldfields Province; 6 - Albany-Fraser Province; 7 - Mount Barren Beds 8 - Albany-Fraser Province; 9 - Perth Basin; 10 - Gascoyne Province; 11 - Bangemall Basin; 12 - Nabberu Basin. MI - Meekatharra; D - Deimals; E - Leonora L - Laverton; K - Kalgoorlie; C - Coolgardie; I - Widgiemooltha; N - Norseman; LJ - Lake Johnson. Framed areas indicate the location of Figs.1b and 1c. Solid - greenstones; dotted - sediments; horizontally ruled - acid volcanics; stipled - granulite and gneiss; blank - granite and granite-gneiss.

ensued regarding granite-greenstone relations in, and the genetic interpretation of, the Kalgoorlie terrain. The first views stratigraphically low greenstone sequences as the relics of an early simatic crust (Glikson, 1970, 1972a; Glikson and Lambert, 1976), whereas the other interprets some granites and gneisses in this region in terms of a pre-greenstone sialic basement (Archibald and Bettenay, 1977; Binns et al. 1976; Oversby, 1975). The first approach emphasizes reconstructions of the primary stratigraphy and geochemistry of the greenstones, whereas the second school uses inferred structure-time correlations and metamorphic domain classifications. Currently a coexistence of simatic and sialic crustal domains in Western Australia during the late Archaean is favoured (Gee, 1976; Hallberg and Glikson, in press). The aim of this paper is to examine basic assumptions and criteria inherent in the interpretation of granite-greenstone relations, with particular attention to the significance of the plutonic rocks.

CLASSIFICATION OF GRANITES

Central to the granite-greenstone problem is an elucidation of time-rock sequences within both supracrustal belts and plutonic terrains. In recent years detailed and regional mapping in the Eastern Goldfields has given rise to consistent stratigraphies in the greenstone belts, indicating the existence of at least two, and possibly a larger number of, volcanic-sedimentary cycles (McCall, 1969; Glikson, 1972b; Williams 1973; Gemuts and Theron, 1976). A similar dichotomy of greenstone cycles has been recognized by Hallberg (1976) in the Murchison Province and Southern Cross Subprovinces of the Yilgarn Block. In contrast, the nature of granite- and gneiss-dominated terrains remains largely unknown. One question is whether the gneisses represent deep-level coeval roots of the granite-greenstone system, or are at least in part of pre-greenstone age.

Gemuts and Theron (1976) and Archibald and Bettenay (1977) classified granites in the Kalgoorlie-Norseman area in terms of a concordant synkinematic group and a discordant postkinematic group. No isotopic or geochemical criteria supporting this scheme have been assessed, though it is known that the second group is dominated by adamellite (e.g. Binns and Marston, 1976, p. 24). On Archibald and Bettenay's (1977) sketch map synkinematic granites invariably border greenstone belts, i.e. the Kalgoorlie-Norseman and Southern Cross belts, whereas postkinematic granites occupy the bulk of granite-dominated terrains as well as marginal positions in relation to the greenstone belts. This distribution pattern gives rise to several questions:

(1) If no systematic isotopic and geochemical distinctions were recognized between synkinematic and postkinematic granites, plutons mapped as postkinematic may in some instances merely represent least-deformed massive equivalents of granites mapped as synkinematic, i.e differential deformation of plutonic bodies renders their correlation on structural basis uncertain.

(2) If granites mapped as synkinematic and postkinematic are distinguished by concordant and discordant relations to the greenstones, respectively, by its nature this criterion cannot be used for the classification of granites which crop out away from greenstone belts.

(3) It has been recognized in several Archaean terrains, including the Eastern Goldfields, that adamellite and quartz monzonite crop out more markedly than granodiorite, tonalite and trondhjemite (Macgregor, 1951; Viljoen and Viljoen, 1969; Glikson and Sheraton, 1972; Hickman and Lipple, 1975). This difference arises from a greater resistance to weathering of K-feldspar and quartz as compared to plagioclase and biotite, resulting in a sampling bias in favour of differentiated granites. Thus, extrapolations from relatively resistant granite tors to wider areas can be misleading, yet such a procedure has been followed by Archibald and Bettenay (1977) in designating the vast Coolgardie-Southern Cross terrain and areas about Menzies as postkinematic.

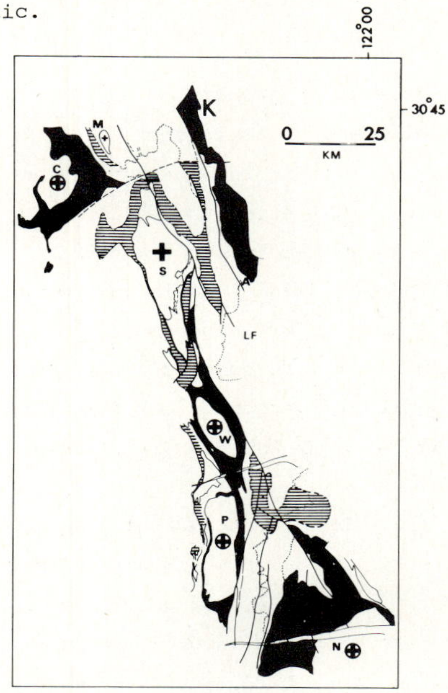

Fig. 2. a sketch map of the Kalgoorlie-Norseman area (after Gemuts and Theron, 1976). Solid - stratigraphically low ultramafic-mafic greenstone sequences; horizontally ruled - stratigraphically high greenstone sequences; circled crosses - early concordant plutons; crosses - late discordant plutons; K - Kalgoorlie; A - Kambalda; M - Mungari granite; C - Coolgardie dome; S - Spargoville granite; W - Widgiemooltha dome; P - Pioneer granite; N - Norseman granite.

The granite-dominated region between Coolgardie and Southern Cross was interpreted by Williams (1973) as an upfaulted block relative to the Kalgoorlie-Norseman greenstone belts. This concept is based on a correlation between greenstones of the Southern Cross Subprovince, and lowermost greenstone associations in the Kalgoorlie Subprovince, and was supported by Glikson and Lambert (1976) on the basis of a proposed eastward tilt of the Yilgarn Block indicated by seismic data (Mathur, 1974). However, Hallberg et al. (1976) have demonstrated an existence of stratigraphically high calc-alkaline volcanics at Marda, Southern Cross Subprovince, which shows that this terraine has not undergone deep level erosion in subsequent times. Thus, it is only possible to regard the Southern Cross terrain as upfaulted in so far as such vertical movement and associated erosion have taken place prior to the calc-alkaline volcanic activity at Marda. This requires an existence of a prolonged hiatus, as possibly represented by the unconformity and basal conglomerate reported by Hallberg et al. (1976) from the base of the calc-alkaline pile.

To date no data have been published which enable a subdivision of components of the continuous granitic terrain between the Kalgoorlie-Norseman and Southern Cross greenstone belts. If the very poor exposure of the granitic rocks which underlie this region is any indication, it may largely consist of deeply weathered plagioclase rich rocks. At least along its easternmost extremity, i.e. near the Kalgoorlie-Norseman belt (Fig. 2), greenstone xenoliths abound in the granites, in places forming discontinuous trains which merge into the Coolgardie greenstones (Gemuts

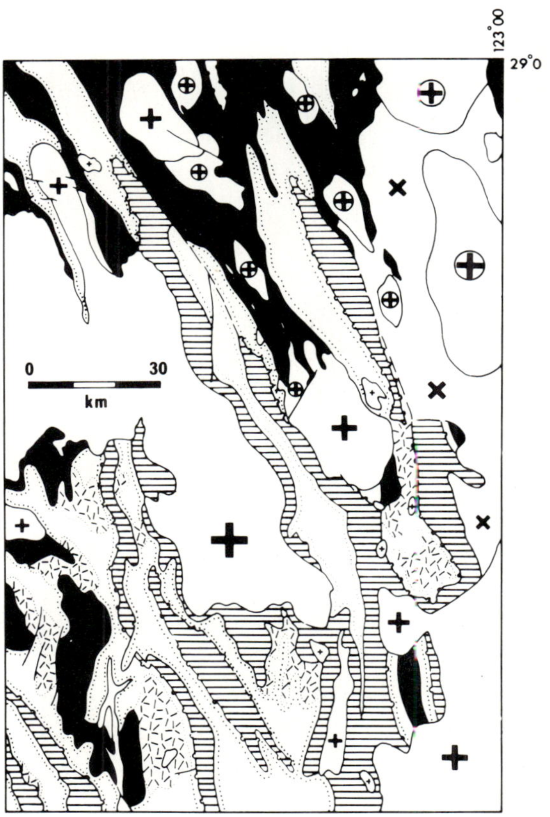

Fig. 3. Sketch map of the Kurnalpi-Edjudina area NE of Kalgoorlie (after Williams, 1970; Williams et al. 1976). Solid - stratigraphically low greenstone sequences; horizontally ruled - stratigraphically high greenstone sequences; dot-bounded areas - sediments; stipled - acid volcanics; circled crosses - early concordant plutons; crosses - late discordant plutons; oblique crosses - magmatites.

and Theron, 1976). In the immediate vicinity of the greenstones, the relations between the acid plutonic bodies and the supracrustal rocks enable a two-fold classification of the intrusions supported by geochemical and isotopic data.

1. Granodiorite-tonalite suite

These intrusions are of oval to eliptical geometry and are concordantly confined to anticlines or domes within stratigraphically low units of the regional greenstone succession. They have gneissose xenolith-rich margins which grade into massive cores. Examples are the Norseman, Pioneer, Widgiemooltha, Coolgardie, and Kambalda intrusions and similar plutons emplaced along the cores of the Carey and Edjudina anticlines (Williams et al., 1976) (Fig. 3). Equivalents are in the north represented by tonalite of the Agnew Dome and the Mount Keith Granodiorite (Cooper et al., 1976; Roddick et al., 1975), and in the west by granodiorite in the Diemal area (Andersen et al., 1976). Available isotopic age data and K_2O/Na_2O ratios are listed in Table 1 and plotted in Figs. 5 and 6.

2. Adamellite suite

These discordantly intrude stratigraphically low to high units of the greenstone belts (Figs. 2,3). The intrusions are little deformed and contain few greenstone xenoliths. Examples are the Mungari and Spargoville plutons and extensive adamelliti[c] and porphyritic granites in the Kurnalpi and Edjudina 1:250.000 Sheet areas (Williams, 1970; Williams et al., 1976). Adamellite dominates among younger intrusion[s] in the Lake Johnson area, west of Norseman (Gower and Bunting, 1976) and in the Diemal area (Andersen et al., 1976). Isotopic age data and K_2O/Na_2O ratios for analyzed samples from this suite are listed in Table 1.

In accordance with the above classification originally proposed on the basis of field relations (Williams, 1970; Glikson and Sheraton, 1972), positive relations pertain between the K_2O/Na_2O ratio of granites and their Pb isochron ages determined by Oversby (1975), but no correlation is observed between this ratio and Rb-Sr ages (Fig. 5). Thus, the Lawlers Leucotonalite (Cooper et al., 1976) yields a comparatively very young Archaean age but has a low K_2O/Na_2O ratio (Table 1). This is a small intrusion and could be possibly compared with minor late-stage tonalites intruded into the Bulawayan and Midlands greenstone belts in Rhodesia, e.g. Sesombi Tonalite (Jahn and Condie, 1976) and several intrusions listed by Phaup (1973).

Significant trace element differences pertain between the Kambalda Granodiorite and the Mungari Adamellite, which are the best knownrepresentatives of the granodiorite-tonalite suite and adamellite suite respectively. Thus the Kambalda intrusion has low Th and U abundances (Th - 5.6ppm; U - 1.9ppm), whereas the Mungari Adamellite has high levels of these elements (Th - 28ppm; U - 7ppm) (Nance and Taylor, 1977). A general increase in the U^{238}/Pb^{204} value of granites with time was reported by Oversby (1975) for granites of the Kalgoorlie-Norseman area

Table 1. Isotopic ages, initial Sr^{87}/Sr^{86}, K_2O/Na_2O and Rb/Sr of granites, granite clasts and acid volcanics in the Kalgoorlie System, Western Australia (pegmatites and aplites are not included).

Geological unit	age (m.y.)	Ri	K_2O/Na_2O	Rb/Sr	Reference
(1) Whole rock Rb-Sr ages					
granite pebbles, Kurrawang conglomerate.	2670±255	0.7025	0.066 (7)	0.052 (11)	Compston & Turek, 1973.
porphyry pebbles, Kurrawang conglomerate	2670±255	0.7025	0.12 (10)	0.1 (3)	Compston & Turek, 1973.
acid volcanics (associations II to IV)	2660±30 to 2595±40		0.40 (50)	0.076 (17)	Turek & Compston, 1971.
Lawlers tonalite	2632±35	0.70145	0.52	0.104	Cooper et al. 1976.
Lawlers leucotonalite	2549±18	0.70223	0.57	0.5	Cooper et al. 1976.
Mt. Keith granodiorite	2689±17	0.70149	0.72 (10)	0.23 (22)	Roddick et al.1975.
Mungari granite	2620±20	0.701 (assumed)	1.2	4.6 (2)	Compston & Turek, 1973.
Deimal granodiorites	2670 to 2610				Andersen et al. 1976.
Deimal adamellite	2560 to 2500				Andersen et al. 1976.
(2) Pb-Pb ages					
Kambalda granodiorite	2760±70		0.42 (4)	0.11	Oversby, 1975 O'Beirne, 1968.
Karonie quarry	2699±75		0.67 (4)		Oversby, 1975.
Stennet rocks	2671±79		0.60 (2)		Oversby, 1975.
Buldania rocks	2655±35		0.87 (3)		Oversby, 1975.
Lake Johnson	2632±28		0.9-1.0 (west) 1.4-2.7 (centre)		Oversby, 1975.
Mungari-Karramindie Soak	2640	Mungari- Karramindie	1.28 (4) 1.27	4.6	Oversby, 1975. O'Beirne, 1968.

Bracketed numbers represent the sample on which the average is based.

Likewise, marked differences pertain to the Rb/Sr ratios of these plutons (Table 1). Rare earth element study by Nance and Taylor (1976) has shown that the Kambalda Granodiorite has a highly fractionated chondrite-normalized pattern (LREE/HREE - 27; La/Yb - 146) and a small positive Eu anomaly (Eu/Eu* - 1.16), suggesting equilibration with garnet and/or amphibole and thus derivation from basic parents. In contrast the Mungari Adamellite shows a highly fractionated LREE curve segment (La/Sm - 8.5) but an almost horizontal HREE curve segment, and a strong negative Eu anomaly (Eu*/Eu - 0.37). The latter pattern can be interpreted in terms of a derivation of the adamellitic melts from a mixed source consisting of about equal amounts of older granodiorites of the Kambalda type and of tholeiitic basalts. Alternatively, the melts could be derived by anatexis of greywacke whose detrital components were derived from such sources. In either case the fractionated LREE part of the pattern is inherited from the Na-rich granites and the flat HREE segment from the basic rocks. The marked negative Eu anomaly indicates extensive fractionation of plagioclase.

The Coolgardie Granodiorite (a likely representative of the granodiorite-tonalite suite) and the Mungari Adamellite are shown on the aerial photomosaic in Fig. 4. There is some evidence that the formation of these suites was separated in time by a period during which volcanic and sedimentary activity persisted in the supracrustal belts - resulting in the accumulation of upper greenstone sequences. This is suggested by the consistent confinement of intrusive margins of the older plutons to lower greenstone sequences (Figs. 2,3), in contrast to the adamellite suite. This view is supported by the occurrence in the Jones Creek area, near Agnew, of a greenstone sequence which possibly overlies the Mount Keith Granodiorite unconformably (Durney, 1972), although structural complications do not allow the confirmation of such relations (Marston and Travis, 1976).

SIGNIFICANCE OF ISOTOPIC DATA

Because of the generally narrow range of granite ages determined in the Eastern Goldfields, i.e. c. 2.75-2.55 b.y. (Table 1), and due to differences between Pb-Pb and Rb-Sr isochron ages, available geochronological data do not as yet furnish an adequate basis for a classification of the granites. Nevertheless, these data provide important constraints for petrogenetic models.

Oversby (1975) determined high μ (U^{238}/Pb^{204}) values in feldspars from granites collected in three localities in the Norseman area, and on the basis of multistage Pb model calculations suggested that these rocks were derived from high μ precursors of adamellitic composition whose age was at least 3300 m.y. This conclusion, however, gives rise to the following questions:

(1) Two of the analyzed granites (at Stennet Rocks and Buldania Rocks) have low K_2O/Na_2O ratios which correspond to granodiorite (Table 1). It is improbable that anatexis of adamellitic parents (K_2O/Na_2O about 1) could give rise to melts characterized by a lower ratio, unless the analyzed granites are regarded as K-depleted residues of the anatectic melt.

(2) Adamellites are comparatively rare in early Archaean terrains, which are dominated by Na-rich acid plutonic rocks, although c. 3.0 b.y. old rocks which show transitional granodiorite-adamellite affinities form a component of Pilbara

Fig. 4. Aerial photomosaic of area between Kalgoorlie and Coolgardie. C = Coolgardie granite dome; M = Mungari granite. These plutons represent the early concordant, and the late discordant, suite.

batholiths (Hickman and Lipple, 1975).

(3) The inferred existence of 3300 m.y. old adamellites is based on a general positive correlation between K_2O/Na_2O and μ values in granites. However, volcanic and sedimentary rocks can have an equally wide range of μ values (e.g. Moorbath et al. 1973), and such rocks could therefore constitute equally suitable parental materials for the granodiorites.

(4) An assumption of a mantle μ values range of 7.4-8.1 (Oversby, 1975) inherent in the Pb model calculations may not be justified if the Archaean mantle was less developed in LIL elements than the modern mantle, as suggested by Hart and Brooks (1974).

An isotopic criterion inherent in which are fewer uncertainties than multistage Pb models and μ - K_2O/Na_2O correlations is provided by the R_i (initial Sr^{87}/Sr^{86}) values of the granites. The generally low R_i of Eastern Goldfields rocks (Table 1, Fig. 6) places limits on the age at which precursors of the granites were derived from material of mantle-type chemistry.* Had the granites originated by the anatexis of parental materials characterized by similar Rb/Sr ratios to those observed in the granites themselves, i.e. higher than 0.1 (Table 1), such precursors must have been separated from mantle-type sources less than 400 m.y., and in most instances less than 200 m.y., prior to the indicated isotopic ages (Fig. 6). However, it is unlikely that the low-R_i granodiorites were derived by anatexis of similar parents, because such a process would have been preceded by partial melting and the segregation of differentiated high-LIL melts - but such igneous rocks characteristically occur only late in the evolution of the Kalgoorlie System (Fig.5). It is far more likely that the granodiorites were derived from precursor materials with low Rb/Sr and K_2O/Na_2O values, such as are characteristic of the greenstone assemblages (see Hallberg, 1972). Partial melting of early, predominantly mafic-ultramafic, volcanic rocks at infracrustal levels is petrogenetically consistent with the generation of tonalitic to granodioritic magmas (Arth and Hanson, 1975; Glikson and Lambert, 1976; Glikson, 1976a). Anatexis of the minor dacitic component of greenstone belt assemblages can account for the more highly fractionated adamellitic component of the early batholiths, e.g. the c. 3.0 b.y. old transitional granodiorite-adamellite rocks in the Pilbara (Hickman and Lipple, 1975). The observed abundance of progressively assimilated mafic-ultramafic xenoliths in high-grade metamorphic Archaean terrains provides direct evidence for the partial melting

* Cooper et al. (1976) measured a pegmatite Rg-Sr age of 2556 ± 18 m.y. with a very high R_i of 0.7624 ± .0068 in the Agnew-Lawlers area, which they interpreted in terms of derivation from an old basement not exposed at the surface. However, because K-rich Rb/Sr high acid volcanics form a minor component of the greenstone sequence (O'Beirne, 1968), derivation from such rocks is an alternative possibility.

process. Assuming an original Rb/Sr ratio of less than 0.05 for mafic parents of the granodiorites, these precursors could have been derived from the mantle at a very early stage of Earth history (Fig. 6).

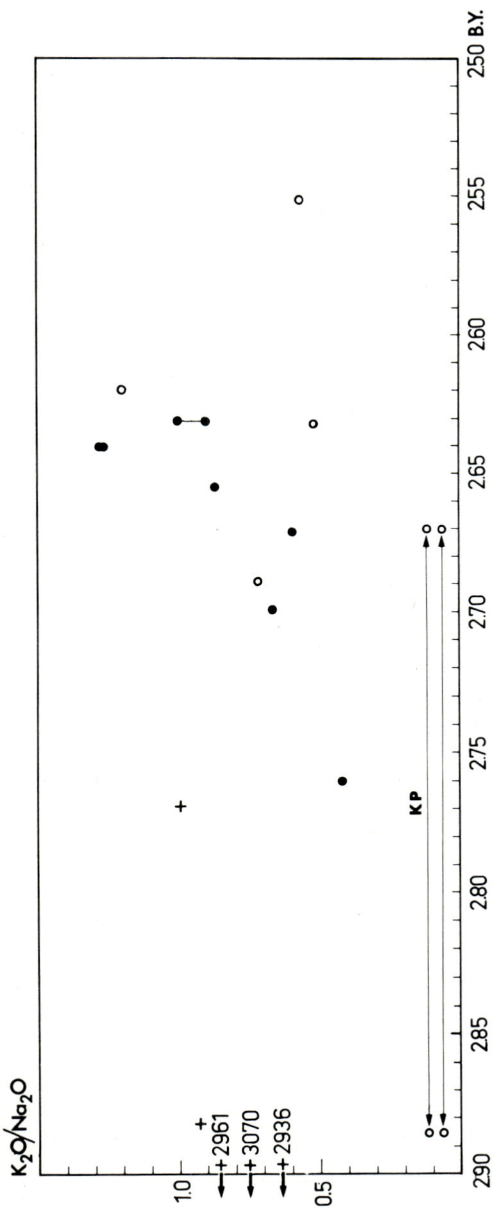

Fig. 5. K_2O/Na_2O of granites from the Eastern Goldfields region plotted against the corresponding isotopic ages (based on data in Table 1). Open circles - Rb-Sr isochron ages; solid circles - Pb-Pb ages; KP - age range of granite and porphyry pebbles from the Kurrawang Conglomerate. Crosses - ages of Pilbara granites.

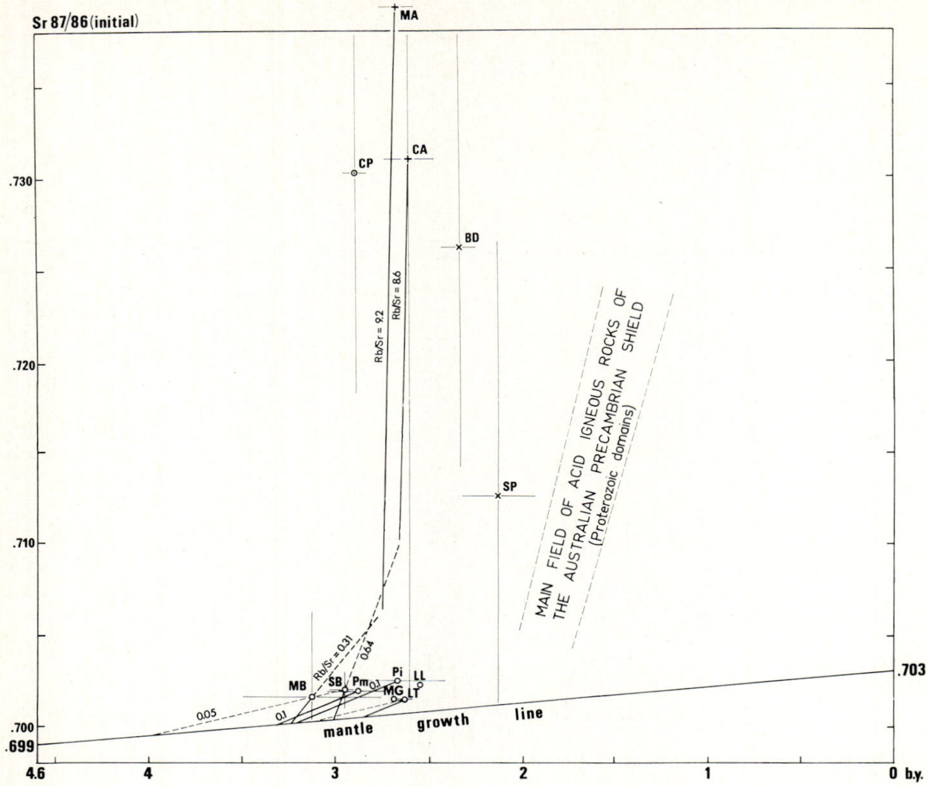

Fig. 6. Initial Sr^{87}/Sr^{86} plotted against corresponding whole rock Rb-Sr ages of granites from the Eastern Goldfields region and the Pilbara Block. LL - Lawlers leucotonalite; LT - Lawlers tonalite; MG - Mount Keith granodiorite; P_i - minimum age of granite and porphyry pebbles from the Kurrawang Conglomerate; P_m - maximum age of granite and porphyry pebbles from the Kurrawang Conglomerate; SB - Shaw Batholith; MB - Mount Edgar Batholith; CP - Copper Hills porphyry; CA - Cooglegong adamellite; BD - Black Range dolerite; SP - Spinaway porphyry. Slopes represent Sr^{87}/Sr^{86} growth lines for the indicated Rb/Sr ratios, and their intersections with the mantle growth line (Rb/Sr = 0.03) represent maximum crustal residence times of precursors of the granites. The approximate limits of the Proterozoic field for acid igneous rocks are after Glikson (1977).

GRANITE-GREENSTONE RELATIONS

In attempting to unravel granite-greenstone relations in the poorly exposed Eastern Goldfields region, it is instructive to refer to the well-documented patterns observed in well-exposed terrains such as the Pilbara and Rhodesia - although it does not necessarily follow that such relations are applicable in the Kalgoorlie terrain. In these terrains trains of xenoliths within the oldest-dated tonalitic and granodioritic gneisses merge imperceptibly into synclinal keels of adjacent greenstone belts. Examples are (1) transitions between xenolith zones in the Mount Edgar Batholith, Shaw Batholith and Tambourah Granodiorite and the Talga Subgroup greenstones (Hickman and Lipple, 1975); (2) A transition between supracrustal xenoliths in the Mashaba tonalite gneiss complex (3580 ±200 m.y., Hawkesworth et al., 1975) and the Victoria greenstone belt (Wilson, 1973); and (3) a transition between the Gwenoro migmatite belt with its abundant supracrustal xenoliths, the Selukwe schist belt, and the Sebakwian Group in the Midlands greenstone belt (Stowe, 1973). Such correlations are uncertain where the batholiths and greenstone belts are intervened by faults, an example being the correlation problem between the Barberton greenstone belt and the Ancient Gneiss Complex in Swaziland (Anhaeusser, 1973; Hunter, 1973). In such instances supracrustal xenoliths in the gneisses have been interpreted as relics of volcanic sequences which predate those of the main greenstone belts. That more than one greenstone sequence occurs is now established in several Archaean terrains in Western Australia, India, South Africa, Rhodesia and Canada (Glikson, 1976b). Whereas stratigraphically high greenstone sequences are known in places to overlie granites unconformably (Bliss and Stidolph, 1969; Bickle and Nisbet, 1975; Baragar and McGlynn, 1976), basal greenstone sequences such as the Sebakwian Group, Tjakastad Subgroup (eastern Transvaal), Sargur Group (southern India), Malartic Group (Ontario-Quebec) and Talga Subgroup (Pilbara) are neither observed to overlie granite nor do they contain granite-derived detritus (Glikson, 1976a; Glikson and Lambert, 1976).

In the Eastern Goldfields region the distribution of xenoliths is difficult to trace owing to poor outcrop. However, larger xenoliths crop out more strongly than their host gneisses and have been mapped in several areas, for example west and south of Coolgardie (Gemuts and Theron, 1976). Moreover, a delineation of xenolith-rich zones is facilitated by aerial total magnetic intensity maps, on which they manifest themselves as linear to arcuate positive anomalies which allow a subdivision of the batholith into subsidiary intrusions. The observation of internal boundaries defined by synclinal keel zones of greenstones within the batholiths suggests the latter represent deeply eroded levels of the granite-greenstone system. The observed transition between the Coolgardie greenstone succession and such xenoliths trains, and the widespread distribution of the latter in the granite-gneiss terrain, confirm the view of the early greenstones as relics of regional volcanic sheets, as contrasted to narrow trough deposits.

Archibald and Bettenay (1977) observed polyphase coaxial folds in the Pioneer Granite dome (Fig. 2), including (a) tight isoclincal folding in banded gneiss along the eastern margin of this pluton, and (b) open folds related to the emplacement of the granite. Because only the latter structures were observed in the adjacent greenstones, they suggested that these rocks postdate the gneiss. However, the basic assumptions inherent in this relative dating method are questionable. An application of comparative superposition orders to relative time determination, useful within lithologically uniform structural domains, is beset by severe difficulties where comparisons between lithologically contrasted domains are concerned, e.g. between greenstone belts and granitic batholiths. In the Kalgoorlie-Norseman greenstone belt styles of deformation vary from open folding in synclines to tight isoclinal folding accompanied by subparallel faulting in anticlinal hinges. At least three fold axis orientations are recognized, including NNW, NS and ENE-trending fold systems (Williams, 1973). At least two sets of foliation occur in ductile schist units in the Kalgoorlie-Coolgardie area, including a NNW striking flow-cleavage and an ENE-striking strain-slip cleavage - the latter possibly cotectonic with the ENE-trending cross folds (Glikson, 1971). However, these structural elements are only partly represented, or may be altogether absent, in associated less ductile lithologies - e.g., acid porphyries and gabbroic rocks, nor can it be expected that these structures should occur within the adjacent batholiths. Thus, even if the assumption is made that both the greenstones and the granites have been subjected to the very same tectonic stress fields - an unsupported supposition - the rheological contrasts between these rock bodies would have resulted in different structures.

Inherent in the evolution of batholiths is the internal polyphase deformation related to syn and late-magmatic flow, lit-par-lit intrusion, injection of late magma increments into semiconsolidated rock, and faulting and shearing of contact zones. Superposed on these structures is solid-state deformation associated with emplacement of the plutons at late or post-magmatic stages. Thus, concomitant formation of chilled marginal mantles and their deformation due to continuing upward movement of the diapirs are reflected by transitions from marginal gneissose zones into essentially undeformed cores. In contrast, the detailed structural concordance between layered greenstone successionsand the external geometry of intrusive batholiths indicates that the mode of emplacement and geometry of the latter constituted controlling factors with regard to the deformation of the supracrustal rocks. Differences between the essentially endemic intrabatholithic structures and the structures of their greenstone envelopes cannot therefore be utilized for relative age determinations. The banded gneisses described from the eastern periphery of the Pioneer Granite (Archibald and Bettenay, 1977) could be of late and/or post-magmatic origin related to their marginal location. This interpretation is consistent with the parallel orientation of banding, gneissosity and the contact of the Pioneer Granite, whereas had the gneisses represented basement xenoliths (Archibald and Bettenay, 1977) a random distribution could be expected. It is concluded that, to date, no pre-greenstone rocks have been identified in the Eastern Goldfields.

METAMORPHIC AND TECTONIC EFFECTS OF GRANITE EMPLACEMENT

Central to the elucidation of granite-greenstone relations is the metamorphic and tectonic effects of the plutonic activity on the supracrustal successions. Binns et al. (1976) and Archibald and Bettenay (1977) suggested that the greenstone belts include dynamic and static metamorphic domains, and that this division has predated the emplacement of the synkinematic granites. According to their map the dynamic high-grade metamorphic domains are consistently associated with the synkinematic granites - a relationship suggestive of thermal contact effects rather than a pre-granites zonation. As noted earlier, in the Kalgoorlie-Norseman greenstone belt deformation is more intense in anticlinal compared to synclinal zones, and as the position of anticlines and domes is closely related to that of the early oval granodiorites these differences in tectonic style and degree of strain can be directly attributed to the controlling factor of plutonic activity. Degrees of strain, however, while generally higher along anticlinal hinges, are also locally high in the vicinity of shear zones. Deformation analysis of pebbles in the Kurrawang Conglomerate in an open folded synclinal structure between Kalgoorlie and Coolgardie has suggested degrees of strain in excess of 50% (Glikson, 1971). Further strain analyses are required to assess whether a systematic division of the greenstone terrain is possible.

Although the contact metamorphic aureoles associated with the granites are in places well pronounced, to the author's knowledge upper amphibolite and granulite facies rocks are comparatively uncommon. This observation is in contrast to the metamorphic map of Binns et al. (1976), on which mid to upper amphibolite facies zones predominate in wide zones along granite contacts. An examination of the mineralogical criteria used by these authors to define metamorphic grade, however, reveals that calcic plagioclase composition (An content) has been used as a prograde metamorphic indicator in the dominantly mafic rocks with which these metamorphic zones coincide. Previously petrological and geochemical studies (Glikson, 1971; Hallberg, 1972) have established the common preservation of relic igneous calcic plagioclase in the amphibolites - a feature attributed to their low pressure Abukuma type facies series and low partial H_2O pressure. The greenschist facies metamorphism in this terrain is indicated by assemblages containing calcic plagioclase and actinolite, and by chlorite-biotite-muscovite assemblages in sedimentary intercalations. The transition to the amphibolite facies is marked by a prograde conversion of the labradorite-actinolite to the andesine-hornblende assemblage. Similar prograde reactions in amphibolites have been documented in the Cloncurry region by England (1973) by means of an electron probe study, showing a progressive equilibration and narrowing of compositional ranges with rising grade. Thus, at the present state of knowledge, the bulk of the supracrustal sequence is thought to have been metamorphosed according to the greenschist to lower amphibolite facies, whereas narrow mid to upper amphibolite zones may be developed in close proximity to intrusive granites, clearly constituting thermal contact aureoles. No evidence for a control of the metamorphic zonation pattern by any factor other than these intrusions is apparent.

NATURE OF THE EARLY CRUST

The very old isotopic ages yielded by Archaean gneisses in several shield areas (e.g. Black et al., 1971; Goldich and Hedge, 1974; Hawkesworth et al., 1975), when compared with the commonly younger metamorphically reset ages of greenstones, has resulted in the impression as if: "the majority of recent structural and isotopic evidence from Archaean greenstone terrains in Africa and America suggest that greenstone belts formed on a sialic basement" (Archibald and Bettenay, 1977, p.370). An important intuitive factor in this regard is the impression of antiquity imparted by the structurally complex and more highly metamorphosed gneiss suites. Thus, although the tonalitic and granodioritic gneisses regarded as basement contain abundant xenoliths of older supracrustal rocks, their presence is explained by proponents of the early sial hypothesis in terms of basement "remobilization" or "reactivation". Such models, however, take no account of the fact that any anatectic processes within a sial basement would have initially produced large volumes of LIL-enriched eutectic adamellite and quartz monzonite melts, nor are they consistent with the low R_i values of the gneisses.

In the bid to prove a sial basement beneath the Kalgoorlie greenstones a comparison has been suggested between the Archaean ironstones and those of the ensialic lower Proterozoic Hamersley Basin (Gole, 1976; Binns et al., 1976). Further, occurrences of acid volcanic rocks and their sedimentary derivatives at low stratigraphic levels of the greenstone succession are often cited as evidence for continental environment. However, the validity of such criteria is questionable. The bathymetric significance of banded iron formation is unknown, and their common occurrence as intercalations in thick successions of pillow lava requires their deposition in either deep water or rapidly subsiding basins, i.e. in order to account for the persistence of subaqueous conditions as indicated by the pillows. In considering the significance of the acid volcanics, attention is drawn to their predominantly dacitic to Na-rich rhyolitic chemistry and the attendant LIL-depleted nature (O'Beirne, 1968). Because such magmas are petrogenetically incompatible with ensialic anatexis, but are consistent with the partial melting of basic source materials, the occurrence of this type of acid volcanics - and possible hypabyssal and plutonic counterparts - requires the existence of earlier greenstone-type parental rocks. A case in point is the abundance of dacitic to Na-rich rhyolitic lavas in modern island arc-trench domains, as well as the occurrence of oceanic plagiogranites in abyssal and ridge environments (Coleman and Peterman, 1975).

It is instructive to compare some principal characteristics of Archaean and Proterozoic terrains in the Australian Precambrian Shield with reference to the crustal

Table 2. A comparison between some lithological and geochemical characteristics of Archaean and Proterozoic rocks of the Australian Precambrian Shield.

Characteristic	Archaean terrains	Proterozoic terrains
Mafic volcanics	low-LIL element tholeiites, high-Mg to peridotitic komatiites.	Low- to high-LIL element tholeiites. Komatiites are very rare. Andesite is rare.
Intermediate volcanics	Andesite may be very common, or bimodal. Mafic-felsic volcanics predominate.	
Felsic volcanics	Mostly Na-rich K-poor dacite and rhyolite porphyries. Some K-high types.	Mostly K-rich rhyolites.
Granites	Prior to ca 2.7 b.y. mainly granodiorite, tonalite and trondhjemite with low R_i (mostly below 0.705) and highly fractionated LREE-enriched types. Adamellites appear mainly about 2.6 b.y.	Mostly adamellite, quartz monzonite, some granodiorite; R_i values show a wide range, and are mostly above 0.710. Negative Eu anomalies common.
Clastic sediments	Greywackes and siltstones very common; orthoquartzite and aluminous shales are rare. Volcanogenic sediments and polymictic conglomerates are common.	Cross-bedded quartzites and shales are very common. Greywacke and siltstone are common in some terrains.
Chemical sediments	Banded ironstone and chert are are common. Carbonates are rare.	Banded ironstones abound in the lower Proterozoic. Carbonates are very common.
Chemical features of sediments	Low K/Na of clastic sediments; REE curves with positive Eu anomalies, or flat. Low Sr^{87}/Sr^{86} ratios of carbonates and, by inference, of sea water. High ferromagnesian trace element contents.	High K/Na of clastic sediments; REE curves with negative Eu anomalies. High Sr^{87}/Sr^{86} ratios of carbonates and, by inference, of sea water. Lower ferromagnesian trace element contents.

environment in which they have evolved. The comparison outlined in Table 2 includes significant differences which point to the development of Archaean and Proterozoic basins in distinct domains. Conversely, had both Archaean and Proterozoic volcanic-sedimentary assemblages developed above sial, their differences remain inexplicable. The geochemical and isotopic data of the Archaean igneous rocks are mostly indicative of mantle melting and basic crust fusion, i.e. one- or two-stage melting processes (Ringwood and Green, 1966). In contrast, differentiated products of 3-stage melting, i.e. extensive fusion of sialic rocks, abound and commonly predominate in Proterozoic terrains (Fig. 6). It is impossible to envisage a formation of thick komatiite-low LIL tholeiite-dacite successions above granitic basement which would not result in anatexis of the latter under the high heat flow conditions in the Archaean (Lambert,

1976). It is equally difficult to conceive a rise of LIL-depleted tonalitic to granodioritic diapirs through an older sial which would not ensue in extensive contamination and elevation of R_i values. Most significantly, differentiated K-rich acid magmas make appearance toward the close of the Archaean cycle in every terrain (e.g. Fig. 5), whereas a deposition of greenstone sequences above subsiding sial should have been manifested by an abundance of eutectic magmas at early stages of these cycles.

Whereas it is highly unlikely, for the above reasons, that stratigraphically low greenstone successions in the Eastern Goldfields were deposited above sial, it is suggested that the development of late greenstone cycles has taken place in a crustal environment consisting of both the early greenstones and the ca 2.7 b.y. old granodiorites and tonalites intruded into them. These plutonic events have clearly post-dated those recorded from the Pilbara and Wheat Belt (southwestern Yilgarn) terrains at ca 3.0 b.y. ago (Arriens, 1971; LeLaeter and Blockley, 1972; Oversby, 1976). In this interpretation, following the cratonization of the early sima in the Pilbara to the north and the Wheat Belt to the west, simatic environments persisted over most of the central and eastern parts of the Yilgarn Block until about 2.7 b.y. ago (Fig. 7). No older age limits have to date been placed on the stratigraphically lowermost greenstone units in the Yilgarn and the Pilbara, and recent zircon age (Pidgeon, 1978) and Pb model age (Sangster, 1977) determinations indicate an age in excess of 3.5 b.y. for the oldest known greenstones (Talga Subgroup, Hickman and Lipple, 1975) in the latter terrain. The possibility that small sialic nuclei may have evolved in pre-2.7 b.y. times in the central and eastern parts of the Yilgarn Block must be examined by further isotopic studies of the granites. To date, all the information at hand is consistent with a view of Archaean granite-greenstone systems as the product of progressive and diachronous transformation of sima into sial by means of two-stage - and toward the close of the Archaean three-stage - anatectic processes.

SUMMARY

(1) Granites associated with greenstones of the Kalgoorlie terrain are classified in terms of a ca 2.7 b.y. old granodiorite-tonalite suite and a ca 2.6 b.y. old adamellite suite. Minor late-stage tonalites are known.

(2) The geochemical and isotopic characteristics of the 2.7 b.y. old suite are consistent with its derivation by partial melting of greenstone assemblages but militate against models inherent in which is the anatectic reactivation of older sial.

(3) The region-wide distribution of outliers and xenoliths of early greenstone units suggests their interpretation as relics of a once-continuous volcanic crust. To date, no granites or gneisses which predate this crust have been reported from the central and eastern parts of the Yilgarn Block.

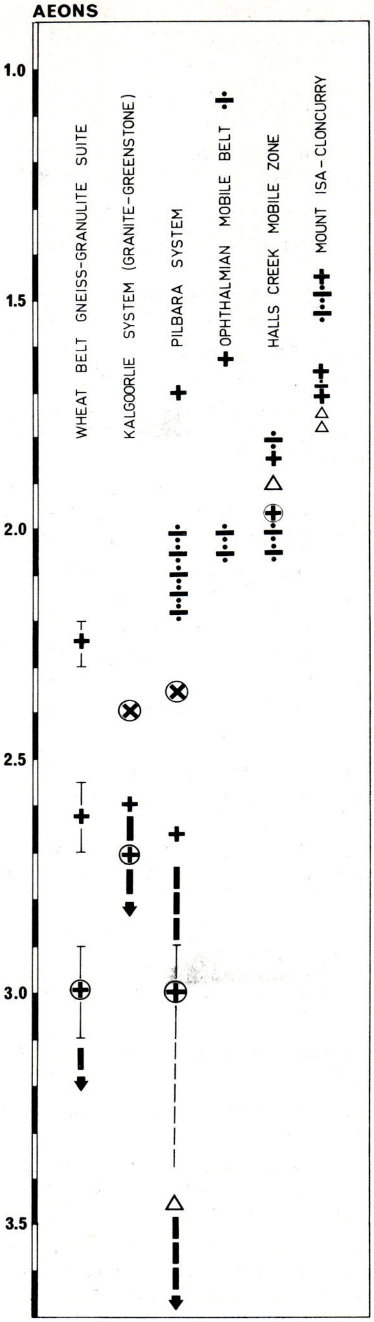

Fig. 7. Interpreted temporal distribution of major events in the evolution of several Precambrian terrains in Australia. Black lines and arrows - minimum ages of early greenstone successions; circled crosses - tonalitic and granodioritic plutonism; dashed lines - late greenstone successions; crosses - late differentiated granites; horizontal lines and dots - continental supracrustal deposits; open triangles - acid volcanic rocks; circled oblique crosses - basic dykes.

(4) Metamorphism and deformation within the greenstone belts can be largely attributed to thermal and dynamic effects of batholith emplacement.

(5) All the field, geochemical and isotopic evidence to date point to the evolution of the Kalgoorlie granite-greenstone system by a progressive and diachronous transformation of simatic into sialic crustal environments.

ACKNOWLEDGEMENTS

Grateful acknowledgement is due to R. Davy, J.A. Hallberg, A.H. Hickman, I.B. Lambert, D. Lewis, W. Libby and R. Pidgeon for their detailed and constructive comments and criticism. This paper is published with the permission of the Director of Mineral Resources, Geology and Geophysics.

REFERENCES

Andersen, L.S., Bettenay, L.F., Binns, R.A., Delaeter, J.R., Gorton, M.P. and Groves, D.I., 1976. Archaean crustal history of the central Yilgarn Block, Western Australia (abstract). Inter. Geol. Congr. 25th, 1: 3-4.

Anhaeusser, C.R., 1973. The evolution of the early Precambrian crust of southern Africa. Phil. Trans. R. Soc. London, A273: 359-388.

Archibald, N.J. and Bettenay, L.F., 1977. Indirect evidence for tectonic reactivation of a pre-greenstone sialic basement in Western Australia. Earth Planet. Sci. Lett., 33: 370-378.

Arriens, P.A., 1971. The Archaean geochronology of Australia. Geol. Soc. Aust. sp. Publ., 3: 11-24.

Arth, J.G. and Hanson, G.N., 1975. Geochemistry and origin of the early Precambrian crust of northeastern Minnesota. Geochim. Cosmochim. Acta, 39: 325-362.

Baragar, W.R.A. and McGlynn, J.C., 1976. Early Archaean basement in the Canadian shield: a review of the evidence. Geol. Surv. Can. Pap. 76-14.

Bickle, M.J., Martin, A. and Nisbet, E.G., 1975. Basaltic and peridotitic komatiites and stromatolites above a basal unconformity in the Belingwe greenstone belt, Rhodesia. Earth Planet. Sci. Lett., 27: 155-162.

Binns, R.A., Gunthorpe, R.J. and Groves, D.I., 1976. Metamorphic patterns and development of greenstone belts in the eastern Yilgarn Block, Western Australia. In: Early History of the Earth, B.F. Windley (Editor), Wiley, London, pp. 331-350.

Binns, R.A. and Marston, R.J., 1976. Archaean geology of the Yilgarn Block, Western Australia. Inter. Geol. Congr. 25th Excursion 40A guidebook.

Black, L.P., Gale, N.H. Moorbath, S., Pankhurst, R.T. and McGregor, V.R., 1971. Isotopic dating of very old Precambrian amphibolite facies gneisses from the Godthåb District, West Greenland. Earth Planet. Sci. Lett., 12: 245-259.

Coleman, R.G. and Peterman, Z.E., 1975. Oceanic plagiogranites. J. Geophys. Res. 80: 1099-1108.

Compston, W. and Turek, A., 1973.Radiometric age limits for the provenance and deposition of the Kurrawang Beds, Coolgardie Goldfield, Western Australia. J. Geol. Soc. Aust., 20: 217-222.

Cooper, J.A., Nesbitt, R.W. and Mortimer, G., 1976. Crustal development in the Agnew-Lawlers area, Western Australia, as shown by Rb-Sr isotopic and chemical studies (abstract). Inter. Geol. Congr. 25th, 1: 7.

DeLaeter, J.R. and Blockley, J.G., 1972. Granite ages within the Pilbara Block, Western Australia. J. Geol. Soc. Aust., 19: 363-370.

Doepel, J.J.G., 1970. Norseman 1:250,000 Geological Sheet with explanatory notes. Geol. Surv. W. Aust.

Durney, D.W., 1972. A major unconformity in the Archaean, Jones Creek, Western Australia. J. Geol. Soc. Aust., 19: 251-259.

England, R.N., 1973. Progressive metamorphism of amphibolites from the Cloncurry and Petermann Ranges areas. Bur. Miner. Resour. Rec. 1971/109.

Gee, R.D., 1976. Regional geology of the Archaean nuclei of the Western Australian Shield. in: Economic Geology of Australia and Papua-New Guinea, C.L. Knight (Editor), Aust. Instit. Min. Met., Melbourne. pp. 43-55.

Gemuts, I. and Theron, A., 1976. The stratigraphy and mineralization of the Archean between Kalgoorlie and Norseman, Western Australia. in: Economic Geology of Australia and Papua-New Guinea, C.L. Knight (Editor), Aust. Instit. Min. Met., Melbourne. pp. 66-74.

Glikson, A.Y., 1970. Geosynclinal evolution and geochemical affinities of early Precambrian systems. Tectonophysics, 9: 397-433.

Glikson, A.Y., 1971. Structure and metamorphism of the Kalgoorlie System southwest of Kalgoorlie, Western Australia. Geol. Soc. Aust. Sp. Publ., 3:121-132.

Glikson, A.Y., 1972. Petrology and geochemistry of metamorphosed Archaean ophiolites, Kalgoorlie-Coolgardie, Western Australia. Bull. Bur. Miner. Resour. Aust., 125: 121-189.

Glikson, A.Y., 1972. Early Precambrian evidence of a primitive ocean crust and island nuclei of sodic granite. Bull. Geol. Soc. Am., 83: 3323-3344.

Glikson, A.Y., 1976a, Stratigraphy and evolution of primary and secondary greenstones: significance of data from shields of the southern hemisphere. In: Early History of the Earth.B.F. Windley (Editor), Wiley, London, pp. 257-278.

Glikson, A.Y., 1976b. Trace element geochemistry and origin of early Precambrian acid igneous series, Barberton Mountain Land, Transvaal. Geochim. Cosmochim. Acta, 40: 1261-1280.

Glikson, A.Y., 1977. Evidence on the radius of the Precambrian Earth. Bur. Miner. Resour. J. Aust. Geol. Geophys., 2: 229-232.

Glikson, A.Y. and Sheraton, J.W., 1972. Early Precambrian trondhjemitic suites in Western Australia and Northwestern Scotland and the geochemical evolution of shields. Earth Planet. Sci. Lett., 17: 227-242.

Glikson, A.Y. and Lambert, I.B., 1976. Vertical zonation and petrogenesis of the early Precambrian crust in Western Australia. Tectonophysics, 30: 55-89.

Gole, M.J., 1976. Metamorphosed banded iron formations in the Yilgarn Block, Western Australia (abstract). Inter. Geol. Congr. 25th, Vol. 1: 11-12.

Goldich, S.S. and Hedge, C., 1975. 3800 Myr granitic gneiss in southwestern Minnesota. Nature, 252: 467-468.

Gower, C.F. and Bunting, J.A., 1972. Lake Johnstone 1:250.000 Geological Sheet with explanatory notes. Geol. Surv. W. Aust.

Hallberg, J.A., 1972. Geochemistry of Archaean volcanic belts in the Eastern Goldfields region of Western Australia. J. Petrol., 13: 45-56.

Hallberg, J.A., 1976. A petrochemical study of a portion of the western Yilgarn Block. Com. Sci. Indust. Res. Organ. Min. Res. Lab. Rep. FP13.

Hallberg, J.A., Johnston, C. and Bye, S.M., 1976. The Archaean Marda igneous complex, Western Australia. Precambrian Res., 3: 111-136.

Hallberg, J.A. and Glikson, A.Y. (in press). Archaean granite-greenstone terrains of Western Australia. In: Precambrian shields of the southern hemisphere, D.R. Hunter (Editor).

Hart, S.R. and Brooks, C., 1974. The geochemistry and evolution of early Precambrian mantle. Carnegie Instit. Wash. Yearbook 73: 967-977.

Hawkesworth, C.J., Moorbath, S. and O'Nions, R.K., 1975. Age relationships between greenstone belts and granites in the Rhodesian Archaean craton. Earth Planet. Sci. Lett., 25: 251-262.

Hickman, A.H., 1975. Precambrian structural geology of part of the Pilbara region. Geol. Surv. W. Aust. Ann. Rep. for 1974: 68-73.

Hickman, A.H. and Lipple, S.L., 1975. Marble Bar 1:250.000 Geological Sheet with explanatory notes. Geol. Surv. W. Aust.

Hunter, D.R., 1970. The ancient gneiss complex in Swaziland. Trans. Geol. Soc. S. Africa., 73: 107-150.

Jahn, B. and Condie, K.C., 1976. On the age of Rhodesian greenstone belts. Contr. Miner. Petrol., 57: 317-330.

Lambert, R. St. J., 1976. Archaean thermal regimes, crustal and upper mantle temperatures and progressive evolutionary model for the Earth. In: Early History of the Earth, B.F. Windley (Editor), Wiley, London. pp. 363-376.

Macgregor, A.M., 1951. Some milestones in the Precambrian of southern Rhodesia. Proc. Geol. Soc. S. Africa, 54: 27-34.

Marston, R.J. and Travis, G.A., 1976. Stratigraphic implications of heterogeneous deformation in the Jones Creek Conglomerate, Kathleen Valley, Western Australia. J. Geol. Soc. Aust., 23: 141-156.

Mathur, S.P., 1974. Crustal structure in southwestern Australia from seismic and gravity data. Tectonophysics, 24: 151-182.

McCall, G.J.H., 1969. The Archaean succession west of Lake Lefroy. J. Proc. R. Soc. W. Aust., 52: 119-128.

Moorbath, S., O'Nions, R.K. and Pankhurst, R.J., 1973. An early Archean age for the Isua iron formation. Nature, 245: 138-139.

Nance, W.B. and Taylor, S.R., 1977. Rare earth element patterns and crustal evolution - II. Archaean sedimentary rocks from Kalgoorlie, Australia. Geochim. Cosmochim. Acta, 41: 225-231.

O'Beirne, W.R., 1968. The acid porphyries and porphyroid rocks of the Kalgoorlie area. Ph.D. Thesis, Univ. W. Aust. (unpubl.).

Oversby, V.M., 1975. Lead isotopic systematics and ages of Archaean acid intrusives in the Kalgoorlie-Norseman area, Western Australia. Geochim. Cosmochim. Acta, 39: 1107-1125.

Oversby, V.M., 1976. Isotopic ages and geochemistry of Archaean acid igneous rocks from the Pilbara, Western Australia. Geochim. Cosmochim. Acta. 40: 817-829.

Pidgeon, R.T. (in press). Isotopic dating of early Precambrian volcanism in the layered greenstone succession of the Pilbara, Western Australia. Earth Planet. Sci. Lett.

Phaup, A.E., 1973. The granitic rocks of the Rhodesian craton. in: Geol. Soc. S. Africa sp. Publ., 3: 59-67.

Ringwood, A.E. and Green, D.H., 1966. An experimental investigation of the gabbro-eclogite transformation and some geophysical implications. Tectonophysics, 3: 383-427.

Roddick, J.C., Compston, W. and Durney, D.W., 1975. The radiometric age of the Mount Keith granodiorite - a maximum age estimate for an Archean greenstone sequence in the Yilgarn Block, Western Australia. Precambrian Res., 3: 55-78.

Sangster, D. and Brook, W.A., 1977. Primitive lead in an Australian Zn-Pb-Ba deposit. Nature (in press).

Stowe, C.W., 1973. The older tonalite gneiss complex in the Selukwe area, Rhodesia. Geol. Soc. S. Africa Sp. Publ., 3: 85-96.

Viljoen, M.J. and Viljoen, R.P., 1969. A proposed new classification of granitic rocks of the Barberton region. Geol. Soc. S. Africa Sp. Publ., 2: 153-188.

Williams, I.R., 1970. Kurnalpi 1:250.000 Geological Sheet with explanatory notes. Geol. Surv. W. Aust.

Williams, I.R., 1974. Structural subdivision of the Eastern Goldfields Province, Yilgarn Block. Geol. Surv. W. Aust. Ann. Rep. for 1973: 53-59.

Williams, I.R., Gower, C.F. and Thom, R., 1976. Edjudina 1:25C.000 Geological Sheet with explanatory notes. Geol. Surv. W. Aust.

Wilson, J.F., 1973. Granites and gneisses of the area around Mashaba, Rhodesia. Geol. Soc. S. Africa Sp. Publ., 3: 79-84.

GRANULITES

MADRAS GRANULITES: GEOCHEMISTRY AND P-T CONDITIONS OF CRYSTALLISATION

B.L. Weaver and J. Tarney
Department of Geological Sciences, University of Birmingham, U.K.

B.F. Windley
Department of Geology, University of Leicester, U.K.

E.B. Sugavanam and V. Venkata Rao
Geological Survey of India, Tamil Nadu Circle, Madras, India

Abstract

A study of the major and trace element geochemistry of Madras granulites has been made using 64 representative samples from the Pallavaram area. They can be divided into four geochemically distinct rock groups: meta-igneous pyroxene granulites, metagabbros and charnockites, and metasedimentary khondalites.

The meta-igneous rocks, taken together, display a tholeiitic trend with marked Fe-enrichment, contrasting with the general calc-alkaline trends shown by Scottish Lewisian granulites. Geochemical differences between the pyroxene granulites and metagabbros suggest that whereas the former may represent original basaltic liquid compositions, the latter approximate more closely to cumulates. The charnockites mostly correspond to adamellite in composition, but the paucity of intermediate compositions emphasises the bimodality of the meta-igneous association, a common feature in other Archaean terrains. In terms of lithophile element concentrations of Ba, Rb, Sr and the light rare earths, the charnockitic rocks are broadly similar to modern Andean equivalents. However the concentration of yttrium is generally low, indicating significant though variable heavy rare-earth depletion and the influence of garnet or hornblende in their genesis.

In common with other granulite terrains there is a trend towards increasing K/Rb rations with decreasing K_2O, and K/Rb ratios of 1700 are recorded in some rock types. However there is a strong dependence of K/Rb ratio on mineralogy, those rocks with potassic phases having lower K/Rb ratios. Nevertheless, even the K-feldspar-rich charnockites have K/Rb ratios higher than the average crustal value.

Microprobe analyses of coexisting pyroxenes and garnets in different mineralogical assemblages provide consistent P-T data (by several methods) defining the metamorphic crystallisation conditions as 720-840°C and 9-10 kb. Feldspar thermometry suggests crystallisation at about 680°C. These P-T conditions agree well with those recorded from other Precambrian granulite facies terrains.

INTRODUCTION

The aim of the present study was to re-examine in detail the geochemistry of rocks from the type-charnockite area of Pallavaram, Madras, with emphasis on trace element geochemistry, and to apply the now standard methods of geothermometry/geobarometry to elucidate the P-T conditions of crystallisation. To this end 64 samples were analysed for major and trace elements and the relevant mineral phases analysed in some samples.

The occurrence of 'hypersthene granulite' in the Pallavaram area was originally described by Holland (1893, 1900) who termed it charnockite. He considered the charnockite suite (ranging in composition from ultrabasic to acid) to represent a differentiated igneous sequence. The geochemistry and mineral chemistry of the type-charnockite area was subsequently examined by Howie (1955). Howie and Subramaniam (1957), Subramaniam (1959) and Ray and Sen in a series of papers (Ray, 1970; Ray and Sen, 1970; Sen, 1970, 1973; Sen and Ray, 1971a, b) all investigated various aspects of the field relations, petrography and mineral chemistry. The charnockite suite was largely considered by these authors to be meta-igneous.

The essential petrography of granulites from the Madras area was well surveyed by Howie (1955), Subramaniam (1959) and Sen and Ray (1971b). 100 thin sections were examined in the present study. The mineralogy of the main petrographic rock groups can be summarised as follows:

(a) khondalite suite

quartz-perthite-garnet-(plagioclase-sillimanite-biotite-spinel-opaque oxide)

(b) charnockite suite

Acid division: microcline (perthite)-plagioclase-quartz-hypersthene-(garnet-opaque oxide)

Intermediate division: plagioclase (antiperthite)-microcline-quartz-hypersthene-(diopside-hornblende-opaque oxide)

Basic division: plagioclase-hypersthene-diopside-hornblende-(opaque oxide-quartz)

Common accessory minerals: apatite, biotite and zircon.

Within the charnockite suite garnet is developed only in the charnockites sensu stricto and appears to have formed by the reaction described by McLelland and Whitney (1977) in which garnet and clinopyroxene are produced in acid granulites by the reaction of anorthite, orthopyroxene, Fe-oxides and quartz. Quartz may be a product or reactant in this reaction, but appears here to be a reactant.

Hornblende in the basic granulites was considered to be primary, from both petrographic and chemical evidence, by Sen (1970, 1973), Ray (1970) and Sen and Ray (1971a, b), and we have no evidence to refute this conclusion. Clearly the assemblages present in the charnockite suite are those of intermediate pressure granulites (Green and Ringwood, 1967). Petrographically there is very little evidence of retrogressive activity in the type area; all the samples appear unusually fresh.

The terminology adopted here is simple and in line with that of Holland (1900) and Howie (1955). Khondalites are those granulites which are recognised as being metasedimentary in origin. Rocks recognised as being meta-igneous are termed the charnockite suite (Holland, 1900). The acid division of the charnockite suite includes the charnockites sensu stricto. There are some enderbites (Tilley, 1932) amongst the acid members of the charnockite suite but this group is here referred to in bulk as the charnockites for simplicity. Intermediate and basic members of the charnockite suite are termed simply intermediate and basic granulites.

In the field there is a conformable intercalated sequence of acid and basic charnockites, khondalites, meta-norites and meta-gabbros. The khondalites contain high alumina pelitic, psammitic and calcareous assemblages. The basic granulites are presumed to have been basic volcanic rocks whilst pink acid granulites rich in K_2O and low in CaO (not analysed by us) may represent acid volcanics. Intruded into the supracrustal rocks were norites in sills, and gabbros and leucogabbros forming layered complexes (outside the type area similar complexes contain calcic anorthosites). Widespread acid charnockites of tonalitic-trondjhemitic-granodioritic composition (hypersthene gneisses) are thought by E.B.S. and V.V.R. to belong to a pre-supracrustal basement. Indeed, whereas on fresh surfaces the charnockitic rocks appear dark and homogeneous, on weathered surfaces they are distinctly gneissic and migmatitic with an appearance similar to that of the Peninsular gneisses.

The basic bodies occur as boudinaged lenses. Late stage potassic pegmatoid lenses show well-developed pinch and swell structures. It is possible that these pegmatoid bodies may be associated with a phase of granitic injection which migmatised and metasomatised the gneisses to their present K-rich state immediately before the granulite facies maximum. The Pallavaram rocks are atypical in being so potash rich and, while this may be a primary characteristic, it is to be noted that elsewhere in the Peninsular Shield tonalitic gneisses predominate.

GEOCHEMISTRY

A total of 64 samples were analysed for both major and trace elements by X-ray florescence spectrometry (a full list of the analyses is available upon request). Trace elements were determined on powder pellets with calibrations produced from international standards; major element determinations were made on powder pellets, but calibrations were based on results from a range of fused samples calibrated against international standards. This method gives enhanced precision for major element results.

Average analyses of the geochemically and petrographically distinct units are

Table 1 Average analyses of Madras granulites

	Khondalites	Basic granulites I	Basic granulites II	Intermediate granulites	Charnockites
n*	10	7	10	4	28
SiO_2%	75.05 (1.21)	50.14 (1.86)	49.43 (0.77)	58.98 (1.55)	72.60 (3.22)
TiO_2	0.37 (0.25)	1.80 (0.20)	0.93 (0.28)	1.59 (0.30)	0.37 (0.18)
Al_2O_3	11.53 (2.04)	13.24 (0.78)	13.11 (0.53)	13.65 (0.75)	13.52 (1.30)
Fe_2O_3	7.24 (4.41)	18.52 (0.89)	12.68 (1.49)	12.64 (1.60)	3.62 (1.11)
MnO	0.09 (0.09)	0.225(0.04)	0.181(0.02)	0.199(0.052)	0.050(0.02)
MgO	1.82 (1.21)	5.13 (1.12)	8.36 (1.10)	2.62 (0.38)	1.16 (0.50)
CaO	0.38 (0.30)	7.36 (1.96)	11.82 (1.75)	5.96 (0.95)	1.67 (1.03)
Na_2O	1.11 (1.35)	2.85 (0.42)	2.00 (0.42)	3.20 (0.54)	2.88 (0.47)
K_2O	2.11 (2.40)	0.87 (0.35)	0.41 (0.43)	0.75 (0.13)	4.08 (1.66)
P_2O_5	0.040(0.010)	0.285(0.14)	0.091(0.028)	0.402(0.207)	0.070(0.040)
Trace elements in p.p.m.					
Cr	5 (3)	38 (30)	351(126)	20 (14)	16 (20)
Ni	< 1 (0)	33 (41)	152(25)	4 (5)	3 (5)
Zn	29 (28)	136 (33)	83 (13)	124(17)	43 (20)
Rb	56 (67)	17 (22)	4 (4)	5 (1)	100(45)
Sr	47 (58)	150(23)	113(29)	180(35)	166(124)
Zr	361(207)	112(25)	56 (18)	158(22)	277(133)
Nb	8 (4)	13 (7)	4 (1)	13 (1)	9 (7)
Y	77 (79)	41 (11)	20 (5)	30 (9)	23 (26)

Table 1 continued.

La	39 (33)	28 (9)	9 (4)	39 (6)	36 (4)
Ce	71 (62)	51 (20)	19 (7)	71 (12)	67 (31)
Ba	560(542)	319(235)	67 (36)	308(186)	879(368)
Pb	14 (12)	7 (4)	5 (2)	9 (3)	19 (7)
Th	23 (27)	< 3 (3)	< 3 (1)	7 (5)	19 (3)
Ga	18 (6)	23 (5)	17 (2)	22 (3)	19 (3)
K/Rb	274	832	730	1187	373
Rb/Sr	0.95	0.11	0.04	0.03	0.89
Ba/Rb	19.83	30.71	21.97	65.23	11.19
Ba/Sr	20.56	2.14	0.65	1.62	7.08

n* = Number of samples on which mean is based.

Figures in brackets are standard deviations.

presented in Table 1. The subdivision of the basic granulites into units I and II is a geochemical distinction.

The khondalites are distinctive both in their geochemistry (Table 1) and petrography. They are all silicic (SiO_2 > 72%), but exhibit a wide range in content of other major and trace elements (as shown by standard deviations, Table 1), though Cr and Ni are low. All are high in normative corundum and quartz. Their metasedimentary origin, as opposed to the meta-igneous charnockite suite, can be demonstrated by the reference to discrimination diagrams such as those of Leake (1964).

Normatively the basic granulites are quartz and olivine tholeiites. They show inter-element correlations (all significant at the 99.5% confidence level) typical of tholeiitic rocks, i.e. Zr shows positive correlations with Zn, Sr, Y, Nb, Ba, La, Ce, Pb, Si, Ti, Na, K, P and negative correlations with Mg and Ca. In this respect Zr is behaving as an incompatible element in the basic granulites, suggesting that they represent a fractionated igneous suite. Geochemically it is possible to divide the basic granulites into two well-defined populations, here termed Groups I and II (Table 1), corresponding loosely to the pyroxene granulites and meta-gabbros in the field descriptions of the Indian Geological Survey. From comparison of Groups I and II (Table 1) it is obvious that the rocks of Group II have a cumulate chemistry, while those of Group I have more the character of residual liquids. This is demonstrated by the higher contents of Cr, Ni, Ca and Mg in Group II, and higher levels of relatively incompatible elements such as Zr, Nb, Y, La, Ce, Na, P, etc. in Group I. Moreover, the Group I granulites have a higher content of Fe, favouring separation of phases with an Fe/Mg ratio less than that of the liquid. Thus Group I mafic rocks appear to be differentiated products and Group II to be cumulates from an original tholeiitic liquid. Higher contents of Ni and Cr in Group II rocks could indicate a higher proportion of cumulus olivine and pyroxene, and the behaviour of Sr (cf. Ba/Sr ratios, Table I) indicates that plagioclase may also have been a cumulus phase. Normatively, Group II rocks have greater amounts of olivine, pyroxene and a more calcic plagioclase, suggesting that these are indeed likely to have been cumulus phases. This will be further discussed below.

The intermediate granulites are restricted in number, and thought to be hybrids formed by interbanding between acid and basic granulites (Holland, 1900; Howie, 1955; Subramaniam, 1959). The charnockite suite then appears to be a bimodal suite with acid and basic members well represented but not necessarily genetically related. This will be discussed below.

The charnockites are by definition acid, all having more than 64% SiO_2, and are highly potassic with c. 4% K_2O (Table 1) and high K/Na ratios. Values for trace elements show considerable dispersion (see standard deviations, Table 1) but Ba, Zr, Rb and Sr are all consistently high. The charnockites are granodioritic to quartz monzonitic in composition. Representative analyses are given in Table 2.

An AFM diagram (Fig. 1) for the whole meta-igneous suite reveals a strongly tholeiitic trend, the more fractionated basic and intermediate granulites showing marked iron enrichment. The compositional gap between the charnockites and basic-intermediate granulites again illustrates the bimodal nature of the charnockite

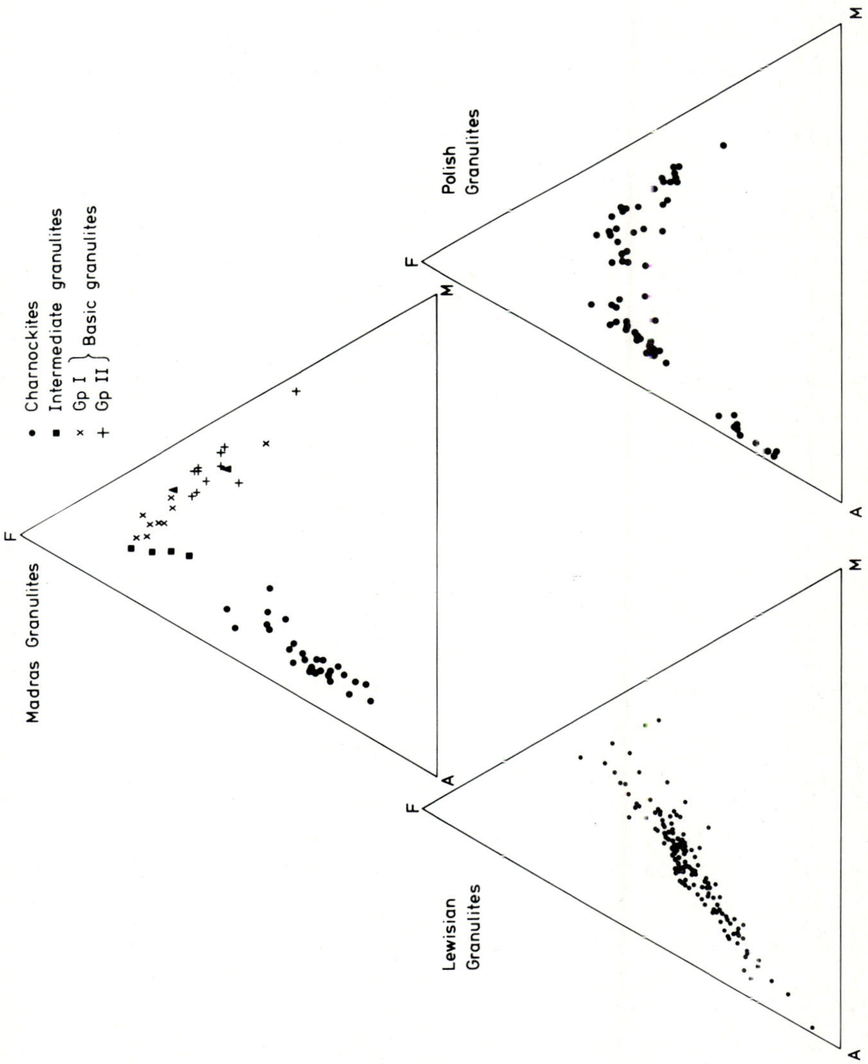

Fig. 1. AFM diagrams for Madras, Lewisian and Polish granulites showing tholeiitic trends for the Madras and Polish granulites and a calc-alkali trend for the Lewisian granulites.

suite. The trend for Scottish Lewisian granulites and retrogressed granulites (Sheraton et al., 1973) is distinctly calc-alkaline (Fig. 1), but this trend may be partially due to tectonic and retrogressive activity (Tarney, 1976) 'smearing' an originally bimodal suite. Polish granulites also display a tholeiitic trend (Tarney, unpublished data).

Although full rare earth element (REE) data were not obtained in the present study, simulated REE plots can be obtained from measured values for La, Ce and Y, using Y as a monitor of heavy rare earth (HREE))behaviour. Such simulated REE plots are presented in Fig. 2 for the khondalites, basic granulites and charnockites, respectively. That the REE are relatively immobile during prograde metamorphism has been demonstrated by Green et al. (1972), O'Nions and Pankhurst (1974) and many other workers, and thus these simulated patterns may be used to make inferences about the original unmetamorphosed nature of the rocks.

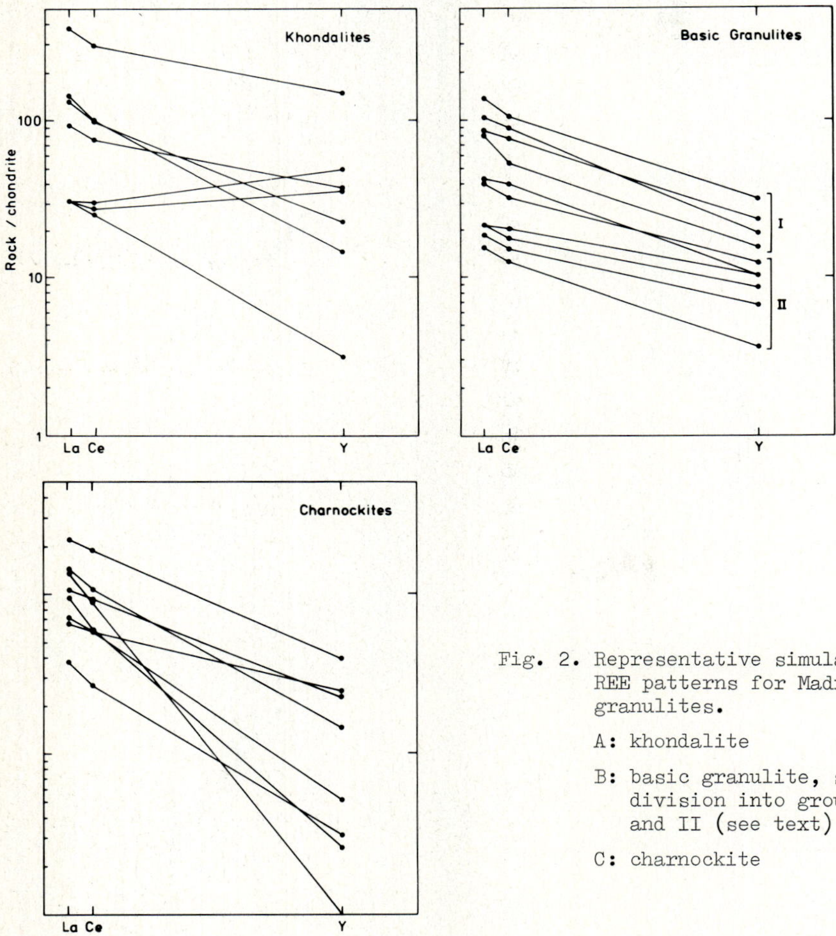

Fig. 2. Representative simulated REE patterns for Madras granulites.

A: khondalite

B: basic granulite, showing division into groups I and II (see text)

C: charnockite

The khondalites show considerable variability in their REE patterns (Fig. 2) though the majority have rare earth distributions enriched in light rare earths (LREE); two samples, however, show LREE-depleted patterns. Quartz-rich sediments rich in heavy mineral concentrates such as garnet could well have bulk REE distributions showing HREE enrichment.

The distinction between Groups I and II of the basic granulites is well demonstrated by their REE patterns (Fig. 2), the Group II samples having lower total levels of REE and slightly lower $(Ce/Y)_N$ ratios; this suggests that Group I rocks are cumulates and Group II residual. Furthermore, the implication is that the crystallising phases must have had low partition coefficients for the REE (of less than 1) thus raising total levels of REE in the melt, but with higher partition coefficients for the HREE than the LREE, increasing the Ce/Y ratio in the residual liquid. The most likely phases capable of achieving this are clino- and orthopyroxene and it would seem that these may have been the dominant phases during crystallisation. Consideration of the REE pattern for the parent liquid (by averaging the patterns for groups I and II) indicates that it must have had an original LREE-enriched pattern similar to that of continental tholeiites. This could be achieved by partial melting of a mantle source with residual pyroxene or possibly minor garnet.

Simulated REE plots for the charnockites (Fig. 2) show rather different relationships. No general trend (apart from LREE enrichment) is obvious, and a wide range in Ce/Y ratios is observed, some samples showing extreme Y depletion. Such Y depletion could only be achieved through processes of fractional crystallisation or partial melting involving minerals with high partition coefficients for the heavy REE such as hornblende (Arth and Barker, 1976) or garnet. Comparison of the charnockite REE patterns with those of the basic granulites suggests very strongly that the patterns observed in the charnockites could not be produced by fractionation of the low-pressure mineral assemblages which seem to have governed REE behaviour in the basic granulites. Thus consideration of REE patterns again demonstrates that the basic and acid granulites are not cogenetic.

A plot of Y_N against $(Ce/Y)_N$ also demonstrates features of the REE patterns. As can be seen from Fig. 3 there is a good negative correlation between Y and Ce/Y for the charnockites, with Y increasing markedly as Ce/Y decreases. Such a trend as this again implies involvement of garnet as a residual phase in the source region of the charnockites. On this diagram (Fig. 3D) the basic granulites show minor horizontal scatter in the Ce/Y values, but a general vertical trend in line with the cumulate-residual relation outlined above. Again there appears to be no relation between acid and basic granulites. In comparison with other areas (Fig. 3A,B,C) it can be seen that the negative correlation between Y and Ce/Y is a common feature in Archaean granulites, probably reflecting the major influence of garnet control on the REE's in the genesis of these rocks.

In recent years a distinct relation between K and Rb has been demonstrated for granulite facies rocks (Lambert and Heier, 1968; Sighinolfi, 1971; Sheraton et al., 1973; Cooper and Field, 1977). Shaw (1968) found that the K/Rb ratio in normal upper crustal rocks averages 230, but shows a slight tendency to increase with decreasing K content. In Archaean granulite facies rocks (i.e. Fig. 4) this

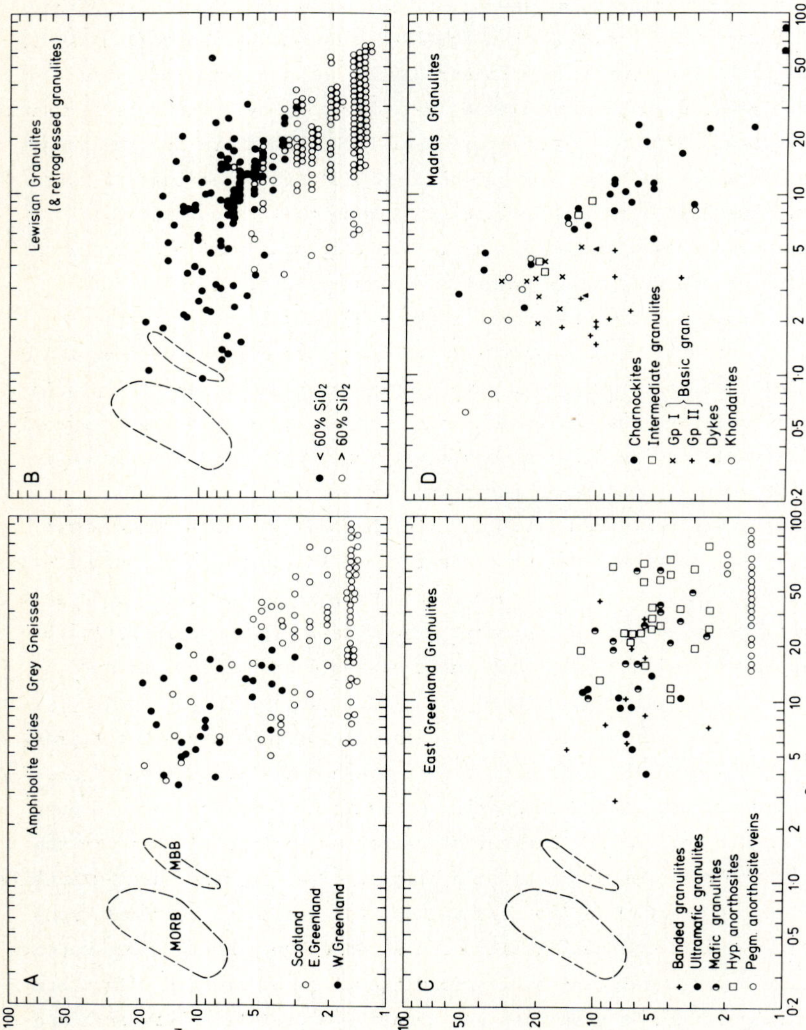

Fig. 3. Y_N against $(Ce/Y)_N$ plot for Scottish and Greenland amphibolite facies grey gneisses (A), Lewisian granulites (B), East Greenland granulites (C), and Madras granulites (D). Modified from Tarney and Windley (1977).

Table 2

Representative analyses of Pallavaram charnockites

	MP 22	MP 72	MP 14	MP 69	MP 97	MP 37
SiO_2%	72.9	71.8	70.6	64.0	64.4	67.3
TiO_2	0.18	0.50	0.28	0.09	0.59	0.41
Al_2O_3	13.8	13.0	14.7	17.0	17.0	15.3
Fe_2O_3	2.39	5.26	1.95	2.66	4.57	3.12
MnO	0.025	0.050	0.024	0.029	0.051	0.038
MgO	0.72	2.42	0.96	1.66	2.74	1.38
CaO	1.15	1.78	1.26	2.10	5.97	2.08
Na_2O	3.23	3.21	2.42	3.60	3.72	2.99
K_2O	5.13	2.21	7.29	7.93	0.37	5.74
P_2O_5	0.140	0.046	0.100	0.218	0.046	0.126
Total	99.67	100.58	99.58	99.29	99.36	98.48
Trace elements in p.p.m.						
Cr	8	91	11	28	54	21
Ni	< 1	18	< 1	9	16	2
Zn	15	63	14	35	57	29
Rb	138	17	179	223	7	104
Sr	96	311	211	369	279	479
Y	48	13	7	12	2	2
Zr	24	189	251	32	237	385
Nb	11	6	4	< 1	4	2
Ba	683	890	1055	1320	352	1708
La	21	30	29	57	29	43
Ce	46	48	48	120	51	69
Pb	24	21	23	47	12	22
Th	3	14	13	76	3	25
Ga	17	18	18	25	28	20
$(Ce/Y)_N$	2.3	9.0	16.3	24.4	62.3	84.3
K/Rb	309	258	338	295	1032	458
Rb/Sr	1.44	0.23	0.35	0.60	0.03	0.22
Ba/Rb	4.95	12.54	5.39	5.92	50.29	16.42
Ba/Sr	7.11	2.86	5.00	3.58	1.26	3.57

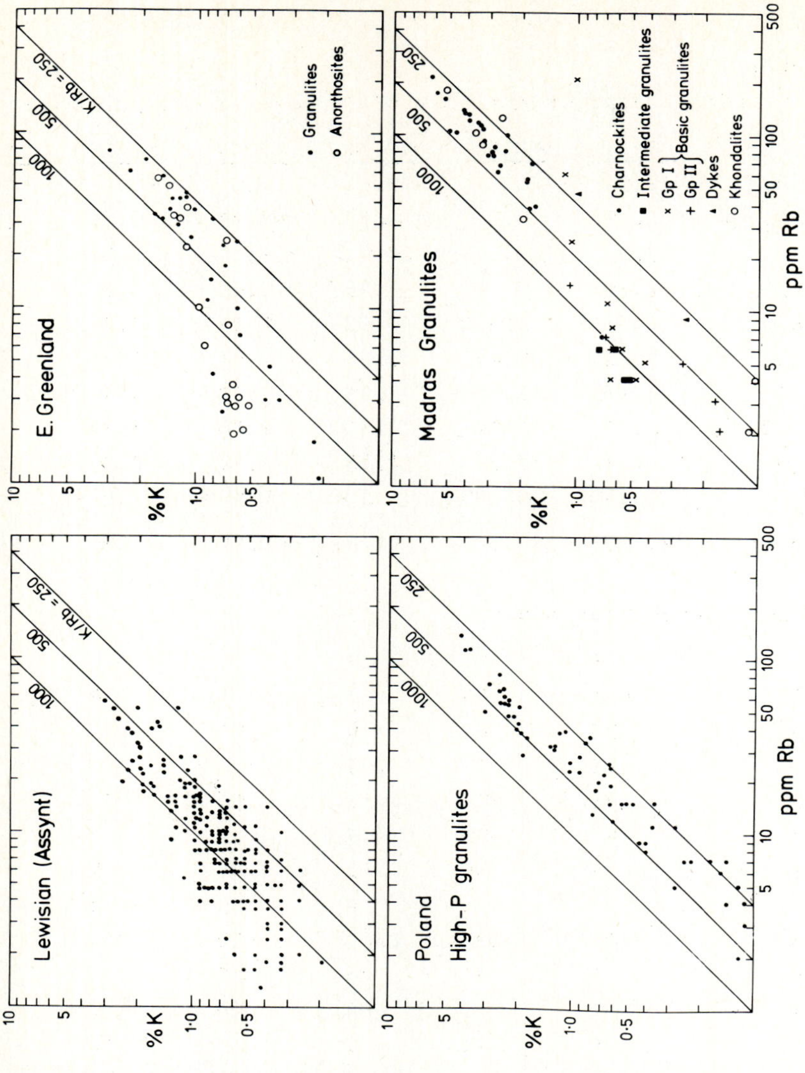

Fig. 4. Plot of K agains Rb for Lewisian, Greenland, Polish and Madras granulites. Modified after Tarney and Windley, 1977.

trend is much more marked, the K/Rb ratio increasing rapidly as K decreases (Tarney and Windley, 1977). In this respect the Madras rocks are typical granulites (Table 2), with the K/Rb ratio reaching 1700.

Development of the K/Rb ratio in the Madras granulites shows a good degree of mineralogical control (Fig. 4), the charnockites having much lower K/Rb ratios (Table 2), the basic granulites developing the higher values. This is due to the presence of K-feldspar as a stable phase in the charnockites and is in marked contrast to Lewisian granulites where there is little mineralogical control over the K/Rb ratio with all samples (whether acid or basic) showing generally high K/Rb ratios. Significantly Lewisian granulites are diopside normative and K-feldspar free.

Bimodal meta-igneous suites are a common feature of many Archaean terrains (Barker and Peterman, 1974; Barker and Arth, 1976), the two components being trondhjemitic or tonalitic and basaltic in composition, with acid material dominant, and with a distinct lack of intermediate (andesitic) material. Petrogenetic models for the evolution of the components of these suites were proposed by Arth and Hanson (1975) and Barker and Arth (1976). The basaltic component was considered by Arth and Hanson (1975) to be the product of partial melting of mantle peridotite at shallow depths, and the basic granulites conform to this pattern, with the implication that pyroxenes (or possibly minor garnet) were residual phases.

Barker and Arth (1976) suggested that low Al_2O_3 (less than 15%) trondhjemite-tonalite liquids could be generated by partial melting of amphibolite or hornblende-bearing gabbro, in which plagioclase is residual but garnet and/or hornblende are not. This was invoked to explain the negative Eu anomalies, enriched LREE and flat HREE patterns observed in some tonalite-trondhjemites. Barker et al. (1976) describe low Al_2O_3 trondhjemitic rocks from Colorado with fractionated REE patterns but with small or absent Eu anomalies. These they consider to have been generated at depths in excess of 50 km by partial melting leaving residual hornblende, garnet, clinopyroxene and plagioclase, or melting at slightly greater depths with residual garnet and clinopyroxene. Similarly, Arth and Hanson (1975) suggest the acid rocks of northeastern Minnesota were largely derived by partial melting of either eclogite or mixed eclogite-peridotite parent material at mantle depths or by partial melting of shortlived greywacke, in order to account for the high K_2O and SiO_2, negative Eu anomalies and low initial Sr ratios of the rocks. The strongly fractionated REE patterns (LREE enriched) observed in trondhjemites, tonalites and granodiorites require residual phases in the source that preferentially retains the HREE at small degrees of partial melting. Hornblende and garnet are the only phases with high enough partition coefficients (for acidic liquid compositions) for the HREE to produce significant HREE depletion. Extreme HREE depletion (as, for example, observed in some of the charnockites, Fig. 2) could only be achieved with residual garnet in the source, with its high partition coefficients for the HREE as compared to hornblende. For the charnockites our favoured model is of partial melting at depths where garnet is a stable phase. Varying degrees of partial melting would be able to produce the range in Ce/Y values observed, early formed melts having high contents of LREE, high Ce/Y ratios with residual garnet, and later melts would have lower contents of LREE, but higher HREE and lower Ce/Y ratios, as garnet is extracted from the source. Such a mechanism would be capable of producing the trend of increasing Ce/Y ratio with falling Y depicted in Fig. 3 for the charnockites.

The Madras charnockites display a negative correlation of $(Ce/Y)_N$ with SiO_2 but a positive correlation of Ce and Y with SiO_2. Residual garnet or hornblende in the source would seem to be necessary to account for the low Y levels, and the negative (Ce/Y)-SiO_2 correlation might indicate the increased dominance of garnet in the residual mineral assemblages at higher degrees of partial melting. An origin of the charnockite suite by partial melting of a garnet-pyroxene-amphibole-plagioclase assemblage at lower crustal depths (or greater) would satisfy many of the geochemical constraints, but complete REE data are necessary to rigorously test any such model, together with an appraisal of the role of metasomatism.

MINERAL CHEMISTRY AND P-T CONDITIONS OF CRYSTALLISATION

The mineral chemistry of Madras granulites was studied by Howie (1955) and much additional data on the mafic granulites in particular were provided by Ray (1970), Ray and Sen (1970), Sen (1970, 1973) and Sen and Ray (1971a,b). Rather than duplicate this work, we have analysed only those coexisting phases (orthopyroxene, clinopyroxene, garnet, plagioclase, K-feldspar) to which recently refined techniques of geothermometry and geobarometry can be applied.

The analyses were carried out on the Cambridge Microscan 5 microprobe in the Department of Geology, University of Leicester. Full ZAF corrections were applied in computing the final results.

Wood (1974) investigated the P-T conditions of the Madras granulites utilising the data of Howie (1955). The new probe data presented here provide further constraints on the P-T field suggested by Wood (1974). Study of the P-T conditions of metamorphism of the Madras granulites would appear to be useful for two reasons. Firstly, the consistently 'fresh' nature of the rocks implies lack of any retrograde event, suggesting that the present assemblages are those that equilibrated under granulite-grade conditions, and might allow quite accurate estimation of the physical conditions of metamorphism. Secondly, rocks such as these may allow evaluation of the relative merits of the standard geothermometers, geobarometers, and some estimation of their accuracy.

In hand specimen and thin section the rocks appear remarkably fresh. By the use of distribution coefficients, Sen (1970) and Ray and Sen (1970) demonstrated that the basic granulites contain equilibrated assemblages. Fig. 5 supports this, all points falling close to the curve of Saxena (1968) for granulite facies rocks, indicating that the pyroxenes equilibrated under granulite facies conditions with minimal subsequent exchange on further cooling. A corollary of these conclusions is that the Madras granulites were excavated relatively rapidly from the region of metamorphism, unlike, for instance, the Scourian complex of Scotland, which had an extended period of excavation over 1000 my (O'Hara, 1977).

Table 3a Chemical analyses of coexisting pyroxenes

	MP-34		MP-61		MP-77		MP-91		MP-94		MP-100	
	Opx	Cpx	Opx	Cpx	Opx	Cpx	Opx	Cpx	Opx	Cpx	Opx	Cpx
SiO_2	50.08	50.01	51.94	50.96	50.31	50.39	54.74	52.98	49.32	49.11	50.30	50.69
TiO_2	0.09	0.42	0.11	0.45	0.07	0.38	0.01	0.09	0.10	0.25	0.14	0.38
Al_2O_3	2.02	3.53	2.06	3.60	1.86	3.32	2.46	2.58	0.66	3.45	1.14	2.32
Cr_2O_3	0.01	0.04	0.02	0.04	0.04	0.10	0.18	0.26	0.01	0.02	0.01	0.00
FeO	30.84	14.22	23.88	9.86	29.27	13.16	12.68	4.05	37.96	18.49	32.40	14.73
MnO	0.72	0.22	0.60	0.29	0.55	0.28	0.29	0.13	0.72	0.40	0.66	0.34
MgO	15.63	10.59	20.61	12.60	17.03	11.53	29.05	15.74	10.42	8.08	14.44	10.38
CaO	0.76	20.29	0.62	21.56	0.56	20.00	0.51	23.97	0.68	19.52	0.76	21.01
Na_2O	0.04	0.49	0.02	0.58	0.04	0.52	0.01	0.38	0.02	0.47	0.09	0.54
K_2O	0.04	0.00	0.00	0.00	0.01	0.00	0.00	0.00	0.01	0.01	0.03	0.02
Total	100.23	99.81	99.86	99.94	99.74	99.68	99.93	100.16	99.90	99.80	99.97	100.41

Table 3b Chemical analyses of coexisting feldspars

	MP-14		MP-20		MP-44		MP-50		MP-72		MP-74		MP-82	
	Plag	Mic	Plag	Perth	Plag	Mic	Plag	Perth	Plag	Mic	Plag	Perth	Plag	Mic
SiO_2	60.22	64.38	61.95	66.71	61.06	63.20	62.33	65.44	62.04	62.90	60.73	63.74	62.73	63.60
Al_2O_3	23.38	17.52	23.48	18.38	24.07	18.34	23.67	19.03	23.69	18.39	23.99	18.94	21.85	17.69
CaO	6.35	0.28	5.24	0.29	6.08	0.20	5.53	0.15	4.77	0.13	5.55	0.19	4.56	0.25
Na_2O	7.87	2.21	8.76	1.98	7.97	1.85	8.53	1.13	8.85	1.21	8.28	1.25	8.84	1.32
K_2O	0.17	13.41	0.17	0.13	0.13	13.56	0.20	14.87	0.21	14.46	0.11	14.73	0.25	14.55
BaO	-	-	0.00	0.27	-	-	0.00	0.55	-	-	0.02	0.37	-	-
Total	97.99	97.80	99.60	101.21	99.31	97.15	100.27	101.17	99.56	97.09	98.68	99.22	98.23	97.42

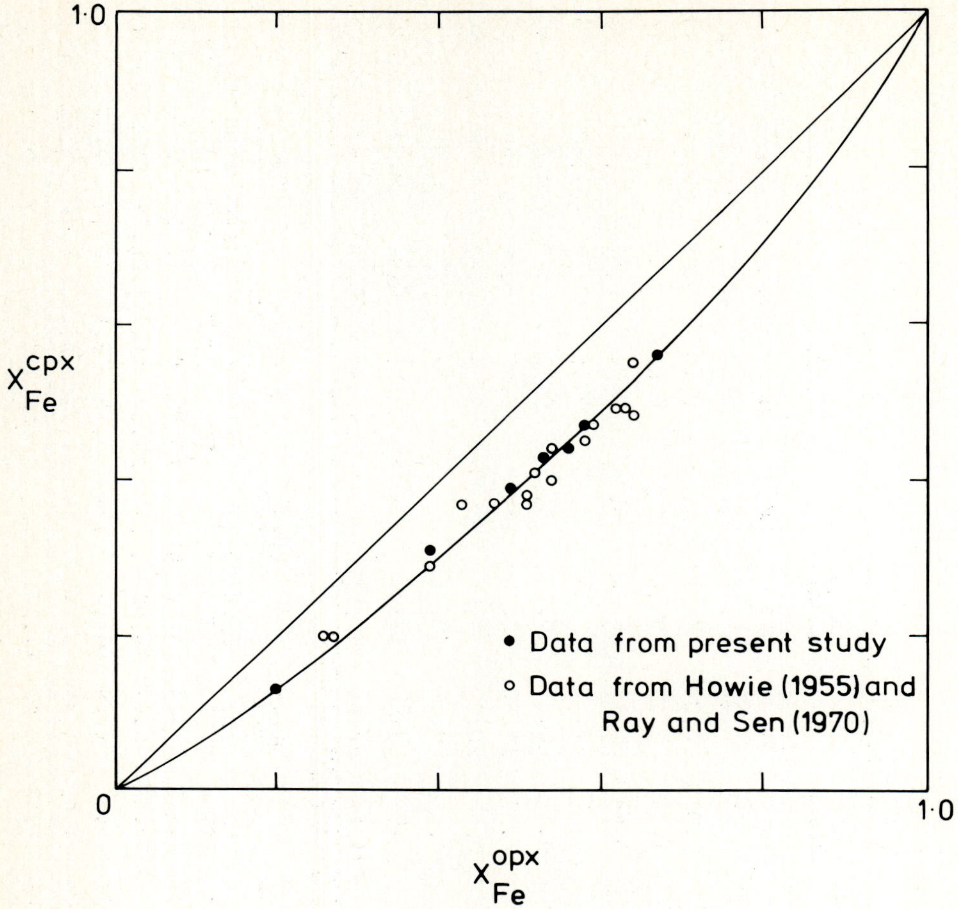

Fig. 5. Plot of X_{Fe}^{CPX} against X_{Fe}^{OPX} for Madras granulite pyroxenes. Diagonal line is the line of equal mole fraction of Fe. Curved line is that of Saxena (1968) for pyroxene compositions that equilibrated under granulite facies conditions.

The standard methods of geothermometry/geobarometry developed in recent years were used to examine the P-T conditions of crystallisation of the Madras granulites and, combined with petrological observations related to experimental work, allow us to define a restricted P-T region for the metamorphism of Madras granulites.

1. Two-pyroxene geothermometer

Seven pairs of coexisting ortho- and clinopyroxenes (Table 3a) and two single orthopyroxenes (Table 7) were probed in the present study. The analysed pyroxenes are from samples covering the full compositional range (48-75% SiO_2)

exhibited by the charnockite suite. The orthopyroxenes are bronzites to ferro-hypersthenes, whereas the clinopyroxenes range from diopside to salite, augite and ferroaugite.

The original formulation of the two-pyroxene geothermometer (based on the diopside-enstatite miscibility gap) was by Wood and Banno, who derived their equation 27 (Wood and Banno, 1973) as a geothermometer. In the light of new experimental results Wells reformulated the data and obtained a different expression (Wells, 1977, eqn. 5). Temperatures obtained from the new analyses by applying the methods of Wood and Banno (1973) and Wells (1977) are given in Table 4, together with the results obtained from the coexisting pyroxene analyses of Howie (1955) and Ray and Sen (1970). Such pyroxene temperatures should represent a peak thermal event (Hewins, 1975), in this case the peak metamorphic (granulite grade) temperature. These temperatures have been calculated using the empirical correction of Bence et al. (1975) for estimation of the ferric iron

TABLE 4: Pyroxene temperatures

	X_{Fe}^{opx}	$a_{Mg_2Si_2O_6}^{cpx}$	$a_{Mg_2Si_2O_6}^{opx}$	$T_1(^{\circ}C)$	$T_2(^{\circ}C)$
MP 16	0.580	0 0.032	0.163	816	872
MP 34	0.522	0.040	0.203	842	902
MP 61	0.394	0.041	0.334	846	870
MP 77	0.486	0.050	0.239	865	929
MP 91	0.197	0.024	0.592	841	770
MP 94	0.672	0.026	0.099	820	885
MP 100	0.557	0.030	0.178	809	858
From data of Howie (1955)					
3709	0.241	0.032	0.532	873	822
4645	0.265	0.030	0.473	873	813
2270	0.383	0.027	0.331	805	807
2941	0.454	0.043	0.259	850	900
4642A	0.568	0.040	0.170	840	913
115	0.631	0.021	0.111	800	846
From data of Ray and Sen (1970)					
M 36	0.594	0.044	0.140	869	958
M 181	0.570	0.033	0.148	837	904
SCH 516	0.529	0.064	0.187	909	1011
M 236	0.502	0.048	0.189	882	961
M 93	0.525	0.036	0.208	823	874
SCH 300	0.431	0.050	0.298	865	912
M 247	0.513	0.029	0.202	807	848
SCH 456	0.630	0.016	0.098	779	821
SCH 537	0.626	0.034	0.096	870	963
SCH 320	0.615	0.020	0.124	782	825
SCH 65	0.522	0.052	0.209	870	947

T_1 = temperature from equation 27 of Wood and Banno (1973)

T_2 = temperature from equation 5 of Wells (1977)

content of pyroxenes which cannot be directly determined by the microprobe. Ignoring possible Fe^{3+} content for these compositions lowers the quoted temperatures by 5-10°C.

Accuracy of the two-pyroxene geothermometer is limited by inexact knowledge of the diopside-enstatite miscibility gap and of the effect of pressure on the limbs of the solvus. It was suggested (Hewins, 1975; Wood, 1975; Wilson, 1976) that the Wood-Banno method produces temperature overestimates, although Hewins (1975) still finds it to be a precise method. Wood (1975) suggested a downward revision of 60°C for temperatures obtained by this method, a figure in agreement with that suggested by Hewins (1975) and Wilson (1976) as the amount by which the thermometer may be overestimating. Temperatures obtained by the Wood-Banno method (Table 4) fall in the range 780-900°C, with a mean of 840°C which, upon revision downwards by 60°C, becomes 780°C. Wood and Banno (1973) considered the accuracy of this method to be ± 60°C.

Temperatures obtained by the Wells method (Table 4) are rather higher, ranging from 770-1010°C with a mean of 885°C. Accepting that the Wood-Banno method does produce temperature overestimates of c. 60°C, it would appear that for the compositions considered here the method of Wells (1977) also produces temperature overestimates of the order of 100°C. Wells (1977) considers the accuracy of his method to be ± 70°C.

Bearing in mind the above considerations, the best estimate of temperature of crystallisation from the two-pyroxene data seems to be 780 ± 60°C. This is in agreement with a figure of 750°C obtained by Fleet (1974) from the data of Howie (1955) applied to the augite limb of the pyroxene solvus, which Fleet (1974) developed as a possible geothermometer, and also with the two-pyroxene geothermometer of Saxena (1976) which yields temperatures of crystallisation of c. 800°C for Madras granulites.

2. <u>Two-feldspar geothermometer</u>

Seven pairs of coexisting plagioclase and alkali feldspar were probed (Table 3b) and combined with two pairs of analyses from Howie (1955) for the purposes of geothermometry. All probed samples are charnockites. The plagioclases range from An_{22}-An_{31}, and the alkali feldspars from Or_{79}-Or_{89}. In addition to the elements analysed here the feldspars may contain appreciable amounts of iron with $FeO + Fe_2O_3$ up to 0.6% (Howie, 1955). Problems associated with microprobe analysis of perthitic feldspars, such as these, are discussed below.

The original formulation of the two-feldspar geothermometer was by Stormer (1975), whose thermometer (eqn. 18) was based on the thermodynamic parameters for high sanidine. Whitney and Stormer recently reformulated the geothermometer (eqn. 8 of Whitney and Stormer, 1977) on the basis of thermodynamic parameters for microcline-low albite solid solutions. Thus the structural state of the

TABLE 5: Feldspar temperatures

	$X_{Na,AF}$	$X_{Na,PF}$	P (Kb)	T_1(°C)	Av(°C)	T_2(°C)
MP 14	0.198	0.685	9	678	713	747
			11	702	737	772
MP 20	0.179	0.744	9	627	648	689
			11	650	682	714
MP 44	0.170	0.698	9	636	666	696
			11	660	690	720
MP 50	0.103	0.728	9	530	554	577
			11	551	575	598
MP 72	0.112	0.761	9	535	559	583
			11	555	580	605
MP 74	0.113	0.725	9	547	571	595
			11	568	593	617
MP 82	0.120	0.767	9	544	569	594
			11	565	590	615
2270	0.135	0.620	9	620	645	670
			11	643	668	693
6436	0.173	0.659	9	659	689	719
			11	683	714	744

T_1 = temperature from equation 18 of Stormer (1975)

T_2 = temperature from equation 8 of Whitney and Stormer (1977)

Av = average of T_1 and T_2 (approximate temperature for the orthoclase structural state)

Samples 2270 and 6436 from Howie (1955)

alkali feldspar becomes of importance in assessing temperatures obtained from these methods. Orthoclase is likely to be the stable structural state during granulite-grade metamorphism and, although the observed alkali feldspar in the charnockites is generally microcline, there is evidence (Howie, 1955) that the microcline has inverted from orthoclase. Unfortunately there are no currently available thermodynamic parameters for the orthoclase structural state, and thus no geothermometric expression for this structural state. Whitney and Stormer (1977) do suggest that reasonably accurate results may be obtained for the orthoclase structural state by averaging temperatures obtained from the methods of Stormer (1975) and Stormer and Whitney (1977). This has been done for the present analyses, and the results are presented in Table 5. Such an average should give a good approximation to the true temperature. The modification to the equation of Stormer (1975, eqn. 18) suggested by Powell and Powell (1977), who considered the effect of Ca on the thermodynamics of alkali feldspar, has little effect (c. 10°C difference) on the temperatures generated for the Madras granulites by this method, due to the low content of Ca ($X_{Ca,AF}$ = 0.02) in the alkali feldspar.

In assessing these results (Table 5) the possible effect of analytical uncertainty in the alkali feldspar analyses needs to be considered. The alkali feldspar in these rocks invariably displays exsolution (perthites and microperthites), and the difficulties in obtaining homogeneous bulk analyses from the microprobe are large (i.e. Bohlen and Essene, 1977). The methods adopted here to overcome these difficulties are outlined in the Appendix, but it may be that these methods have not fully overcome the analytical difficulties. In this respect the analyses of Howie (1955) are relevant as they are based on mineral separates and should represent truly bulk analyses. They give temperatures (Table 5) higher than the mean temperature from probed samples (611°C and 8 kb and 635°C at 11 kb), although they fall within the same temperature range (Table 7). It may be that the mean temperatures for probed samples given here are low by c. 40°C. Adoption of a more rigorous analytical procedure for the alkali feldspars (such as that used by Bohlen and Essene, 1977) should produce more reliable temperatures.

Nevertheless, taking the mean of all temperatures in Table 5, the temperatures indicated by the two-feldspar method are 624 ± 50°C at 9 kb and 648 ± 63°C at 11 kb.

3. Fe-Ti oxide geothermometer

Fe-Ti oxides were not analysed in the present study, due to the analytical difficulties involved in obtaining homogeneous magnetite analyses from the microprobe (i.e. Bohlen and Essene, 1977), and the probability that the Fe-Ti oxides re-equilibrate to low temperatures (i.e. Hewins, 1975) and at best record minimum temperatures of metamorphism (Sen and Ray, 1971b). Relevant data were extracted from the literature and the temperatures thus obtained are given in Table 6. This geothermometer/oxygen barometer is calibrated experimentally from the work of Buddington and Lindsley (1964).

TABLE 6: Fe-Ti oxide temperatures and oxygen pressures

	A		B	C	
	2270	2941	4642A	3	6
T (°C)	560 ± 50	600 ± 50	670 ± 50	605 ± 50	555 ± 50
log f_{O_2}	-20.0 ± 1	-18.7 ± 1	-18.0 ± 1	-19.2 ± 1	-20.1 ± 1

A: data from Howie (1955) quoted in Buddington and Lindsley (1964)
B: data from Howie (1955) quoted in Sen and Ray (1971)
C: data of Sen and Ray (1971)

Temperatures indicated by this method (Table 6) average 580 ± 50°C. The result (Sen and Ray, 1971) for analysis of Fe-Ti oxides from sample 4642A of Howie (1955) may be of dubious value due to the presence of 7.3% sulphur in the magnetite analysis (Howie, 1955) indicating a significant amount of pyrrhotite.

The above data indicate that the closure of the various pairs of phases to intercrystalline diffusion with falling temperature was in the order: orthopyroxene-clinopyroxene, plagioclase-alkali-feldspar, ilmenite-magnetite. The two-pyroxene method indicates peak metamorphic temperatures of 780 ± 60°C, the two-feldspar method 640 ± 50°C, and the Fe-Ti oxide method 580 ± 50°C. Consideration of only feldspar or Fe-Ti oxide data can only yield a minimum estimate of temperature of metamorphism. The above sequence is equally well demonstrated for a single sample 2270 of Howie (1955) for which all three temperatures are available (Tables 4-6): two-pyroxene 745 ± 60°C, two-feldspar 657 ± 50°C, and Fe-Ti oxide 560 ± 50°C.

That Fe-Ti oxides re-equilibrate to a lower temperature than coexisting feldspars in an igneous plutonic environment was demonstrated by Whitney and Stormer (1976). It would seem that reliable estimates of peak metamorphic temperature can only be obtained from the two-pyroxene geothermometer.

4. Orthopyroxene-garnet geobarometer

Two pairs of coexisting orthopyroxene-garnet were analysed in the present study (Table 7) and are combined with two pairs of analyses from Howie and Subramaniam (1957) to obtain P-T lines by the method of Wood (1974), formulated on the basis of the solubility of Al_2O_3 in coexisting orthopyroxene and garnet. All the garnets (Table 7) are pyrope-almandine in composition, with greater than 60% of the almandine end member.

TABLE 7: Chemical analyses of coexisting orthopyroxene-garnet and garnet-plagioclase

	MP-44		MP-72		MP-74	
	Opx	Gt	Opx	Gt	Gt	Plag
SiO_2	48.88	37.40	49.74	39.28	38.63	60.73
TiO_2	0.10	0.04	0.05	0.04	0.03	-
Al_2O_3	2.54	21.31	5.46	22.29	22.00	23.99
Cr_2O_3	0.02	0.02	0.07	0.07	0.02	-
FeO	34.88	32.16	25.91	28.58	30.90	-
MnO	0.53	1.12	0.15	0.62	0.72	-
MgO	12.97	3.99	18.61	8.95	7.01	-
CaO	0.34	3.23	0.07	1.26	1.85	5.55
Na_2O	0.02	-	0.03	-	-	8.28
K_2O	0.02	-	0.02	-	-	0.11
Total	100.30	99.27	100.11	101.09	101.17	98.66

A possible disadvantage of applying this method to microprobe analyses of coexisting phases is the uncertainty introduced by the non-determination of ferric iron. Ignoring the Fe^{3+} content (especially of orthopyroxene) can have a dramatic effect on the pressures obtained, especially from the corrected method of Wood (1974), where Fe^{3+} is combined with Al in the correction procedure.

Here the empirical correction of Bence et al. (1975) was used for estimating Fe^{3+} in orthopyroxene. The possible Fe^{3+} content of garnet was ignored, the analyses of Howie and Subramaniam showing low Fe_2O_3. The validity of these assumptions can be tested by comparing P-T curves thus generated with those from the analyses of Howie and Subramaniam (1957) in which Fe^{3+} is available for both phases. Results from application of this geobarometer are given in Table 8 and the P-T lines generated (calculation following the Wood, 1974, corrected method), illustrated in Fig. 6. The P-T curves from the probe-analysed samples bracket the curves from the data of Howie and Subramaniam (1957), indicating that the above assumptions regarding Fe^{3+} may be valid for these mineral compositions within the overall error of the method (i.e. validity of linear extrapolation from high temperature experimental system to lower temperature, assumptions regarding Al site occupancy in orthopyroxenes, and effect of analytical uncertainty, etc.). Wood (1974) considers the accuracy of the method to be 2-3 kb.

TABLE 8: Orthopyroxene-garnet pressures

	X_{Fe}^{opx}	$X_{Al}^{M1,opx}$	$(1-y)^{Gt}$	ΔV_r	P (Kb) at 800°C	P (Kb) at 900°C
A						
Ch 113	0.483	0.092	0.917	-8.29	5.5	9.7
Ch 199	0.574	0.049	0.765	-8.05	4.9	9.3
MP 44	0.599	0.060	0.882	-8.11	7.3	11.9
MP 72	0.439	0.122	0.952	-8.44	4.5	8.5
B						
Ch 113	0.483	0.077	0.876	-8.20	5.7	10.0
Ch 199	0.574	0.032	0.736	-7.96	8.1	12.8
MP 44	0.599	0.051	0.881	-8.06	9.0	13.7
MP 72	0.439	0.121	0.950	-8.44	4.5	8.5

A = uncorrected method of Wood (1974)
B = corrected method of Wood (1974)

5. Method of Ghent (1976)

This method is based on the distribution of Ca between coexisting garnet and plagioclase according to the reaction:

anorthite = grossular + aluminium silicate + quartz

Aluminium silicate (sillimanite) was observed in one thin section of charnockite (MP 74). Its composition was verified by analysis and the coexisting garnet and plagioclase were also probed (Table 7). The P-T line produced is depicted as 'g' on Fig. 6. The grossular component of garnet was calculated allotting all Ca to grossular; thus this method generates maximum pressures (Ghent, 1976).

6. Method of Morse (1971)

The occurrence of mesoperthite in khondalites indicates that the P-T conditions

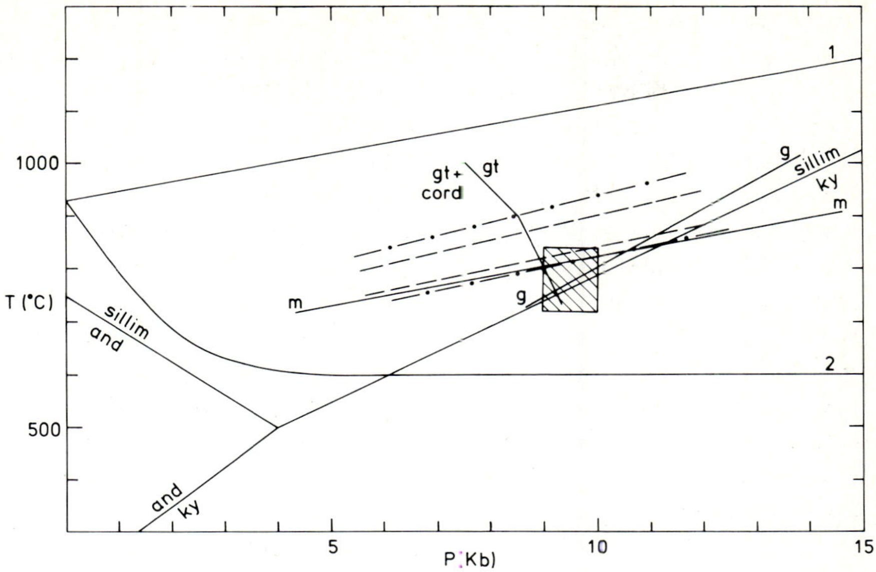

Fig. 6: P-T diagram for Madras granulites. Aluminium silicate phase boundaries from Holdaway (1971). Lines 1 and 2 are the dry and wet granite solidii. m and g are the P-T lines generated by the methods of Morse (1971) and Ghent (1976). Dashed and dot-dash lines are the P-T lines generated by the garnet-orthopyroxene geobarometer (Wood, 1974) for compositions from Howie and Subramaniam (1959) and from the present data. Also shown is the garnet + cordierite-garnet boundary for pelitic compositions with Mg/(Mg+Fe) = 0.4 (from Hensen and Green, 1971). The hatched area is the probable P-T field for crystallisation of the Madras granulites.

of metamorphism may have been close to the critical line of Morse (1971) for the feldspar solvus. This line is indicated as 'm' on Fig. 6.

7. Petrographic observations

The aluminium silicate consistently observed in the khondalites is sillimanite; no kyanite was seen in this study or by Howie and Subrmaniam (1957), although Sen and Ray (1971b) report the occurrence of both sillimanite and kyanite in khondalites from Madras. Appropriate conditions of metamorphism are restricted to the low P/high T region of the sillimanite-kyanite boundary in the aluminium-silicate phase diagram of Holdaway (1971), or, accepting the observations of Sen and Ray (1971b), to the vicinity of this boundary. The boundaries of the aluminium-silicate phase diagram of Holdaway (1971) are given in Fig. 6.

There is a total absence of cordierite in rocks of the khondalite suite, although garnet is common, allowing the experimental data of Hensen and Green (1971) to be applied to these rocks. The $Mg/(Mg+Fe^{2+})$ ratio for garnets from khondalites is c. 0.4 (Howie and Subramaniam, 1957), enabling the boundary for the upper stability of garnet + cordierite for this composition to be taken from diagrams 8 and 9 of Hensen and Green (1971). This boundary is plotted on Fig. 6 and conditions of metamorphic recrystallisation must have lain to the high-pressure side of this line.

All the above data are compiled on Fig. 6, from which the P-T conditions of metamorphism can be deduced. The hatched area represents the indicated two-pyroxene temperature of $780 \pm 60^{\circ}$C, lying to the high-temperature side of the sillimanite-kyanite boundary, and to the high-pressure side of the garnet-garnet + cordierite boundary. Agreement with the methods which generate P-T lines is generally good, lines g, m and two of the P-T lines from the orthopyroxene-garnet method passing through this area, the remaining two lines generated by this method occurring at higher temperatures (for a given pressure) than that indicated; but this is not surprising considering the accuracy of the method. The convergence of all methods is encouraging and suggests that physical conditions of metamorphic recrystallisation of the Madras granulites can be confined to the region 720-840°C and 9-10 kb (Fig. 6). Wood (1974), using only the data

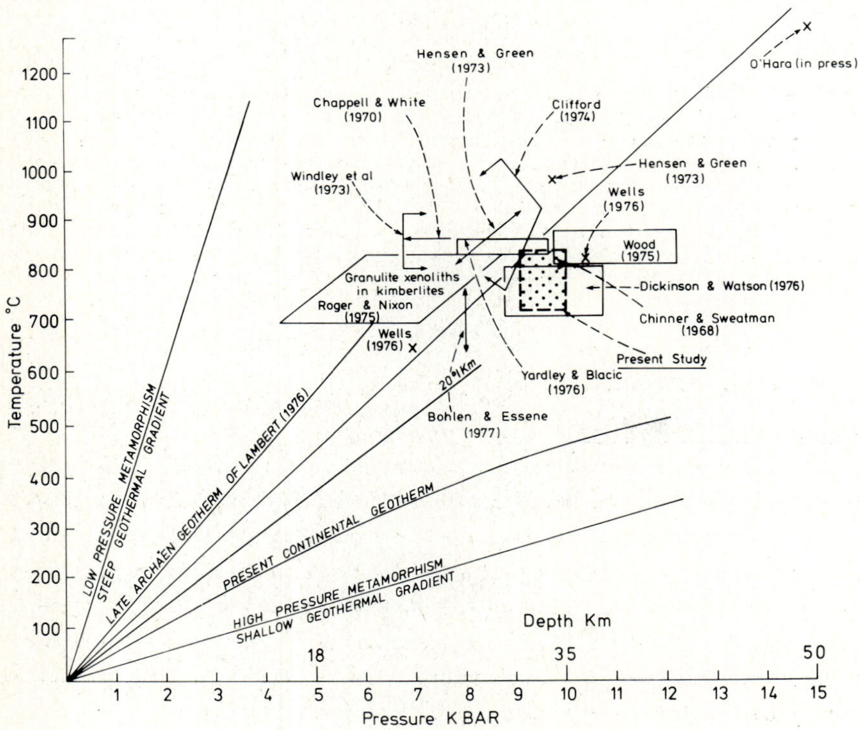

Fig. 7: General P-T diagram for Precambrian granulite facies terrains. Modified from Tarney and Windley (1977).

of Howie (1955), arrived at figures of 800-900°C and 8-12 kb. The lower temperature estimate suggested here comes from the revision of the Wood-Banno method (see above), and the more restricted range of pressure from a consideration of other geobarometric methods.

The P-T estimate for the Madras rocks is in good agreement with that of other Archaean granulites (cf. Fig. 7, modified from Tarney and Windley, 1977). Also indicated on Fig. 7 is the recent calculation (Bohlen and Essene, 1977) of the P-T conditions for the Adirondack highland granulites, which is based on feldspar and Fe-Ti oxide temperatures and which, in view of the earlier discussion, is probably a low value. Pyroxene temperatures for the Adirondacks (McLelland and Whitney, 1977) do indicate higher temperatures (c. 780°C).

Fig. 7 indicates that the P-T conditions for Archaean granulite facies metamorphism generally lie in a field embracing 6-12 kb and 700-1000°C, implying depths of roughly 20-40 km. They are thus intermediate pressure granulites with geothermal gradients of 18-35°C/km. The indications are that the Archaean crust may have been at least 30-35 km thick 2600 my ago. If the Madras charnockites are the product of melting of crustal material at greater depths, then the crust may have been substantially thicker at this time. It follows that although granulites represent deeper crustal material, they do not necessarily represent lower crustal rocks (see Tarney and Windley, 1977).

ACKNOWLEDGEMENTS

We should like to thank R.N. Wilson for much help with the microprobe analyses and Dr G.L. Hendry for assistance with the X.R.F. analyses. BLW gratefully acknowledges the receipt of a NERC studentship.

APPENDIX: MICROPROBE TECHNIQUE

Pyroxenes:

Only optically homogeneous (unexsolved) grains were selected for analysis. The grains were probed along a traverse from core to rim to test for zoning. All analysed grains were found to be unzoned and unexsolved. A bulk analysis for the grain was then obtained by averaging the individual point analyses.

Garnets:

Probe analysis showed the garnets to be unzoned. Individual point analyses were averaged to obtain a bulk analysis for the grain.

Feldspars:

The plagioclase feldspars were probed from core to rim, but found to be unzoned. A bulk analysis for the grain was obtained from an average of the individual point analyses.

The alkali feldspars show varying degrees of exsolution (microperthites and perthites). Two different approaches were adopted in order to obtain bulk probe analyses: for the microperthitic grains a large number of randomly selected points per grain were analysed, and then averaged to obtain a total mean analysis; for the perthitic grains a number of areas per grain were scanned, employing a 30 μ scan, and these results then averaged. This second method appears to give better precision. It is felt that the methods largely overcome the problem of exsolution in the feldspars, but obviously a more exacting method (i.e. Bohlen and Essene, 1977) would give improved precision.

REFERENCES

Arth, J.G. and Hanson, G.N., 1975. Geochemistry and origin of the early Precambrian crust of northeastern Minnesota. Geochim. Cosmoch. Acta, 39: 325-362.

Barker, F. and Peterman, Z.E., 1974. Bimodal tholeiitic-dacitic magmatism and the early Precambrian crust. Precambrian Res., 1: 1-12.

Barker, F. and Arth, J.G., 1976. Generation of trondhjemitic-tonalitic liquids and Archaean bimodal trondhjemite-basalt suites. Geology, 4: 596-600.

Barker, F., Arth, J.G., Peterman, Z.E. and Friedman, T., 1976. The 1.7- to 1.8-b.y. old trondhjemites of southwestern Colorado and northern New Mexico: Geochemistry and depths of genesis. Bull. geol. Soc. Am., 87: 189-198.

Bence, A.E., Papike, J.J. and Aynso, R.A., 1975. Petrology of submarine basalts from the central Caribbean: DSDP Leg 15. J. geophys. Res. 80: 4775-4804.

Bohlen, S.R. and Essene, E.J., 1977. Feldspar and oxide thermometry of granulites in the Adirondack Highlands. Contrib. Mineral. Petrol., 62: 153-169.

Buddington, A.F. and Lindsley, D.H., 1964. Iron-titanium oxide minerals and synthetic equivalents. J. Petrol., 5: 310-357.

Chappel, B.W. and White, A.J.R., 1970. Further data on an "eclogite" from the Sittampundi complex, India. Mineral. Mag., 37: 555-560.

Chinner, G.A. and Sweatman, T.R., 1968. A former association of enstatite and kyanite. Mineral. Mag., 36: 1052-1060.

Clifford, T.N., 1974. Review of African granulites and related rocks. Sp. Pap. geol. Soc. Am., 156: 49 pp.

Cooper, D.C. and Field, D., 1977. The chemistry and origins of Proterozoic low-potash, high-iron charnockitic gneisses from Tromøy, South Norway. Earth Planet. Sci. Lett., 35: 105-115.

Dickenson, B.B. and Watson, J., 1976. Variations in crustal level and geothermal gradient during the evolution of the Lewisian complex of northwest Scotland. Precambrian Res., 3: 363-374.

Fleet, M.E., 1974. Mg, Fe^{2+} site occupancies in coexisting pyroxenes. Contrib. Mineral. Petrol., 47: 207-214.

Ghent, E.D., 1976. Plagioclase-garnet-Al_2SiO_5-quartz: a potential geobarometer-geothermometer. Am. Mineral., 61: 710-714.

Green, D.H. and Ringwood, A.E., 1967. An experimental investigation of the gabbro to eclogite transformation and its petrological implications. Geochim. Cosmoch. Acta, 31: 767-833.

Hensen, B.J. and Green, D.H., 1971. Experimental study of the stability of cordierite and garnet in pelitic compositions at high pressures and temperatures. Contrib. Mineral. Petrol., 33: 309-330.

Hensen, B.J. and Green, D.H., 1973. Experimental study of the stability of cordierite and garnet in pelitic compositions at high pressures and temperatures. Contrib. Mineral. Petrol., 38: 151-166.

Hewins, R.H., 1975. Pyroxene geothermometry of some granulite facies rocks. Contrib. Mineral. Petrol., 50: 205-209.

Holdaway, M.J., 1971. Stability of andalusite and the aluminium silicate phase diagram. Am. J. Sci., 271: 97-131.

Holland, T.H., 1893. The petrology of Job Charnocks tombstone. J. Asiatic Soc. Bengal, 62: 162-164.

Holland, T.H., 1900. The charnockite series, a group of Archean hypersthene rocks in peninsular India. Mem. geol. Surv. India, 28, pt. 2.

Howie, R.A., 1955. The geochemistry of the charnockite series of Madras, India. Trans. R. Soc. Edinb., 62: 725-768.

Howie, R.A. and Subramaniam, A.P., 1957. The paragenesis of garnet in charnockite, enderbite and related granulites. Mineral. Mag., 31: 565-586.

Lambert, I.B. and Heier, K.S., 1968. Geochemical investigations of deep-seated rocks in the Australian Shield. Lithos, 1: 30-53.

Leyreloup, A., Dupuy, C. and Andriambololona, R., 1977. Catazonal xenoliths in French Neogene volcanic rocks: constitution of the lower crust. 2. Chemical compositions and consequences of the evolution of the French Massif Central Precambrian crust. Contrib. Mineral. Petrol., 62: 283-300.

McLelland, J.M. and Whitney, P.R., 1977. The origin of garnet in the anorthosite-charnockite suite of the Adirondacks. Contrib. Mineral. Petrol., 60: 161-182.

O'Hara, M.J., 1977a. Thermometry and barometry of Scourie gneisses. J. geol. Soc. Lond. (in press).

O'Hara, M.J., 1977b. Thermal history of excavation of Archean gneisses from the base of the continental crust. J. geol. Soc. Lond., 134: 185-200.

O'Nions, R.K. and Pankhurst, R.J., 1974. Rare earth element distributions in Archean gneisses and anorthosites, Godthåb area, West Greenland. Earth Planet. Sci. Lett., 22: 328-338.

Powell, M. and Powell, R., 1977. Plagioclase-alkali-feldspar geothermometry revisited. Mineral. Mag., 41: 253-256.

Ray, S., 1970. Significance of hornblende compositions from basic granulites of the type charnockite area near Madras. N. Jb. Miner. Mh., 10: 456-466.

Ray, S. and Sen, S.K., 1970. Partitioning of major exchangeable cations among orthopyroxene, calcic pyroxene and hornblende in basic granulites from Madras. N. Jb. Miner. Mh., 114: 61-88.

Saxena, S.K., 1968. Chemical study of phase equilibria in charnockite, Varberg, Sweden. Am. Mineral., 53: 1674-1695.

Sen, S.K., 1970. Magnesium-iron compositional variance in hornblende-pyroxene granulites. Contrib. Mineral. Petrol., 29: 76-88.

Sen, S.K., 1973. Compositional relations among hornblende and pyroxenes in basic granulites and an application to the origin of garnets. Contrib. Mineral. Petrol., 38: 299-306.

Sen, S.K. and Ray, S., 1971a. Breakdown reactions for natural hornblendes in granulite facies. N. Jb. Miner. Mh., 114: 301-319.

Sen, S.K. and Ray, S., 1971b. Hornblende-pyroxene granulites versus pyroxene granulites: a study from the type charnockite area. N. Jb. Miner. Mh., 115: 291-314.

Shaw, D.M., 1968. A review of K-Rb fractionation trends by covariance analysis. Geochim. Cosmoch. Acta, 32: 573-601.

Sheraton, J.W., Skinner, A.C. and Tarney, J., 1973. The geochemistry of the Scourian gneisses of the Assynt district. In: R.G. Park and J. Tarney (Editors), The Early Precambrian of Scotland and Related Rocks of Greenland. Univ. Keele, pp. 13-30.

Sighinolfi, G.P., 1971. Investigations into deep crustal levels: fractionating effects and geochemical trends related to high-grade metamorphism. Geochim. Cosmoch. Acta, 35: 1005-1021.

Stormer, J.C., 1975. A practical two-feldspar geothermometer. Am. Mineral., 60: 667-674.

Subramaniam, A.P., 1959. Charnockites of the type area near Madras - a reinterpretation. Am. J. Sci., 257: 321-353.

Tarney, J., 1976. Geochemistry of Archean high-grade gneisses, with implications as to the origin and evolution of the Precambrian crust. In: B.F. Windley (Editor), The Early History of the Earth. Wiley, Lond., pp. 405-417.

Tarney, J. and Windley, B.F., 1977. Chemistry, thermal gradients and evolution of the lower continental crust. J. geol. Soc. Lond., 134: 153-172.

Tilley, C.E., 1936. Enderbite, a new member of the charnockite series. Geol. Mag., 73: 312-316.

Wells, P.R.A., 1976. Late Archean metamorphism in the Buksefjord region, southwest Greenland. Contrib. Mineral. Petrol., 56: 229-242.

Wells, P.R.A., 1977. Pyroxene thermometry in simple and complex systems. Contrib. Mineral. Petrol., 62: 129-139.

Whitney, J.A. and Stormer, J.C., 1976. Geothermometry and geobarometry in epizonal granite intrusions: a comparison of iron-titanium oxides and coexisting feldspars. Am. Mineral., 61: 751-761.

Whitney, J.A. and Stormer, J.C., 1977. The distribution of $NaAlSi_3O_8$ between coexisting microcline and plagioclase and its effect on geothermometric calculations. Am. Mineral., 62: 687-691.

Wilson, A.F., 1976. Aluminium in coexisting pyroxenes as a sensitive indicator of changes in metamorphic grade within the mafic granulite terrane of the Fraser Range, Western Australia. Contrib. Mineral. Petrol., 56: 255-277.

Windley, B.F., Herd, R.K. and Bowden, A.A., 1973. The Fiskenaesset complex, West Greenland. Pt. 1: A preliminary study of stratigraphy, petrology and whole-rock chemistry from Qeqertarssuatsiaq. Meddr. Grønland, 196 (2), 80 pp.

Wood, B.J., 1974. The solubility of alumina in orthopyroxene coexisting with garnet. Contrib. Mineral. Petrol., 46: 1-15.

Wood, B.J., 1975. Influence of pressure, temperature and bulk composition on the appearance of garnet in orthogneiss: an example from South Harris, Scotland. Earth Planet. Sci. Lett., 26: 299-311.

Wood, B.J. and Banno, S., 1973. Garnet-orthopyroxene and orthopyroxene-clinopyroxene relationships in simple and complex systems. Contrib. Mineral. Petrol., 42: 109-124.

Yardley, B.W.D. and Blacic, J.D., 1976. Sapphirine in the Sittampundi complex: a discussion. Mineral. Mag., 40: 523-524.

RADIOACTIVE ELEMENTAL DISTRIBUTION IN THE GRANULITE TERRAINS AND DHARWAR SCHIST BELTS OF PENINSULAR INDIA

B.S. Atal, N.S. Bhalla, Y. Lall, T.M. Mahadevan and G.R. Udas

Atomic Minerals Division, Atomic Energy Dept, Hyderabad-500016, India

Abstract

Whole-rock gamma-ray spectrometric data for U, Th and K contents of 133 samples belonging to the granulite belts of North Arcot-Pallavaram (NAP), Eastern Ghats (EG) and Kondapalli and the greenstone gneiss belts of Sargur, Holenarsipur and Kolar are presented. The data reveal that the granulite and greenstone belt rocks have distinctive radioelement abundances and characteristic enrichment patterns for U and Th in relation to K.

The main rocks analysed from the granulite belts include pyroxene granulites, charnockites, associated anorthosites and migmatite gneisses. They are divided into two broad groups on the basis of their silica content: (a) intermediate and basic rocks with SiO_2 content less than 65%, and (b) more acidic rocks with SiO_2 content greater than 65%. Compared with normal magmatic rocks the more basic granulites and charnockites are depleted in U and Th in relation to K. The more acidic rocks, however, have higher concentrations of all three elements and high Th/U ratios, indicating enrichment of Th over U. Later quartzo-feldspathic rocks intrusive into the charnockite-pyroxene granulite group show higher concentrations of all three radioelements and a higher Th/U ratio. These show depletion in U but enrichment in Th in relation to K.

The main rocks from the Dharwar schist belts include amphibolites and associated gneisses, granites and the "Champion gneiss". These show a relative enrichment of U and Th in relation to K, in marked contrast to the patterns of the radioelemental ratios in the granulite belt.

The significance is discussed of the contrasting trends of the radioelement distribution in the granulite and greenstone belts.

INTRODUCTION

The radioelement distribution in granulites has been of special interest to Archaean geochemistry ever since Heier and Adams (1965) found that there is no linear correlation between U, Th and K contents, as exists for magmatic rocks, and that the granulites show a characteristic depletion of U and Th in relation to K. 133 samples from the schist and granulite belts were analysed in the laboratories of the Atomic Minerals Division for U and Th in the ppm ranges and for the major element K. In view of the relevance of this data to problems of Archaean geochemistry, a synthesis of the radioelement content and ratios in granulite and greenstone belts is presented in this paper. The data largely confirm earlier findings (Heier, 1962; Heier and Adams, 1965; Lambert and Heier, 1967, 1968; Narayanswamy and Venkatasubramanian, 1969; Heier and Thoresen, 1971; Zhukova et al., 1974) of depletion of U and Th in relation to K in granulite belts and enrichment of U and Th in relation to K in low-grade rocks, as in greenstone belts. The present work has additional significance because it concerns the first systematic study of both the granulite and greenstone belts in India. The data further point to the fact that remobilisation of radioelements is restricted to the later intrusives within the granulite belts and that no major processes can be envisaged leading to mobilisation of U and Th from the granulite belts into the greenstone belts. The abundances and pattern of enrichment of U and Th in relation to K in each belt seem to be an intrinsic characteristic inherited from early periods of evolution of the two types of belt.

METHOD OF ANALYSIS

U, Th and K estimates were made by using the gamma-ray spectrometric technique. The relevant details regarding the methodology are summarised below:

Crystal size	:	3" dia x 3" thick NaI(Tl) with low K (less than 1 ppm) content
Resolution of the detector	:	7.5% for Cs-137
Energy channels	:	1.46 Mev for K-40; window width 200 Kev 1.76 Mev for U; window width 200 Kev 2.62 Mev for Th; window width 400 Kev
Sample weight	:	c. 800 gm in annular container
Limits of detection	:	0.2 ppm U, 0.5 ppm Th and 0.05% K (for a counting time of 5000 sec)
Accuracy of determination	:	K: 2% at 1% K concentration and 10% near the limit of detection of 0.05% Th: 5% at few ppm level with Th/U 3 and higher for lower concentration, going up to 20% at 0.5 ppm level U: 5-25%, depending upon U content (in the range of U concentrations measured)

A large number of samples were analysed for U by DFN technique to check the accuracy of the spectrometric estimation and to confirm secular equilibrium in the samples. K estimates were confirmed by flame-photometry.

REGIONAL GEOLOGICAL SETTING

The southern peninsular shield of India is mainly made up of rocks of the greenstone-gneiss complex which occupies the central part of the shield, covering most of the Mysore-Deccan plateau, and of the pyroxene granulite-charnockite group which fringes the greenstones to the south and east and forms a major part of the western and eastern Ghats (Fig. 1).

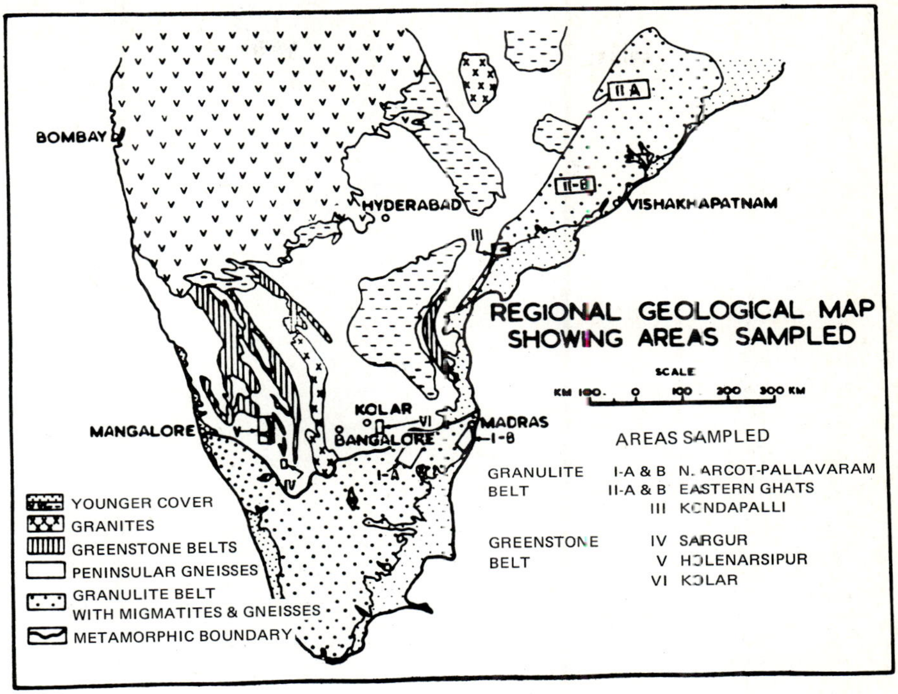

Fig. 1. Geological map of southern India showing areas sampled.

Both groups belong predominantly to the intermediate-pressure metamorphic series, although locally low-pressure assemblages are reported. The anorthosite-granulite association is a characteristic feature of granulite belts. The granulite facies rocks are intruded by later granites and gneisses resulting locally in migmatitic zones. Retrograde metamorphic effects are common.

The greenstone-gneiss complex has a predominant greenschist-to-amphibolite facies mineralogy, the latter especially along its borders with the granulite facies rocks and in association with the Peninsular gneisses and granites. A recent attempt has been made to separate the rocks of the amphibolite facies into the Sargur group (Radhakrishna and Vasudev, 1977).

AREAL DISTRIBUTION OF SAMPLES

The samples analysed were collected from three areas within the granulite belt: (1) the North Arcot-Pallavaram area near Madras, (2) the Eastern Ghats between the Mahanadi and Godavari rivers, and (3) the Kondapalli area near Vijayawada. The samples in all except the second area were collected at random during traverses, whilst those from the Eastern Ghats (collected by Dr V. Divakara Rao of NGRI) are from four traverse lines across the NE-SW strike of the formation, viz. (a) from the Salur-Jeypore profile, (b) from the Narsipatnam-Malakanagiri profile, (c) from the Visakapatanam-Araku profile, and (d) from the Guntur area. The samples from greenstone belts are confined to the Sargur, Holenarsipur and Kolar schist belts.

The areal distribution of the samples are as follows (Fig. 1):

Granulite Belts	:	North Arcot-Pallavaram (NAP)	- 31
		Eastern Ghats (EG)	- 26
		Kondapalli	- 24
Dharwar Greenstone Belts	:	Sargur	- 16
		Holenarsipur	- 15
		Kolar	- 21

PRESENTATION OF DATA

The rocks of the granulite belts are grouped into comparable silica ranges, in view of the fact that all the 21 samples analysed from the Eastern Ghats (Locs. II A and B, Fig. 1) fall within a percent silica range of 50.07 - 66.81. The average radioelement content of samples and their elemental ratios from the Eastern Ghats and NAP are presented in Table 1. Only three samples have been analysed from the Kondapalli area and the elemental averages for these are shown separately.

Table 1. Average radioelement abundances and ratios* in pyroxene granulites (PG) and charnockites (C)

Locality (No. of samples)	Rock types	U (ppm)	Th (ppm)	K (%)	Th/U	K K/U x 10^{-4}	K/Th x 10^{-4}	K/Na
EG (21)	PG, C SiO_2% range 50.07-66.81 av. 58.50	0.45	1.2	1.25	2.7	2.78	1.04	0.56
NAP (9)	PG, C SiO_2% range 45.29-66.29 av. 56.58	0.50	1.1	1.11	2.2	2.22	1.01	0.36
NAP (8)	PG, C SiO_2% range 67.98-75.0 av. 71.78	1.95	14.5	2.86	7.5	1.47	0.20	1.23
Konda-palli (3)	PG, C	0.25	0.55	1.83	2.2	7.32	3.32	0.56

[*Elemental ratios given in the tables in this paper are the ratios of the average element contents and not the average of the observed ratios.]

The radioelement content in the migmatites, granite-gneisses and anorthosites in the granulite belts is summarised in Table 2. This table also contains data on an anorthosite sample from the greenstone belt of Holenarsipur, the radio-element content of amphibolites (metabasaltic rocks) of the Kolar, Sargur and Holenarsipur greenstone belts as well as on the associated gneisses and granites, including the "Champion gneiss". Table 2 also contains data on 5 samples of dolerites intrusive into the granulite belt of NAP.

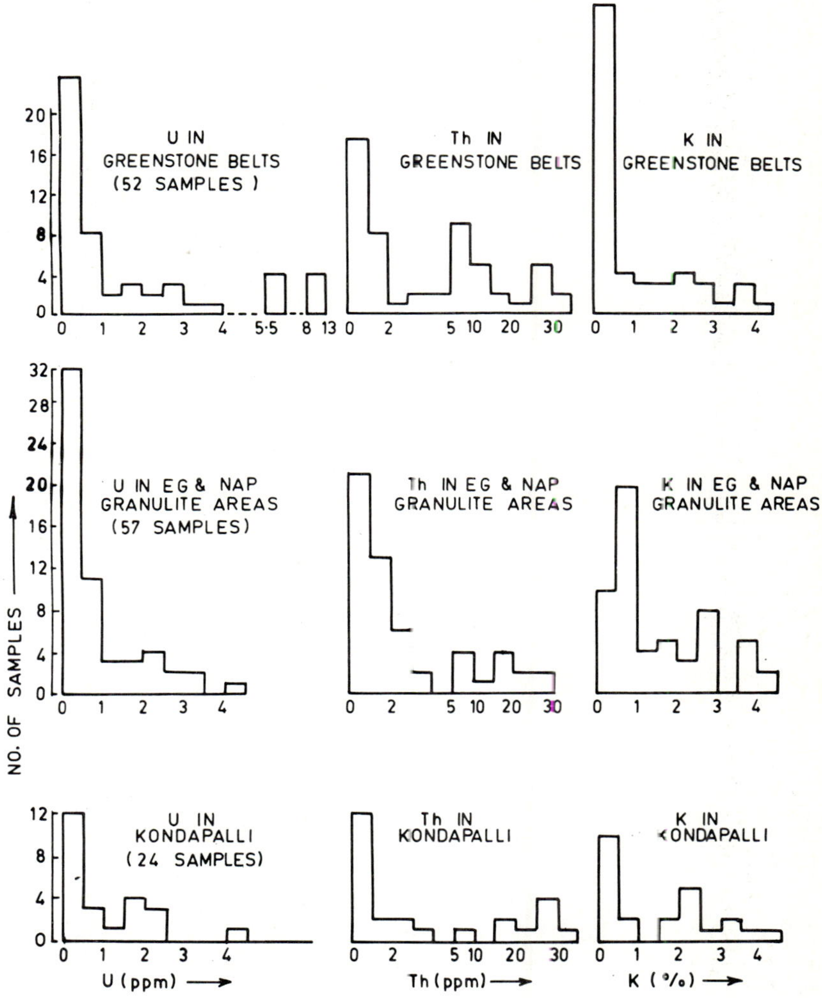

Fig. 2. Histograms of U, Th and K concentrations for different areas.

Table 2. Average radioelement abundances and ratios in anorthosites (An), migmatites (M), granitic gneisses (GG), amphibolites (Am), granites (Gr), gneisses (Gn), the "Champion gneiss" (Ch) and dolerites (meta?) (D)

Locality	Rock type	U (ppm)	Th (ppm)	K (%)	Th/U	K/U x 10^{-4}	K/Th x 10^{-4}
Kondapalli (6 samples)	An	0.35	1.2	0.31	3.4	0.89	0.26
NAP (1 sample)	An	1.6	5.9	0.14	3.7	0.09	0.024
Holenarsipur (1 sample)	An	0.2	0.5	0.06	2.5	0.30	0.12
NAP (7 samples)	M, GG	1.45	22.9	2.84	15.9	1.97	0.12
Kondapalli (6 samples)	M, GG	1.95	40.1	3.47	20.6	1.78	0.086
Sargur schist (9 samples)	Am (mainly)	0.25	0.8	0.20	3.2	0.77	0.24
Kolar schist (4 samples)	Am (mainly)	0.4	1.3	0.26	3.0	0.62	0.21
Holenarsipur (6 samples)	Am	0.4	1.1	0.19	2.7	0.47	0.17
Kolar schist (5 samples)	Gr, Gn	5.3	20.5	2.99	3.9	0.56	0.15
Holenarsipur (5 samples)	Gr, Gn	2.8	7.5	1.26	2.7	0.45	0.17
Kolar schist (3 samples)	Ch	5.5	16.0	2.41	2.9	0.44	0.15
NAP (5 samples)	D	1.1	3.2	0.73	2.9	0.66	0.23

RESULTS

1. General

U, Th and K concentrations for the different areas are shown in histograms in Fig. 2. 75 samples have U below 1 ppm and 35 of these have Th also below 1 ppm. Only 9 samples have their K content below 0.1%. Among the three greenstone belts, Sargur has the lowest concentration for all the three radioelements and Kolar the highest. K contents in the Holenarsipur belt are quite low but are, in general, higher in the granulite belts of the Eastern Ghats, North Arcot and Kondapalli areas as compared to those in the schist belts.

The frequency distribution of U in the greenstone belts (Fig. 3) has a near-lognormal distribution, which is not the case in the granulite belts where U shows a bimodal distribution, suggesting possibly two different events in U mineralisation. Th distribution in all three areas shows a complex pattern (Fig. 4) possibly related to two or more genetic events.

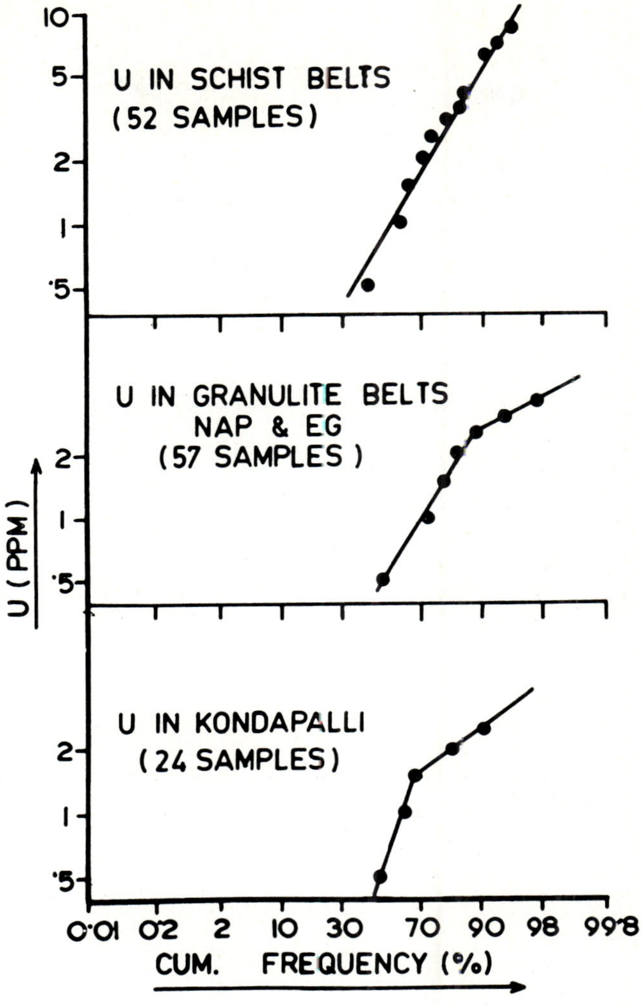

Fig. 3. Frequency distribution of U in schist and granulite belts.

Th/U ratios in all the areas show a wide dispersion (Fig. 5). Greenstone-gneiss areas show a peak between 2.5 and 3.5. Out of 48 samples from those areas for which ratios are available, 31 show a value below, and 17 above, the normal value of 3.5. Three of these samples have a ratio of about 10. For granulite belts of NAP and EG, out of 52 ratios available, 33 have less than the normal, while the rest have above average ratios, 7 having a ratio greater than 10. For the Kondapalli area, out of 18 ratios available, 6 are below and 12 above average, 6 samples having ratios above 10.

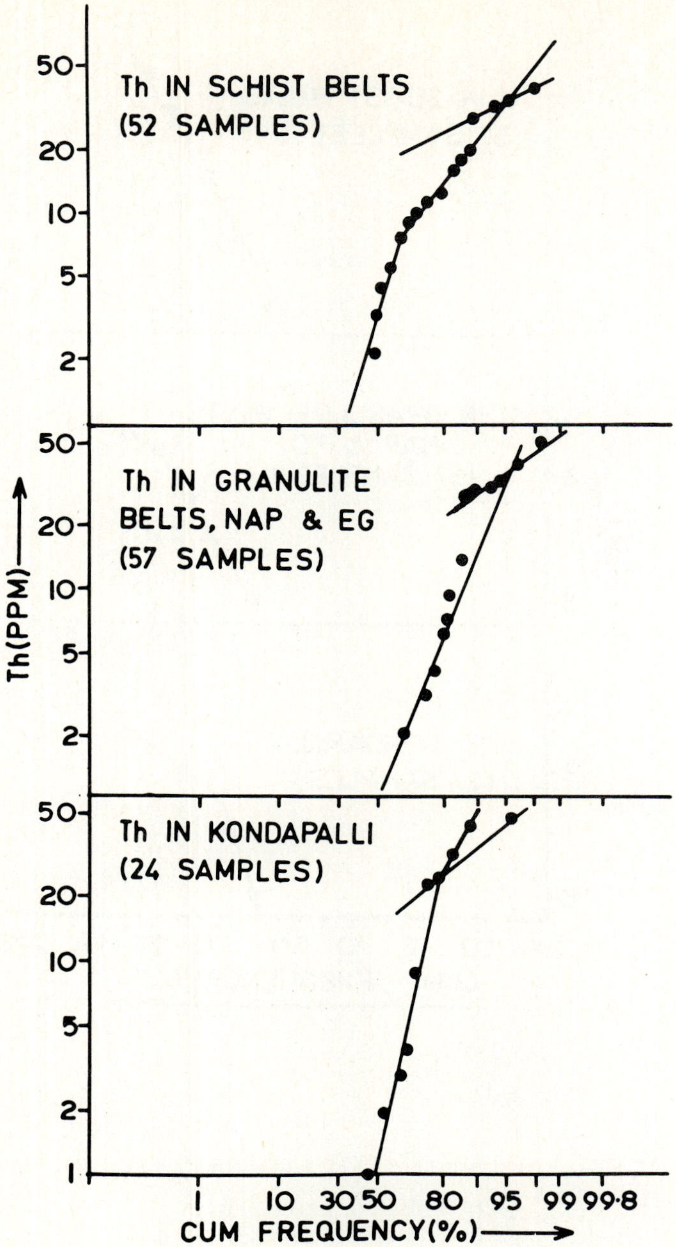

Fig. 4. Frequency distribution of Th in schist and granulite belts.

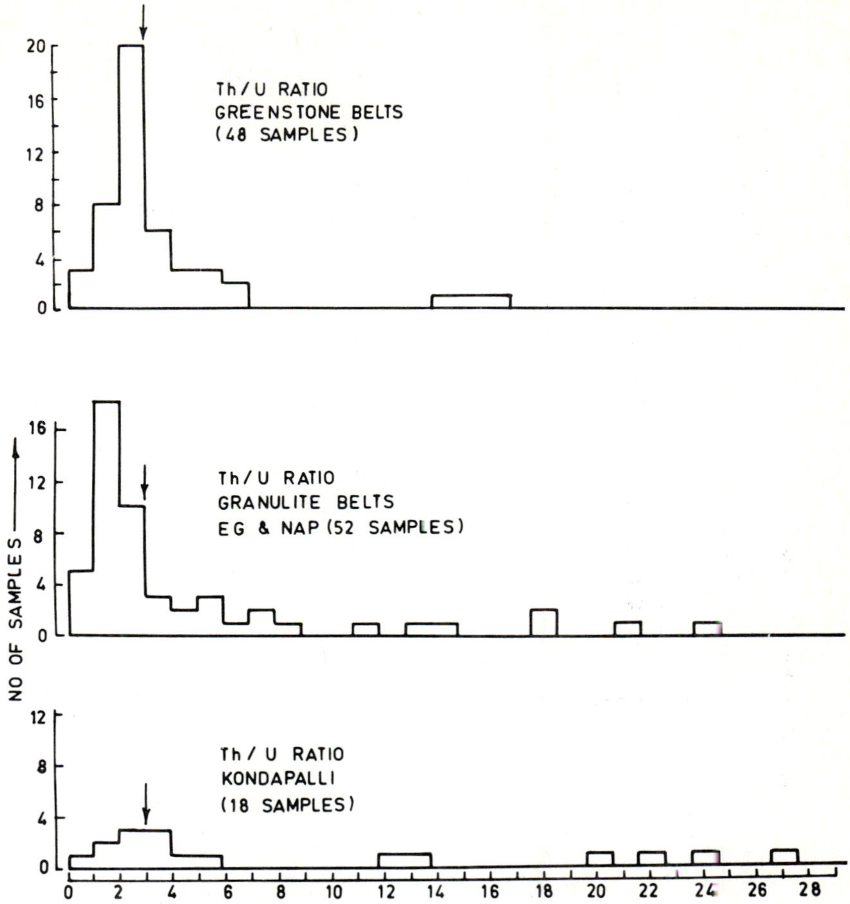

Fig. 5. Histogram showing Th/U ratios for different areas.

For K/U and K/Th ratios the granulite belts show greater scatter towards above-normal values than the greenstone belts (Fig. 6) (normal values for magmatic rocks are taken as K/U x 10^{-4} = 1 and K/Th x 10^{-4} = 0.25).

The plots of Th/U vs U and Th/U vs Th are given in Figs. 7 and 8. These indicate that the higher Th/U ratio in the higher silica ranges is possibly due to the enrichment of thorium in the granulite belt rocks. In the greenstone belts no such characteristic enrichment of thorium is indicated.

Plots of K vs Th, K vs U and Th vs U in the greenstone and granulite belt rocks in Fig. 9 indicate that there is no strict linear relationship between these constituents and the plots show wide scatter. Variation of U and Th with respect

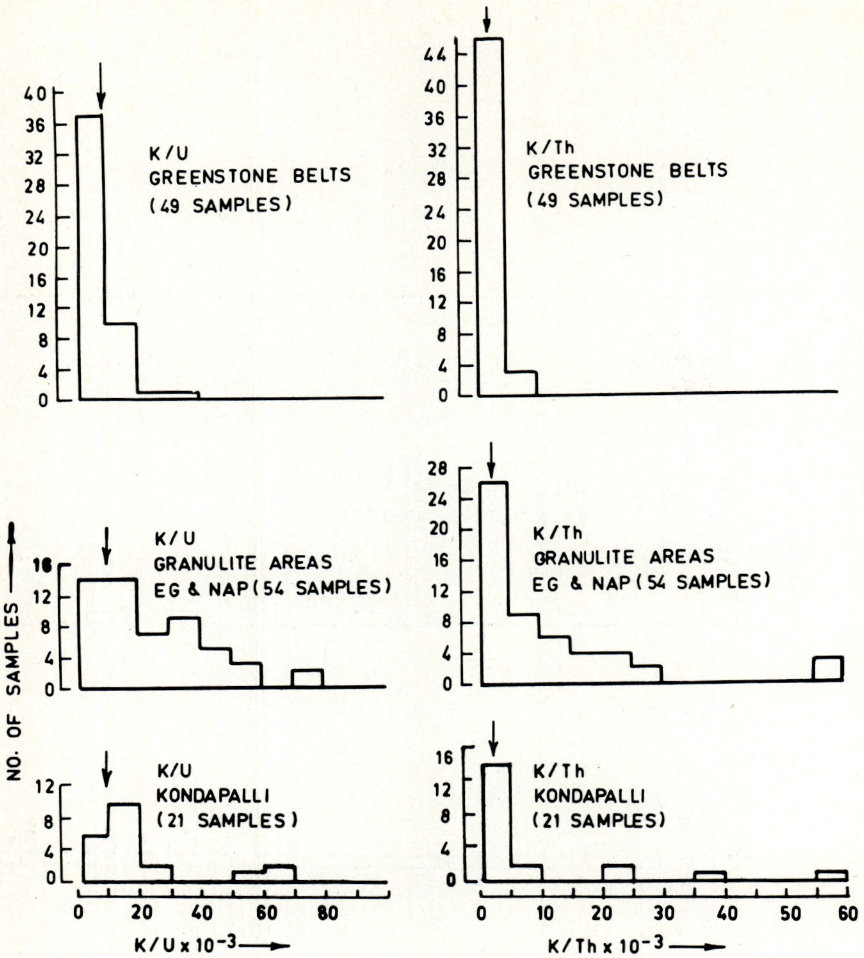

Fig. 6. Histogram showing K/U and K/Th ratios for different areas.

to SiO_2 and Na also shows no linear trend. This confirms that the rocks under study are involved in a complex process of metasomatic evolution.

2. Rocks of the Granulite Belts

(a) Pyroxene granulites and charnockites

These rocks are characterised by low U, Th and Th/U ratios and high K/U and K/Th ratios, where the silica content is less than 65%, indicating depletion of U and Th in relation to K, a feature shared with all reported pyroxene granulites of the world. In the more acidic groups an enrichment of Th in relation to both U

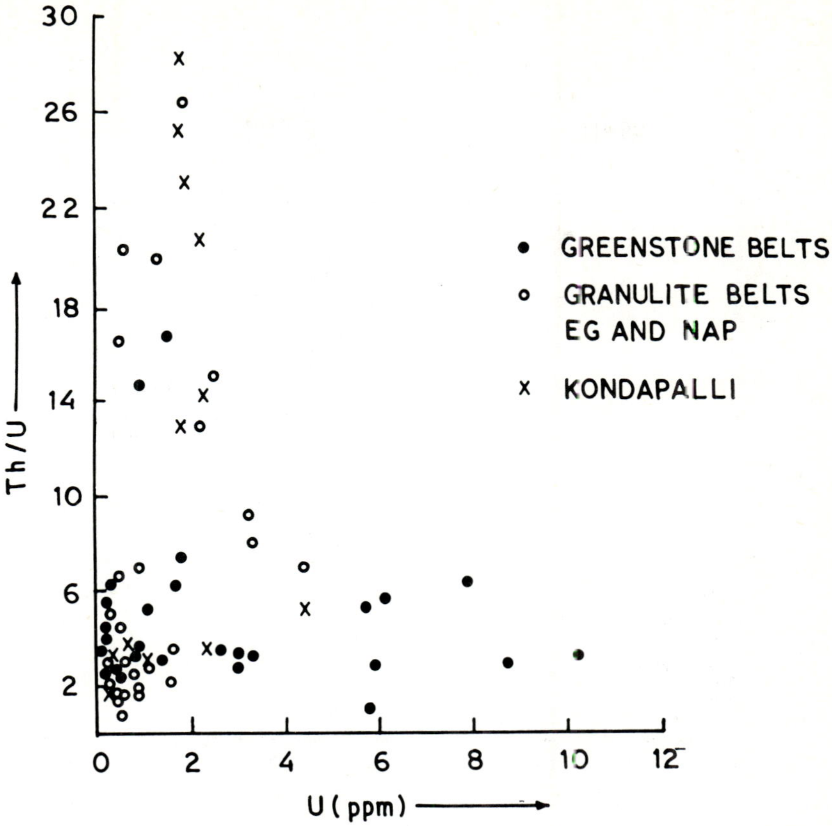

Fig. 7. Th/U vs U distribution.

and K is noted and consequently these rocks have high Th/U ratios.

(b) Gneisses and granites

Radioelement calculations in the quartzo-feldspathic gneisses intrusive into the charnockites of the NAP and Kondapalli areas show relatively far higher concentrations as well as high Th/U ratios. The limited number of samples analysed reveals a depletion in U and enrichment of Th in relation to K in these rocks compared to magmatic rocks.

(c) Anorthosites

Th and U contents of anorthosites (6 samples) from Kondapalli, occurring as layered differentiates in the charnockites, range from 0.4 to 2.3 ppm and 0.2 to 0.6 ppm respectively. In the limited number of samples analysed there is an indication of two types: one contains 0.2 ppm U or less and 0.5 ppm Th, the other has higher values for both elements. Only one sample from North Arcot is depleted in K with respect to U and Th, which is in contrast to the charnockite samples from that area.

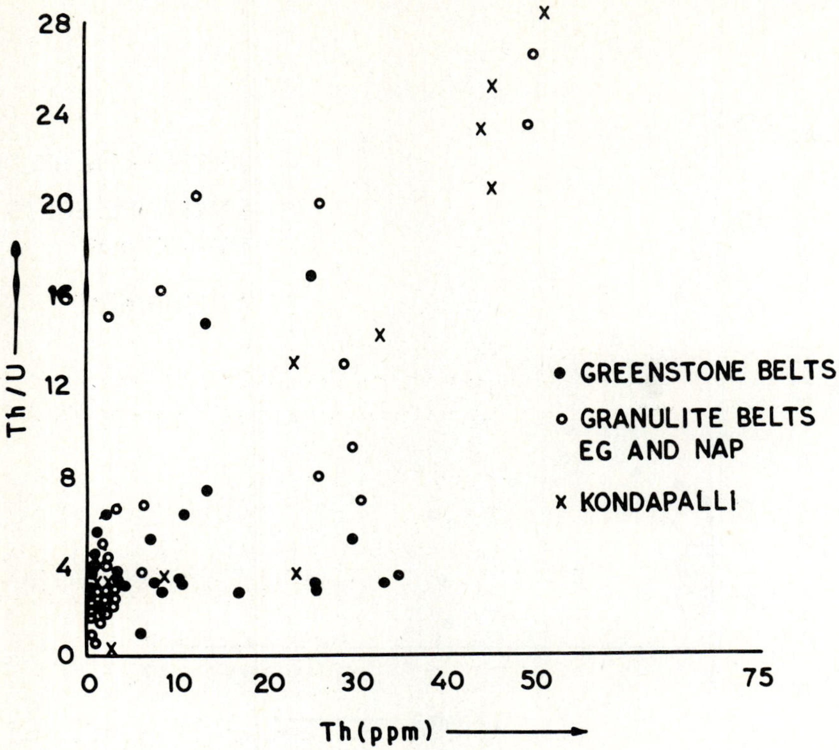

Fig. 8. Th/U vs Th distribution.

(d) Dolerites

In the dolerite dykes of the NAP area, U and Th show some variation whereas K remains nearly constant. Average K/U, K/Th and Th/U ratios are significantly comparable with early basic rocks of the greenstone belts (Table 2), although the content of U, Th and K itself is higher in these dykes.

3. Rocks of the Greenstone Belts

(a) Meta-basic rocks (mainly amphibolites)

The meta-basic rocks in the Sargur, Kolar and Holenarsipur schist belts show comparable U, Th and elemental ratios, a feature in keeping with their possible consanguinity in origin. Compared with pyroxene-granulites of similar silica ranges from the granulite belts, the meta-basic rocks of the greenstone belts have a tendency towards lower K/U and K/Th and slightly higher Th/U ratios.

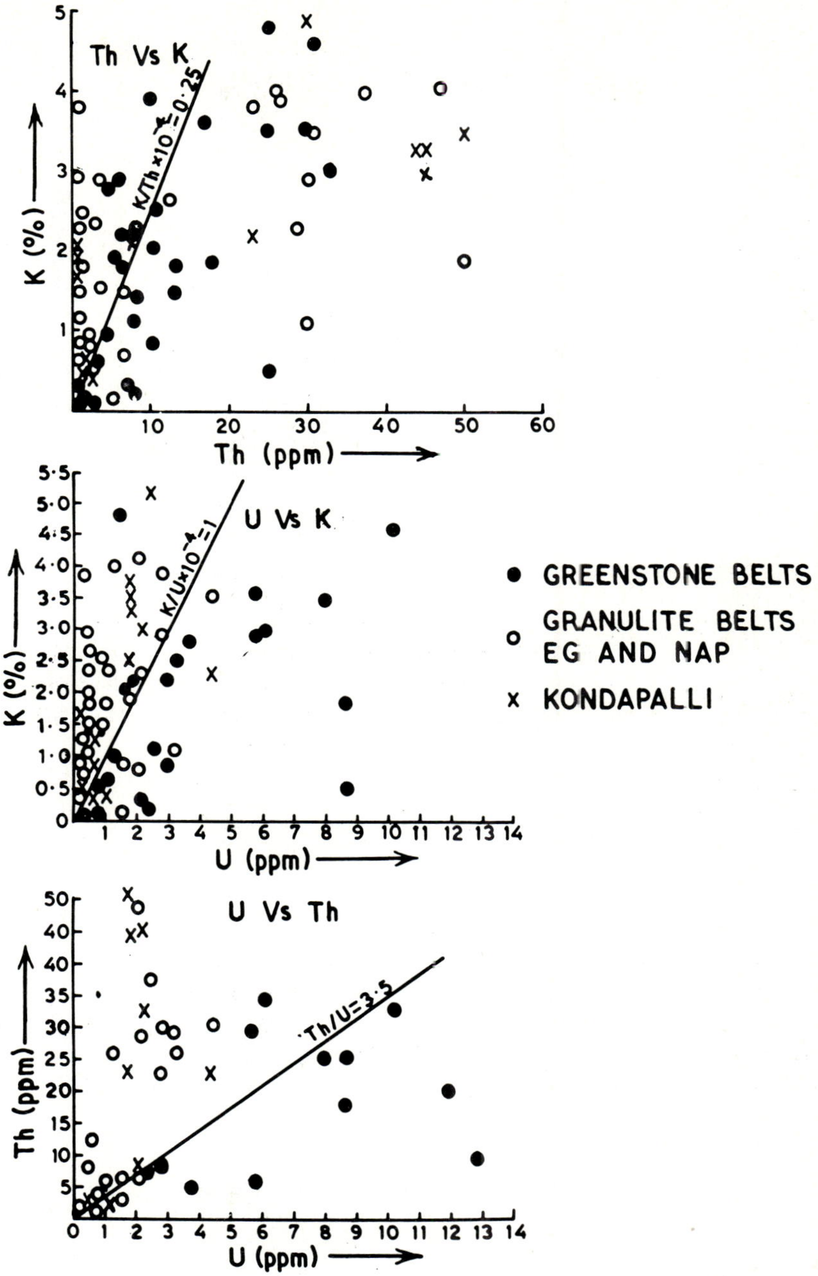

Fig. 9. Th vs K, U vs K and U vs Th distributions.

(b) Gneisses and granites bordering the greenschist belts and the "Champion gneiss"

Gneisses and granites bordering the Kolar and Holenarsipur schist belts have a radioelement distribution distinctive in relation to the charnockites and to the migmatites and granitic rocks intrusive into the charnockites. Like the amphibolites of the greenstone belts, these rocks show relative enrichment of U and Th in relation to K.

Summing up, there is a deficiency of U and Th in relation to K in the pyroxene granulites and charnockites with a tendency to enrichment of both K and Th in relation to U in the later intrusive phases. In the greenstone belts, on the other hand, there is a deficiency of K in relation to U and Th in both the basic rocks as well as the gneisses and granites fringing them (Fig. 10).

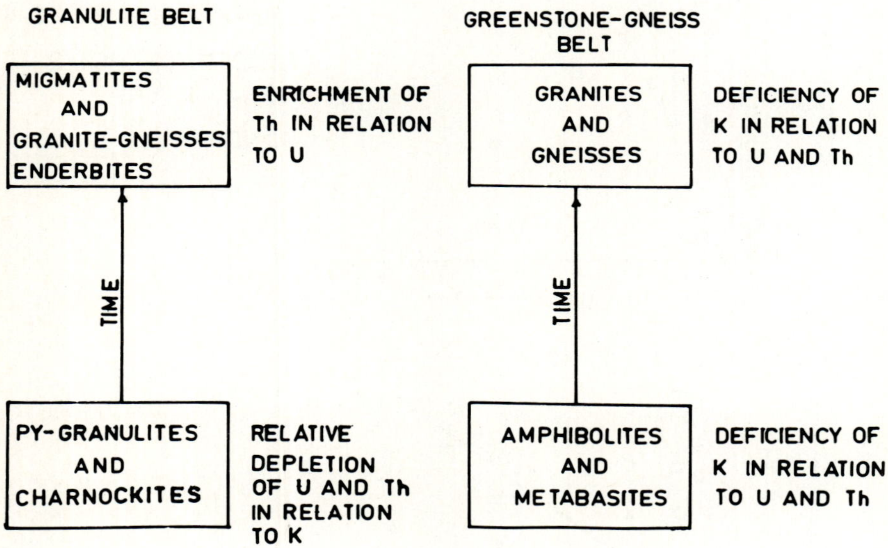

Fig. 10. Trends in geochemical evolution in greenstone-gneiss and granulite belts.

DISCUSSION

The contrasting patterns of radioelement distribution in the rocks of the granulite belts in comparison with those of lower grade metamorphic complexes have been reported by many earlier workers. The relative radioelement enrichment in low-grade metamorphic rocks and their depletion in granulite facies rocks have been attributed to the upward migration of these elements. Zhukova et al. (1973) have specifically proposed extraction of uranium during metamorphism and transfer to upper crustal zones.

The results reported in this paper indicate that remobilisation of radioelements, especially thorium, takes place during the emplacement of younger intrusives in the granulite facies rocks. No evidence is available for a large-scale migration of these elements from the granulite facies rocks to the greenstone belts of lower metamorphic grade. Because U and Th are fixed largely in the lattice of such minerals as zircon, monazite and apatite in the granulite facies rocks, it would be difficult to mobilise these elements without destroying the structure of the minerals. Further, enrichment of U and Th in relation to K seems to be a feature shared by all the rocks of the greenstone belts, both the amphibolites and associated granites and gneisses. It would be difficult to arrive at such a uniform enrichment in all the rocks on a regional scale by a process of upward migration of these elements on their transfer to upper zones, as proposed by various authors, because of the many physical barriers, such as faults and shear zones, and chemical barriers, such as rocks and minerals in which U and Th can be preferentially fixed. In view of these considerations the contrasting trends in enrichment of radioelements appear to be an intrinsic feature of the two belts, the geological reasons for which are still not well understood.

ACKNOWLEDGEMENTS

Thanks are due to members of the Working Groups who collected these samples and made them available for analytical studies. Thanks are also due to S.G. Vasudeva for his useful suggestions during the course of the work.

REFERENCES

Heier, K.S., 1962. Spectrometric uranium and thorium determinations on some high-grade metamorphic rocks of Langoy, Northern Norway. Norsk geol. Tidsskr., 42: 143-156.

Heier, K.S. and Adams, J.A.S., 1965. Concentration of radioactive elements in deep crustal material. Geochim. Cosmoch. Acta, 29: 53-61.

Heier, K.S. and Thorsen, K., 1971. Geochemistry of high-grade metamorphic rocks, Lofoten-Vesteralen, North Norway. Geochim. Cosmoch. Acta, 35: 89-99.

Lambert, I.B. and Heier, K.S., 1967. The vertical distribution of uranium, thorium and potassium in the continental crust. Geochim. Cosmoch. Acta, 31: 377-390.

Lambert, I.B. and Heier, K.S., 1968. Geochemical investigations of deep-seated rocks in the Australian Shield. Lithos, 1: 30-53.

Narayanaswamy, R. and Venkatasubramanian, V.S., 1969. Uranium and thorium contents of coexisting gneisses, granites and pegmatites. Geochim. Cosmoch. Acta, 33: 1007-1009.

Radhakrishna, B.P. and Vasudev, V.N., 1977. The early Precambrian of the southern Indian Shield. J. geol. Soc. India, 18(10).

Zhukova, A.M., Venidiktov, V.M. and Zhukov, G.V., 1974. Pecularities of uranium distribution in highly metamorphic and ultrametamorphic rocks. Geol. Zh., 34(1): 118-122.

EXPERIMENTAL AND THERMODYNAMIC EVIDENCE FOR THE OPERATION OF HIGH PRESSURES IN ARCHAEAN METAMORPHISM

R. C. Newton
Department of the Geophysical Sciences, University of Chicago, Chicago 60637 Illinois, U.S.A.

Abstract

Quantitative analysis of the pressures of granulite-grade metamorphism is now possible as a result of recent calorimetric, experimental and theoretical work. Three lines of approach are used to estimate the apparent recrystallization pressures of well-described samples from Precambrian terranes:

1) The assemblage garnet-sillimanite-quartz. If cordierite is absent, as in the khondalites of southern India, the $Mg_3Al_2Si_3O_{12}$ content of the garnets yields minimum pressures of metamorphism.

Estimates for two khondalites from the Pallavaram area, Tamilnadu, give minimum pressures of 6-9 kilobars in the range 700°-$900^{\circ}C$ from the experimental and thermodynamic data.

2) The assemblages hypersthene-sillimanite-quartz and sapphirine-hypersthene-quartz. These assemblages, chemically equivalent to magnesian cordierite, apparently require minimum pressures of 6-10 kilobars, based on experimental work. The lower estimates are for the completely anhydrous condition.

3) The occurrence of garnet in basic and intermediate-composition granulites. Experimental work apparently requires a minimum of 8 kilobars for garnet in basic compositions at temperatures above $700^{\circ}C$. Thermodynamic calculations pertaining to the assemblage garnet-plagioclase-hypersthene-quartz yield pressures of 8-10 kilobars for two well-described granulite specimens from Uganda and Sweden.

Thus we have good evidence of a thick continental crust even in Archaean times. If there has been continued accumulation of crustal material throughout geologic time, some mechanisms of strong local crustal thickening seem to be required to explain the high pressures of Archaean metamorphism.

INTRODUCTION

The concept that the continental crust has grown thicker and stronger through time is an old and pervasive one that has its origins in the idea of continental accretion, which, for North America at least, is clearly revealed in the distribution of radiometric dates (Gastil, 1960; Goldich et al., 1966). The earliest continental nucleii may have accreted on a thin, weak, primitive oceanic crust (Anhaeusser et al., 1969) and continents may have been increasing in thickness by addition of material transferred to the base of the crust throughout geologic time (Shackleton, 1967). The character of recrystallization of this crust, i.e. metamorphism, may have changed correspondingly, from a type characterized by high-temperature, lower pressure condition in the most ancient times to cooler and often more deep-seated conditions later in geologic time (De Roever, 1956).

This concept has been modified in recent years by a growing recognition of very deep-seated conditions of metamorphism manifest in some Precambrian terranes. The indication of high temperature (granulite facies) conditions remains, but indications of very high pressures, of ten kilobars or more, have been found (Lutts and Kopaneva, 1968; Rabkin, 1968; Morse and Talley, 1971). Most of these are based on relatively recent experimental and theoretical work. It is the purpose of this paper to discuss some of the newer, experimentally based deductions of high pressure conditions operating in the early Precambrian crust, and to apply very recent thermodynamic data to reconsider some of the calculations.

MINERAL ASSEMBLAGES IMPLYING HIGH PRESSURES

The single most diagnostic mineral indicating high pressures is garnet, which can occur in a wide variety of acid to basic rock compositions. At high grades the mineral cordierite competes with garnet for the available aluminium in nearly anhydrous rocks, i.e. the granulites. Cordierite without garnet is usually taken to indicate lower pressures of metamorphism; cordierite and garnet commonly coexist in rocks which are interpreted as progressively more deep-seated as the two minerals become more magnesian (Hensen, 1971). Relatively magnesian garnet without cordierite in acid compositions is considered a high pressure assemblage. In rocks of high Mg/Fe^{2+} ratio, hypersthene plus aluminum silicate plus quartz may appear instead of garnet as denser assemblages replacing cordierite. Garnet can appear in intermediate and basic granulite assemblages (Howie and Subramaniam, 1957; Manna and Sen, 1974;) if the crystallization pressures were sufficiently high. The quantitative aspects of these mineralogic pressure indicators will be considered in specific examples.

1. Garnet in high-grade aluminous felsic compositions.

If an almandine-pyrope garnet is crystallized in a certain

temperature-pressure range along with cordierite, aluminum silicate and quartz, the following reaction, divariant in the system $MgO-FeO-Al_2O_3-SiO_2$, applies:

$$\text{(A)}\quad 3\ (Fe,Mg)_2Al_4Si_5O_{18} \leftrightarrows 2\ (Fe,Mg)_3Al_2Si_3O_{12} + 4\ Al_2SiO_5 + 5\ SiO_2$$

cordierite garnet sillimanite quartz

In P-T ranges where cordierite is stable, the aluminum silicate is almost always sillimanite. In general, the $Mg/Mg+Fe^{2+}$ ratios of coexisting cordierite and garnet are different and, for a dry system, depend on temperature and pressure only, if the garnet is fairly low in other components. It has been shown theoretically and experimentally that, at least for situations where water pressure is low, the isopleths, or curves of constant $Mg/Mg+Fe^{2+}$ ratio, for the minerals have flat dP/dT slopes; that is, they are nearly parallel to the temperature axis. The set of isocompositional curves of cordierite intersect those of garnet at small angles. The situation is shown schematically in Fig. 1. In theory, the compositions of coexisting cordierite and garnet, if they are an equilibrium pair, yield both the temperature and the pressure of crystallization once these scales, implicit in reaction (A), are calibrated experimentally or thermodynamically. In practice, the relatively small temperature dependence of the isopleths renders them good pressure-determination curves, but the small intersection angles can cause large errors in determinations of the associated temperatures. The most practical application at present is for geobarometric estimates in conjunction with independent temperature estimates. If the assemblage garnet-sillimanite-quartz is present without cordierite, as in the khondalites of southern India, the composition of the garnet yields a minimum pressure for the crystallization, subject to an independent temperature estimate. If cordierite is present but not garnet, the relevant isopleth is an upper pressure limit.

The geobarometer scale has been calibrated experimentally by Currie (1971) and Hensen and Green (1971), and discussed theoretically by Lepezin (1969), Wood (1973) and Hutcheon et al. (1974). The work of Hensen and Green (1971) differs significantly from that of Currie (1971). Currie's data show slight positive dP/dT slopes for the garnet isopleths, and yield pressures of crystallization slightly lower for a given garnet composition in the range 700°-900°C than do the data of Hensen and Green (1971). This difference was ascribed by

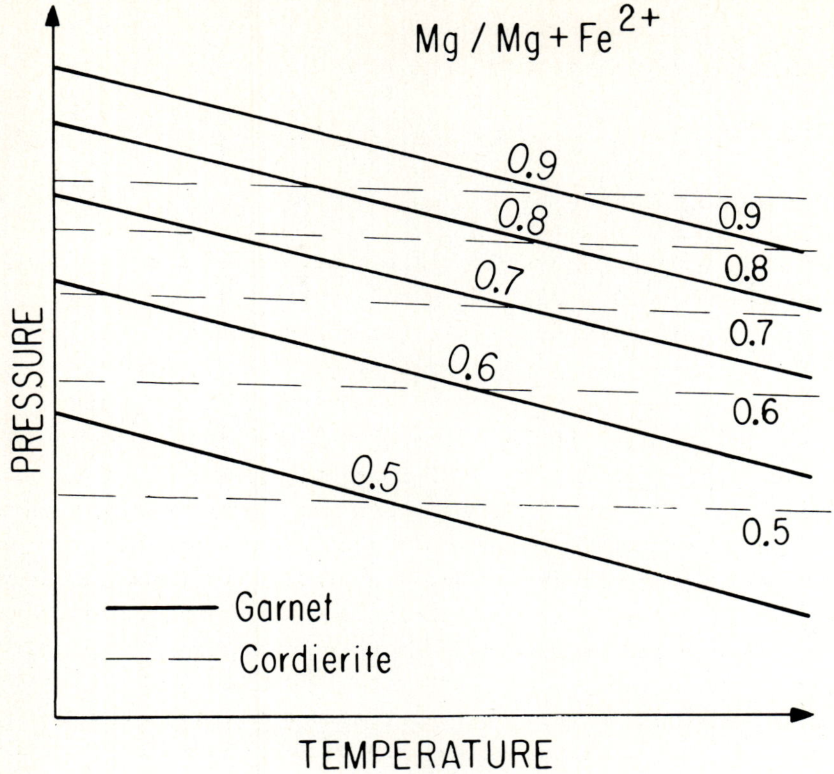

Fig. 1. Schematic representation of $Mg/Mg+Fe^{2+}$ isopleths of coexisting cordierite and garnet.

Wood (1973) to the effect of water pressure. It was first shown by Lepezin (1969) that the "humidity", or content of molecular water in the cordierite framework channels (Smith and Schreyer, 1960), could affect the dP/dT slopes of the distribution isopleths with garnet. Wood's (1973) calculations, based on the experimentally determined dependence of cordierite humidity on water pressure and temperature (Schreyer and Yoder, 1964), showed that Currie's experiments, which were done under hydrothermal conditions of P_{H_2O} equal to P_{total}, could be rectified with those of Hensen and Green (1971), which were done under conditions of $P_{H_2O} \ll P_{total}$. Other possible reasons for the discrepancy include differences in analytic procedure, type of high pressure apparatus (the Hensen and Green, 1971, experiments were done in the solid-pressure-medium piston-cylinder apparatus, with absolute pressure calibration somewhat inferior to Currie's gas-pressure apparatus), and to the fact that Hensen and Green (1971) used a "model" pelite" system with certain amounts of CaO and K_2O present along with the ferromagnesian components. Their synthetic garnets contained a nearly constant 6 mole percent of grossular component.

Specific application may now be made to estimate the crystallization pressures of khondalites. Temperatures in the range 700°-900°C are usually assumed for granulite-grade rocks. This is based on a number of criteria, including estimates of the dehydration temperatures of micas and amphiboles at low P_{H_2O} (Touret, 1971), the presence of associated migmatites in some granulite terrains, which requires that temperatures exceed the beginning of hydrous melting in acid compositions (Winkler, 1976, p. 312), and pyroxene-composition geothermometry (Hewins, 1975). The full chemical analyses of two garnets of khondalites from the Pallavaram area, Tamil Nadu, are given by Howie and Subramaniam (1957). The garnet of specimen Ch 119 has the composition: almandine .649, andradite .026, grossular 0.00, pyrope .321 and spessartine .004 in mole fractions, giving a $Mg/Fe^{2+}+Mg$ ratio of .331. Specimen Ch 121 has the respective composition: .585, .027, 0.00, .380 and .008, or $Mg/Fe^{2+}+Mg$ = .394. Departures from the pyrope-almandine join are negligible. According to Hensen and Green (1971), the minimum pressure of crystallization for Ch 119 is 8.9 kilobars at 770°C and 7.8 kilobars at 900°C. The respective pressures for Ch 121 are 9.3 and 8.3 kilobars. The lowest possible temperature in this pressure range is about 750°, since the specimens contain sillimanite rather than kyanite (Holdaway, 1971). The data of Currie (1971) yield minimum pressures of 7.1 - 7.5 kilobars for the two specimens at 800°C and 7.4 - 8.0 kilobars at 900°C. The data of Hensen and Green may perhaps be preferred, since their experiments were carried out at very low water pressures, which seems more in keeping with granulite facies conditions (Winkler, 1976).

Hutcheon et al. (1974) constructed a thermodynamically based set of garnet-cordierite isopleths for the completely anhydrous situation. Their calibration suffers somewhat from the inadequate measured thermochemical data which were available to the authors. Additional thermodynamic information is available now, and it is tempting to check their calculations. For the equilibrium of reaction (A), considering only the Mg-end members, the governing equation applies:

$$-\frac{\Delta H^{\circ}}{RT} + \frac{\Delta S^{\circ}}{R} \cong \ln \frac{(\alpha^{crd}_{Mg})^{6}}{(\alpha^{gt}_{Mg})^{6}} + \frac{P\Delta V^{\circ}}{RT} \qquad (1)$$

where α^{crd}_{Mg} denotes the activity of $MgAl_2Si_{5/2}O_9$ in cordierite and α^{gt}_{Mg} denotes the activity of $MgAl_{2/3}SiO_4$ in garnet, ΔH^o is the enthalpy change of the end-member reaction at 1 bar and T, and S^o is, likewise, the standard entropy change, ΔV^o the standard solid volume change, P the pressure, T the absolute temperature, R the gas constant; the approximation sign is used because of neglect of the differential compressibilities over several kilobars, which is a very accurate approximation. Table 1 gives recent data for the substances at 1000 K, from which we derive $\Delta H^o = -6.14 \pm 0.73$ kcal. ΔS^o comes out to 18.36 ± 1.2 cal/K. This includes new data for pyrope (Newton et al., 1977). ΔV^o is $150.56 \pm .80\ cm^3$.

At this point we diverge from Hutcheon et al. (1974), who considered pyrope-almandine garnet as an ideal solid solution. An improvement on this oversimplification is possible as a result of recent theoretical deductions and phase equilibrium measurements. Following Ganguly and Kennedy (1974) we regard the garnets as "regular" solutions, whose components have non-unit activity coefficients given by:

$$RT\ln\gamma = W(1-X)^2 \qquad (2)$$

where γ is the activity coefficient of a component, X is the mole fraction of that component, and W, the interaction parameter, is a constant. Oka and Matsumoto (1974) have deduced a W for the pyrope-almandine interaction of 2.58 kcal (based on one-atom mixing) and Ganguly and Kennedy (1974) and Wood (pers. comm., 1977) have deduced similar values.

It is likely that cordierite is much closer to an ideal solution of the Mg- and Fe-end members than is garnet, on the basis of very similar unit-cell volumes and stabilities of the end members. Even if cordierite is non-ideal to the same extent as garnet (W = 2.5 kcal), which is improbable, the pressure calculation to be made would only be affected by about 200 bars. Therefore the activity of $MgAl_2Si_{5/2}O_9$ in cordierite may be taken safely as its mole fraction.

Sample number 4 of Froese et al. serves as a test item. The cordierite-garnet-sillimanite-quartz gneiss comes from a transitional granulite terrane adjacent to Hudson Bay in the North West Territory, Canada. The garnet has a mole fraction composition of .596 almandine, .362 pyrope, .032 grossular and .010 spessartine, with $Mg/Mg+Fe^{2+}$ = .378. The coexisting cordierite has $Mg/Mg+Fe^{2+}$ = .770. The pyrope activity coefficient at 680°C, which is the temperature calculated by

TABLE 1

THERMODYNAMIC DATA OF MINERALS AT 727°C USED IN CALCULATIONS

Mineral	Volume* cm^3/gfw	Enthalpy of Solution** kcal/gfw	Entropy*** cal/K
$Mg_2Si_2O_6$ Enstatite	63.94	17.56	92.42
$Mg_3Al_2Si_3O_{12}$ Pyrope	115.58	27.79	185.15
$Mg_2Al_4Si_5O_{12}$ Cordierite	234.06	27.95	270.50
$CaAl_2Si_2O_8$ Anorthite	101.78	15.25	127.28
$Ca_3Al_2Si_3O_{12}$ Grossular	127.35	42.42	184.30
Al_2SiO_5 Sillimanite	50.49	7.07	70.96
SiO_2 Quartz	23.70	-1.23	27.80

Hutcheon et al., is 1.694, using W = 2.58. Equation (1) gives 6.23 ± .54 kilobars for the cordierite-garnet pair with input from Table 1. This compares very well with the pressure of 5.8 kilobars calculated by Hutcheon et al. (1974). At 900°C the pressure would be about 7.0 kilobars.

* Thermal expansions from Skinner (1966). Average uncertainty .02 %.

** Data from Charlu, Newton and Kleppa (1975), except for anorthite and grossular, which are from unpublished data of the author. Average uncertainty 1.0 %.

*** Data mainly from Robie and Waldbaum (1968). Pyrope entropy from Newton, Thompson and Krupka (1977). Grossular entropy from Westrum, Essene and Perkins (1977). 0.24 cal/K added to clinoenstatite entropy of Robie and Waldbaum (1968). 1.22 cal/K added for cordierite disorder. 0.46 cal/K added for sillimanite disorder. Average uncertainly in entropy 0.3 %.

The two khondalite garnets reported by Howie and Subramaniam (1957) are closely similar in $Mg/Mg+Fe^{2+}$ and other aspects to Hutcheon et al.'s (1974) specimen number 4. Ch 121 calculates slightly lower in pressure at a given temperature and Ch 119 slightly higher if a (fictive) cordierite of $Mg/Mg+Fe^{2+} = 0.77$ is assumed. The calculated pressures are minimum ones for these rocks because cordierite is not present. If water pressure is greater than zero, cordierite stability is promoted to higher pressures (Newton, 1972).

On the basis of the foregoing evidence it seems difficult to escape the conclusion that some very ancient felsic granulites have crystallized at pressures of at least 6 - 9 kilobars.

2. High pressure assemblages chemically equivalent to magnesian cordierite.

Certain rare magnesian aluminous rocks are known which contain the assemblages hypersthene-sillimanite-quartz and sapphirine-hypersthene-quartz, which are chemically equivalent to magnesian cordierite, though considerably more dense, and which therefore represent the high-pressure breakdown products of magnesian cordierite. These relationships, which were discovered experimentally by Schreyer and Yoder (1960, 1964) and deduced chemographically by Hess (1969) and Hensen (1971), are modelled in the following simplified reactions:

(B) $2\,(Mg,FeSiO_3) + 2\,Al_2SiO_5 + SiO_2 \rightleftarrows (Mg,Fe)_2Al_4Si_5O_{18}$

enstatite or hypersthene; sillimanite or kyanite; quartz; cordierite

(C) $2(Mg,Fe)_7Al_{18}Si_3O_{40} + 4(Mg,Fe)SiO_3 + 35\,SiO_2 \rightleftarrows 9(Mg,Fe)_2Al_4Si_5O_{18}$

sapphirine; enstatite; quartz; cordierite

These reactions are divariant in the system $FeO\text{-}MgO\text{-}Al_2O_3\text{-}SiO_2$ and the oversimplifications include neglect of a small Al_2O_3 content of enstatite and variable Al_2O_3 content of sapphirine. The sapphirine stoichiometry expressed in reaction (C) is characteristic of many specimens and adheres to the theoretical structural formula (Moore, 1969).

The assemblage sapphirine-hypersthene-quartz is known in granulite facies rocks from Labrador (Morse and Talley, 1971), Uganda (Nixon et al., 1973) and Antarctica (Dallwitz, 1968). The assemblage enstatite or hypersthene-sillimanite-quartz is known from the Aldan Shield,

Siberia (Marakushev and Kudryavtsev, 1965), Labrador (Morse and Talley, 1971), Finnish Lappland (Eskola, 1952), Uganda (Nixon, Reedman and Burns, 1973), Rhodesia (Chinner and Sweatman, 1968), and the Vizagapatam District, Andhra Pradesh, India (Walker and Collins, 1907). Coexistence of the three minerals in the Vizagapatam specimen was verified by R. K. Herd (pers. comm., 1975). The Rhodesian occurrence is particularly interesting because the assemblage is, to a close approximation, confined to the simple system $MgO-Al_2O_3-SiO_2$. The rock has nearly the bulk composition of Mg-cordierite. Chinner and Sweatman (1968) provide textural evidence that kyanite was present in the assemblage at one time rather than sillimanite. All of the above occurrences are Precambrian.

Various authors have noted the high pressures that these assemblages imply. Much of this conjecture is based on the Mg-cordierite stability diagram of Schreyer and Yoder (1964), worked out under conditions of $P_{H_2O} = P_{total}$. This diagram implies pressures in excess of 11 kilobars for the anhydrous high-pressure breakdown products of cordierite to appear. However, Newton (1972) and Newton et al. (1974) have shown, experimentally and calorimetrically, that Mg-cordierite under anhydrous conditions breaks down several kilobars lower in pressure than under hydrous circumstances. The latter authors showed, mainly calorimetrically, that at temperatures above about 800°C low-iron cordierite in nature should break down to a well-ordered sapphirine of approximately the composition $Mg_7Al_{18}Si_3O_{40}$ at a pressure several hundred bars lower than Mg-cordierite reacts in the synthetic system. The synthetic product of high temperature, high pressure cordierite breakdown is a somewhat disordered sapphirine of nearly the composition $Mg_2Al_4SiO_{10}$. In this case no enstatite enters the breakdown assemblage, since cordierite, sapphirine and quartz are co-linear in the system $MgO-Al_2O_3-SiO_2$.

Fig. 2 shows the experimental and calorimetrically inferred natural breakdown conditions of magnesian cordierite under completely anhydrous conditions and under conditions of $P_{H_2O} = P_{total}$. Fig. 2 includes very recent experimental work on the reaction to sapphirine and quartz done in a gas-pressure system with optimal measurement of the pressures. The following information seems conclusive. If temperatures of granulite metamorphism are in the range 700°-900°C, minimum pressures of 6.4 to 7.0 kilobars are required to produce the high pressure breakdown assemblages of very magnesian cordierite. The pressure necessary to break down dry Mg-cordierite to enstatite, sillimanite

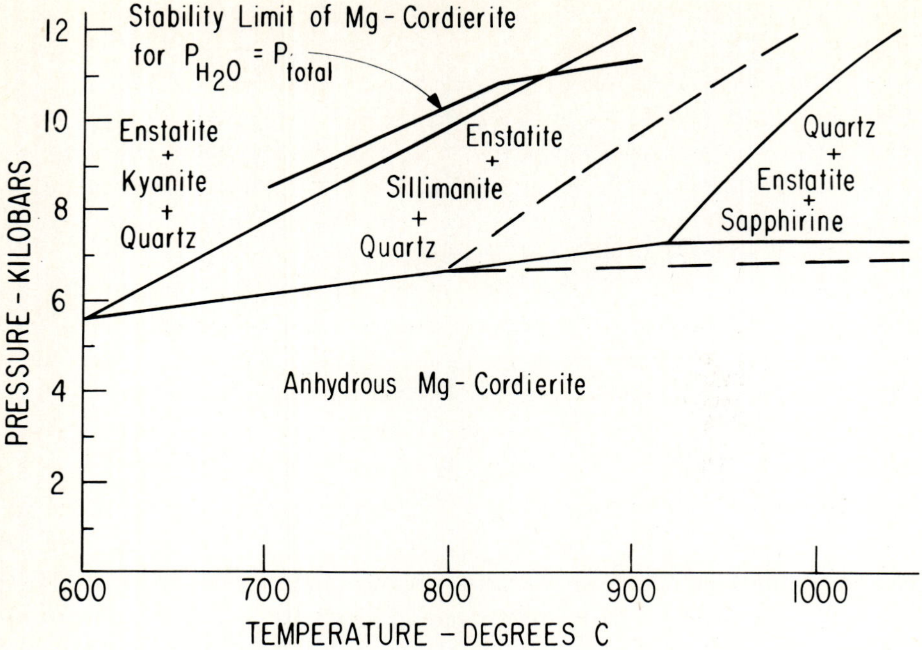

Fig. 2. Equilibrium P,T relations of $Mg_2Al_4Si_5O_{18}$ (Mg-cordierite) bulk composition. Upper pressure stability limits of cordierite, for the anhydrous case (Newton et al., 1974) and case of $P_{H_2O} = P_{total}$ (Newton, 1972) are shown. Experimentally determined field of synthetic sapphirine shown within solid boundaries and calorimetrically estimated field of stability for a natural, well-ordered sapphirine shown within dashed lines. Kyanite-sillimanite boundary from Holdaway (1971).

and quartz at 727°C may be easily calculated by the reader from this thermodynamic data of Table I. Neglect of a small amount of Al_2O_3 in the enstatite at this temperature produces no major uncertainty. The effect of ferrous iron on the equilibrium may be assayed in a manner entirely analogous to the discussion centering around reaction (A) and equation (1). Marakushev and Kudryavtsev (1965) give analyses of coexisting cordierite ($Mg/Mg+Fe^{2+}$ = .83) and hypersthene ($Mg/Mg+Fe^{2+}$ = .67). Using this distribution, and regarding the hypersthene solid solution to be nearly ideal at elevated temperatures (Saxena, 1973, p. 98), shows that the lowering of the pressure requirements of magnesian cordierite breakdown by the presence of ferrous iron at this level to be only about 700 bars.

If kyanite, instead of sillimanite, is among the assemblage alternative to cordierite, as is found in places at Wilson Bay,

Labrador (Morse and Talley, 1971) and inferred for the Rhodesian specimen (Chinner and Sweatman, 1968), minimum pressures of 7.6 kilobars at 700°C and 9.6 at 800°C are required (Fig. 2).

A particularly interesting rock has been described from the Anabar Massif, Siberia, by Lutts and Kopaneva (1968). This is another example of the high-grade, high pressure magnesian-aluminous granulites of dubious origin. It contains pyrope-almandine ($Mg/Mg+Fe^{2+}$ = .61), cordierite ($Mg/Mg+Fe^{2+}$ = .855), hypersthene ($Mg/Mg+Fe^{2+}$ = .821), sapphirine ($Mg/Mg+Fe^{2+}$ = .897), biotite and sillimanite. The hypersthene is very anomalous, containing 10.81 % Al_2O_3 and 7.8 % Fe_2O_3. The cordierite crystals are smaller than the others, and the authors infer a low-pressure back-reaction origin. They estimate conditions of metamorphism of 950°C and 12 kilobars, based mainly on the experimental work at high P_{H_2O} available in 1968. Newton (1972) calculated from thermodynamic data that the assemblage enstatite-sillimanite requires pressures nearly as high for stability as the assemblage enstatite-sillimanite-quartz. At 800°C, this would seem to require a minimum of 6 kilobars. This conclusion is rendered less forceful by the unknown effects of the high Al_2O_3 and Fe_2O_3 contents of the orthopyroxene. Certainly pressures as high as 12 kilobars do not seem necessary in light of more recent experimental and thermodynamic data.

In summary, the evidence from the peculiar ancient magnesian gneisses indicates, again, that minimum pressures of 6 - 9 or more kilobars seem to be necessary. The origin of these odd bulk compositions is quite obscure. Chinner and Sweatman (1968) suggest altered volcanic ash deposits. Lateritic or other weathering and subsequent metasomatic transfer may be an explanation (Meng and Moore, 1972). This combination of surface and deep-seated processes does not seem to have been operative in post-Precambrian times.

3. <u>Garnet in hypersthene-normative basic and intermediate granulites</u>.

Garnetiferous basic charnockites from ancient terranes are widespread; for instance, Howie and Subramaniam (1957) reported a garnetiferous two-pyroxene granulite from Pallavaram (Ch 199) with 12.9 % normative olivine. The composition of the garnet is almandine .645, andradite .057, grossular .146, pyrope .127 and spessartine .025, in mole fractions. This is typical of the complex (polycomponent) garnets characteristic of granulites. Jawardena and Carswell (1976) report similar garnet-bearing basic charnockites from the Highland Series of southeast Sri Lanka. Wood (1975) described and analyzed basic garnet granulites from the Archaean terrane of South Harris, Outer Hebrides.

Ringwood and Green (1966) presented experimental evidence bearing on the occurrence of garnets in basic granulites. Their work was done on various simulated basic compositions. They show (p. 83) minimum pressures of 8 - 9 kilobars required to produce garnet in an alkali-poor olivine-normative tholeiite composition in the temperature range 700°-900°C. This estimate is subject to the uncertainties of relatively poor pressure calibration in the piston-cylinder solid-pressure device, difficulty of proving equilibrium reactions in polycomponent silicate systems, and the long extrapolation necessary from the high temperature experimental range (above 1100°C).

In an alternative approach, Wood (1975) calculated thermodynamically the minimum pressures necessary to produce garnet in the basic granulites of the South Harris terrane. The basis of his calculation is the reaction:

$$\text{(D)} \quad \underset{\text{in olivine}}{Fe_2SiO_4} + \underset{\text{in plagioclase}}{CaAl_2Si_2O_8} \rightleftarrows 2/3\ \underset{\text{in garnet}}{Fe_3Al_2Si_3O_{12}} + 1/3\ \underset{\text{in garnet}}{Ca_3Al_2Si_3O_{12}}$$

This reaction is of high variance in natural basic systems. In a rock containing garnet, plagioclase, and olivine, the pressure of the recrystallization can be calculated by an equation similar to equation (1) if the temperature is independently estimated. The requisite thermodynamic parameters are poorly known for the iron-bearing components of the minerals, and Wood estimated these from available experimental work, especially that of Green and Hibberson (1970). He calculated minimum pressures of 9 - 12 kilobars to stabilize the garnets in the Harris metabasites in the temperature range 700°-900°C. These pressures are minimum estimates because olivine is not actually part of the paragenesis. As additional evidence of the surprisingly high pressure, Wood (1975) points to the presence of kyanite in closely associated metapelites which, in the same temperature range, would call for minimum pressures of 8 - 12 kilobars (Holdaway, 1971).

Garnet in charnockites of intermediate composition is less widespread. Garnet is lacking in metatonalites of the South Harris and was not reported by Howie and Subramaniam (1957) from Pallavaram. They described garnetiferous intermediate charnockites from Uganda, Sweden and Ceylon.

Little experimental work is available to establish the pressures of garnet crystallization in intermediate compositions in the granulite temperature range. Wood (1975) ascribes the lack of garnet in the South Harris metatonalites to greater SiO_2-saturation than in the metabasites, and this is borne out by the experiments of Ringwood and Green (1966) on various basaltic compositions, in which they found a higher pressure necessary to produce garnet in quartz tholeiites than in olivine tholeiites. This general relationship was also found experimentally by Kushiro and Yoder (1966) in the model system $CaO\text{-}MgO\text{-}Al_2O_3\text{-}SiO_2$. An extrapolation from their high temperature data to 800°C on the univariant reaction of anorthite + enstatite to garnet + clinopyroxene + quartz gives about 13 kilobars to stabilize garnet in a simple intermediate system. However, the simple four-component system is too restricted chemically to draw quantitative conclusions.

An ingenious calculation was made by Wood (1975) to provide an upper pressure limit for the South Harris metatonalites. He made use of the equation:

$$\text{(E)} \quad \underset{\text{in orthopyroxene}}{Mg_2Si_2O_6} + \underset{\text{in plagioclase}}{CaAl_2Si_2O_8} \rightleftarrows 2/3\ \underset{\text{in garnet}}{Mg_3Al_2Si_3O_{12}} + 1/3\ \underset{\text{in garnet}}{Ca_3Al_2Si_3O_{12}} + \underset{\text{quartz}}{SiO_2}$$

This reaction will be of high variance in a complex natural system. The governing equation is similar to equation (1):

$$-\frac{\Delta H^o}{RT} + \frac{\Delta S^o}{R} \cong \ln \frac{(\alpha^{Gt}_{MgAl_{2/3}SiO_4})^2(\alpha^{Gt}_{CaAl_{2/3}SiO_4})}{(\alpha^{opx}_{Mg_2Si_2O_6})(\alpha^{plag}_{CaAl_2Si_2O_8})} + \frac{P\Delta V^o}{RT} \qquad (3)$$

where the symbols have the same meanings as in equation (1). The garnet formulae are reduced for convenience. Key thermodynamic data were not available to Wood (1975), so that he was forced to evaluate the parameters of the equation from experimental phase equilibria, including that of Kushiro and Yoder (1966). Since garnet is lacking in the metatonalites, Wood calculated the composition of the garnets which <u>could</u> coexist with the other minerals at various temperatures by making use of empirical data on the distribution of Fe^{2+}, Mg and Ca among pyroxenes and garnets in other rocks, including the South Harris metagabbros. He calculated garnet compositions of about grossular .22, pyrope .27, almandine .51, for the granulite temperature range. He considered the garnets non-ideal, but considered only a Ca-Mg interaction parameter, $W_{Ca\text{-}Mg}$, somewhat temperature dependent, with an average value of about 3 kilocalories. This is in keeping with a calculation of this parameter derived from phase equilibrium measurements subsequently published by Hensen et al. (1975). Wood (1975) found an upper pressure limit for rocks of the South Harris terrane of 11-14 kilobars in the range 700°-900°C. The estimates are maximal because of the absence of garnet in the metatonalites.

Improved thermodynamic data now available encourage a recalculation of the pressures by the Wood method. Table 1 lists the thermodynamic data for the substances, from which we derive $\Delta H^o = 1370 \pm 500$ cal and $\Delta S^o = -7.03 \pm .46$ cal/K at 1000 K. An improvement in the activity coefficients of the garnet components is now possible, owing to the work of Ganguly and Kennedy (1974) and Ganguly (1976). They derived regular solution interaction parameters W for the major garnet components based on a combination of thermodynamic deduction from experiments and natural parageneses involving garnets where temperature could be estimated independently. Following them, an expression for the activity coefficient, γ , of a component in a regular-solution ternary mixture may be written:

$$RT\ln\gamma_1 = W_{1-2} X_2^2 + W_{1-3} X_3^2 + (W_{1-2} + W_{1-3} - W_{2-3}) X_2 X_3 \quad (4)$$

where W_{1-2} is the interaction parameter for components 1 and 2, etc. In granulite garnets, the grossular and andradite components may be lumped together because they share a common cation, and because the mixing of a few percent of Fe^{3+} with Al on the octahedral sites contributes a negligible amount to the garnet free energy. An interaction term for spessartine may be ignored because the specific interaction energy of Mn^{2+} with the other divalent cations, except Mg, is small (Ganguly and Kennedy, 1974) and because Mn is usually very restricted in amount. For the effective Ca, Mg and Fe components, Ganguly and Kennedy give: $W_{Ca-Mg} = 3.83$ kcal, $W_{Ca-Fe} = 0.2$ kcal and $W_{Mg-Fe} = 2.95$ kcal at about 630°C. Hensen, Schmid and Wood (1975) found experimental evidence for a temperature dependence of W_{Ca-Mg}. At 727°C W_{Ca-Mg} would be 3.16 kcal. Oka and Matsumoto (1974) deduced a W_{Mg-Fe} of 2.58 kcal from petrologic evidence, and Wood (pers. comm., 1977) found a value of 2.3 at higher temperatures in experimental Mg, Fe partitioning of garnet with olivine. Wood (1977) found a W_{Ca-Fe} of nearly zero by reduction of phase equilibrium experiments. The values $W_{Ca-Mg} = 3.2$, $W_{Ca-Fe} = 0$ and $W_{Mg-Fe} = 2.5$ will be used here for calculations at 1000 K. Thus, the activity coefficients of the $CaAl_{2/3}SiO_4$ (grossular) and $MgAl_{2/3}SiO_4$ (pyrope) components at 1000 K become:

$$\ln\gamma_{gross} = 1.621\ X_{pyr}^2 + 0.352\ X_{alm}X_{pyr} \tag{5}$$

$$\ln\gamma_{pyr} = 1.621\ X_{gross}^2 + 1.259\ X_{alm}^2 + 2.869\ X_{gross}X_{alm}$$

where $X_{gross} = \dfrac{Ca}{Ca+Mg+Fe^{2+}+Mn}$, etc.

At temperatures above 700°C, Fe-Mg orthopyroxenes are substantially ideal (Saxena, 1973), especially with regard to the $Mg_2Si_2O_6$ component. The presence of other cations (Ca, Al, Ti, etc.) can be allowed for by the ideal two-site activity model (Wood and Banno, 1973), in which the activity of $Mg_2Si_2O_6$ in orthopyroxene is given by $X_{Mg}^{M_2}\ X_{Mg}^{M_1}$, where M_2 is the larger octahedrally coordinated crystallographic site and M_1 is the smaller. The site occupancies of the minor cations can be gotten from the structural formulae, along with rules of occupancy based on relative sizes. Fe^{2+} is considered equipartitioned between M_1 and M_2 by Wood and Banno (1973).

The activity of $CaAl_2Si_2O_8$ in synthetic plagioclase was calculated at 700°C as a function of composition by Orville (1972) from his cation-exchange data in chloride solutions. He found a constant activity coefficient of 1.28 for $CaAl_2Si_2O_8$ in the composition range 0-50 mole percent, and this value will be used here.

As a first calculation the upper pressure limit of Wood's (1975) metatonalite specimen 01 may be calculated. His microanalytic data for the orthopyroxene give $X_{Mg}^{M_1} = .5275$ and $X_{Mg}^{M_2} = .5775$. The activity of $Mg_2Si_2O_6$ is thus .3046. The plagioclase is $An_{44.8}$, with an activity of .573. The activity coefficients of $CaAl_{2/3}SiO_4$ and $MgAl_{2/3}SiO_4$ are, respectively, 1.18 and 2.07, in the model fictive garnet given earlier. These data lead, via equation (3), to a pressure of 12.70 ± 1.77 kilobars at 727°C where the uncertainty is due to uncertainties in ΔH^o and ΔS^o. This compares well with the maximum pressure of about 12 kilobars given by Wood (1975) for the same temperature.

This method may be applied to two intermediate charnockites described and analyzed by Howie and Subramaniam (1957). The specimen S347 from Uganda has plagioclase An_{32}, a garnet of composition almandine .671, andradite .062, grossular .128, pyrope .107 and spessartine .032, an orthopyroxene of $Mg/Mg+Fe^{2+} = .449$, quartz, hornblende, biotite and iron ore. The cation site occupancies of the orthopyroxene are $X_{Mg}^{M_2} = .450$, $X_{Mg}^{M_1} = .411$. The activity coefficients are $\gamma_{gross} = 1.04$

and γ_{pyr} = 2.69. The pressure indicated by the assemblage at 727°C is 9.80 ± 1.8 kilobars. The other intermediate charnockite, specimen V.2 from Varberg, Sweden, has a garnet of composition almandine .713, andradite .020, grossular .173, pyrope .042, and spessartine .052. The $Mg_2Si_2O_6$ activity of the iron-rich orthopyroxenes is .0644. The average plagioclase is $An_{26.5}$. The other minerals are microcline, quartz, hornblende, clinopyroxene, ilmenite and magnetite. The calculated pressure at 727°C is 8.33 ± 1.8 kilobars.

As a result of the foregoing, it can be concluded that minimum pressures of 7 - 10 kilobars were necessary to produce the mineralogy of some garnetiferous charnockites, if the temperatures were around 800°C. Wood's upper pressure limits for the South Harris granulite terrane seem to be valid, but sufficient data to evaluate his calculated lower pressure limits of 9 - 12 kilobars are not yet available. It is possible that these limits are overestimated, in view of the lack of garnet in intermediate compositions in this terrane, whereas a garnet-bearing assemblage has been found in intermediate charnockites from other terranes, with thermodynamically calculated pressures of crystallization of 8.3 - 9.8 kilobars for two examples.

IMPLICATIONS FOR ARCHAEAN METAMORPHISM

Good evidence exists, from three lines of approach, that pressures of 6-10 kilobars were necessary to produce the mineralogy of some Archaean granulites. Pressures of 10-13 kilobars have been estimated, but are less certain. Corresponding depths of burial are 20-33 km for the more moderate estimate. These are depths approaching the base of the present-day normal continental crust. The implication of thick continents at least as far back as 2400 m.y. ago seems unavoidable.

That North America grew in size throughout the Precambrian seems required by the distribution of radiometric dates (Goldich et al., 1966) and the same implication exists for the other continents (Shackleton, 1967). Intuitively, one might tend to correlate the thickness of a continent with area. The foregoing petrologic evidence, however, indicates that some continental masses in the Archaean, if they were relatively small in area, were nevertheless comparable in thickness to the present continents. The widespread occurrence of granulite-facies metamorphic rocks strongly suggests that Archaean geothermal gradients were greater than those of more recent times. Ernst (1972) invokes a secular decrease in average thermal gradient to explain the apparent restriction of high pressure, low temperature blueschist metamorphism to relatively late geologic history.

Modern concepts of lithospheric motions have been employed to explain locally thick crustal sequences in ancient terranes. Bridgwater et al. (1974) suggest that horizontal transport resulted in tectonic thickening and reduplication of protocontinental material in the Archaean terrance of West Greenland. Windley (1976) envisaged microcontinental collisions due to plate motions comparable to modern-day global tectonics, and Burke et al. (1977) found no evidence to invalidate a uniformitarian view of island arc collision and microcontinental and continental collision operating back into Archaean times. If the continental areas were not large enough to sustain deep marginal geosynclines by 2400 m.y. ago, some such horizontal tectonic processes would seem to be needed.

ACKNOWLEDGMENTS

The calorimetric data cited here was mainly produced by Dr. T. V. Charlu under support of a National Science Foundation grant to the author, No. EAR 76-03997. Lelde Kalmite typed the manuscript.

REFERENCES

Anhaeusser, C.R., Mason, R., Viljoen, M.J. and Viljoen, R.R., 1969. A reappraisal of some aspects of Precambrian shield geology. Bull. geol. Soc. Amer., 80: 2175-2200.

Bridgwater, D., McGregor, V.R. and Myers, J.S., 1974. A horizontal tectonic regime in the Archaean of Greenland and its implications for early crustal thickening. Precamb. Res., 1: 179-197.

Burke, K., Dewey, J.F. and Kidd, W.S.F., 1977. World distribution of sutures - the sites of former oceans. Tectonophysics, 40: 69-100.

Charlu, T.V., Newton, R.C. and Kleppa, O.J., 1975. Enthalpies of solution at 970 K of compounds in the system $MgO-Al_2O_3-SiO_2$ by high temperature solution calorimetry. Geochim. Cosmochim. Acta, 39: 1487-1497.

Chinner, G.A. and Sweatman, T.R., 1968. A former association of enstatite and kyanite. Min. Mag., 36: 1052-1060.

Currie, K.L., 1971. The reaction 3 cordierite = 2 garnet + 4 sillimanite + 5 quartz as a geologic thermometer in the Opinicon Lake region, Ontario. Contr. Mineral. Petrol., 33: 215-226.

Dallwitz, W.B., 1968. Coexisting sapphirine and quartz in granulite from Enderby Land, Antarctica. Nature, 219: 476-477.

De Roever, W.P., 1956. Some differences between post-Paleozoic and older regional metamorphism. Geol. en Mijnb., new ser., 18: 123-127.

Ernst, W.G., 1972. Occurrence and mineralogic evolution of blueschist belts with time. Amer. J. Sci., 272: 657-668.

Eskola, P., 1952. On the granulites of Lapland. Amer. J. Sci., Bowen Vol.: 133-172.

Ganguly, J., 1976. The energetics of natural garnet solid solutions. II. Mixing of the calcium silicate end-members. Contr. Min. Pet., 55: 81-90.

Ganguly, J. and Kennedy, G.C., 1974. The energetics of natural garnet solid solution: I. Mixing of the aluminosilicate end-members. Contr. Min. Pet., 48: 137-148.

Gastil, G., 1960. The distribution of mineral dates in time and space. Amer. J. Sci., 258: 1-35, 1960.

Goldich, S., Muehlberger, W.R., Lidiak, E.G. and Hedge, C.E., 1966. Geochronology of the midcontinent region, United States. 1. Scope, methods and principles. J. geophys. Res., 71: 5375-5388.

Green, D.H. and Hibberson, W., 1970. The instability of plagioclase in peridotite at high pressure. Lithos, 3: 209-221.

Hensen, B.J., 1971. Theoretical phase relations involving cordierite and garnet in the system $MgO-FeO-Al_2O_3-SiO_2$. Contr. Min. Pet., 33: 191-214.

Hensen, B.J. and Green, D.H., 1971. Experimental study of the stability of cordierite and garnet in pelitic compositions at high pressures and temperatures. I. Compositions with excess aluminosilicate. Contr. Min. Pet., 33: 309-330.

Hensen, B.J., Schmid, R. and Wood, B.J., 1975. Activity-composition relations for pyrope-grossular garnet. Contr. Min. Pet., 51: 161-166

Hess, P.C., 1969. The metamorphic paragenesis of cordierite in pelitic rocks. Contr. Min. Pet., 24: 191-207.

Hewins, R.H., 1975. Pyroxene geothermometry of some granulite facies rocks. Contr. Min. Pet., 50: 205-209.

Holdaway, M.J., 1971. Stability of andalusite and the aluminum silicat phase diagram. Amer. J. Sci., 271: 97-131.

Howie, R.A. and Subramaniam, A.P., 1957. The paragenesis of garnet in charnockite, enderbite, and related granulites. Min. Mag., 31: 565-586.

Hutcheon, I., Froese, E. and Gordon, T.M., 1974. The assemblage quartz-sillimanite-garnet-cordierite as an indicator of metamorphic conditions in the Daly Bay Complex, N.W.T. Contr. Min. Pet., 44: 29-34.

Jawardena, D.E. and Carswell, D.A., 1976. The geochemistry of charnockites and their constituent ferromagnesian minerals from the Precambrian of south-east Sri Lanka (Ceylon). Min. Mag., 40: 541-554

Kushiro, I. and Yoder, H.S., 1966. Anorthite-forsterite and anorthite-enstatite reactions and their bearing on the basalt-eclogite transformation. J. Petrol., 7: 337-362.

Lutts, B.G. and Kopaneva, L.N., 1968. A pyrope-sapphirine rock from the Anabar massif and its condition of metamorphism. Dokl. Akad. Nauk SSSR, 179: 161-163.

Manna, S.S. and Sen, S.K., 1974. Origin of garnet in basic granulites around Saltora, W. Bengal, India. Contr. Min. Pet., 44: 195-218.

Marakushev, A.A. and Kudryavtsev, V.A., 1965. Hypersthene-sillimanite paragenesis and its petrologic implication. Dokl. Akad. Nauk SSSR, 164: 145-148.

Meng, L.-K. amd Moore, J.M., 1972. Sapphirine-bearing rocks from Wilson Lake, Labrador. Canad. Mineral., 11: 777-790.

Moore, P.B., 1969. The crystal structure of sapphirine. Amer. Mineral., 54: 31-49.

Morse, S.A. and Talley, J.H., 1971. Sapphirine reactions in deep-seated granulites near Wilson Lake, Central Labrador, Canada. Earth Planet. Sci. Lett., 10: 325-328.

Newton, R.C., 1972. An experimental determination of the high pressure stability limits of magnesian cordierite under wet and dry conditions. J. Geol., 80: 398-420.

Newton, R.C., Charlu, T.V. and Kleppa, O.J., 1974. A calorimetric investigation of the stability of anhydrous magnesium cordierite with application to granulite facies metamorphism. Contr. Min. Pet., 44: 295-311.

Newton, R.C., Thompson, A.B. and Krupka, K.M., 1977. Heat capacity of synthetic $Mg_3Al_2Si_3O_{12}$ from 350 to 1000 K and the entropy of pyrope. Eos, 58: 523.

Nixon, P.H., Reedman, A.J. and Burns, L.K., 1973. Sapphirine-bearing granulites from Labwor, Uganda. Min. Mag., 39: 420-428.

Oka, Y. and Matsumoto, T., 1974. Study on the compositional dependence of the apparent partition coefficient of iron and magnesium between coexisting garnet and clinopyroxene solid solution. Contr. Min. Pet., 48: 115-121.

Orville, P.M., 1972. Plagioclase cation exchange equilibria with aqueous chloride solution: Results at 700°C and 2000 bars in the presence of quartz. Amer. Jour. Sci., 272: 234-272.

Rabkin, M.I., 1968. Depth conditions of regional metamorphism of the Anabar Shield. Int. Geol. Congr., 23rd Sess. (Prague), Sect. 4: 215-223.

Ringwood, A.E. and Green, D.H., 1966. An experimental investigation of the gabbro-eclogite transformation and some gecphysical implications. In : Petrology of the Upper Mantle. Aust. Natn'l. Univ. Dept. of Geophys. Geochem. Publ. No. 444, 252 pp.

Robie, R.A. and Waldbaum, D.R., 1968. Thermodynamic properties of minerals and related substances at 298.15 (25.0°C) and one atmosphere (1.013 bars) pressure and at higher temperatures. Bull. U.S. Geol. Surv., 1259: 256 pp.

Saxena, S.K., 1973. Thermodynamics of rock-forming crystalline solutions. Spring-Verlag, 188 pp.

Schreyer, W. and Yoder, H.S., 1960. Instability of anhydrous Mg-cordierite at high pressures. Carn. Inst. Wash. Yrbk., 59: 90-91.

Schreyer, W. and Yoder, H.S., 1964. The system Mg-cordierite-H_2O and related rocks. N. Jahrb. Min. Abh., 101: 271-342.

Shackleton, R.M., 1967. On the origin of some African granites. Proc. Geol. Assoc. Lond., 81: 549-564.

Skinner, B.J., 1966. Thermal expansion. In: S.P. Clark (Editor), Handbook of Physical Constants. Mem. Geol. Soc. Amer., 97: 75-96.

Smith, J.V. and Schreyer, W., 1960. Location of water and argon in cordierite. Min. Mag., 33: 226-236.

Touret, J., 1971. Le facies granulite en Norwege Meridionale. I. Les associations mineralogiques. Lithos, 4: 239-249.

Walker, T.L. and Collins, W.H., 1907. Petrological study of some rocks from the Hill Tracts, Vizagapatam District, Madras Presidency. Geol. Surv. India Rec., 36: 1-18.

Westrum, E.F., Essene, E.J. and Perkins, E., 1977. Thermophysical properties of the garnet, grossularite ($Ca_3Al_2Si_3O_{12}$). J. Chem.

Thermo. (in press).

Winkler, H.G.F., 1976. Petrogenesis of metamorphic rocks. Springer-Verlag, 334 p.

Wood, B.J., 1973. Fe^{2+}-Mg^{2+} partition between coexisting cordierite and garnet - a discussion of the experimental data. Cont. Mineral. Petrol., 40: 253-258.

Wood, B.J., 1975. The influence of pressure, temperature and bulk composition on the appearance of garnet in orthogneisses - an example from South Harris, Scotland. Earth Planet. Sci. Lett., 26: 299-311.

Wood, B.J., 1977. Experimental determination of the mixing properties of solid solutions with particular reference to garnet and clinopyroxene solutions. In: D.G. Fraser (Editor), Thermodynamics in Geology. Oxford Univ. Press, 11-27.

Wood, B.J. and Banno, S., 1973. Garnet-orthopyroxene and orthopyroxene-clinopyroxene relationships in simple and complex systems. Contr. Mineral. Petrol., 42: 109-124.

COMPARISON OF SOME OF THE GEOCHEMICAL FEATURES AND TECTONIC SETTING OF ARCHAEAN AND PROTEROZOIC GRANULITES, WITH PARTICULAR REFERENCE TO AUSTRALIA

Allan F. Wilson

Department of Geology and Mineralogy, University of Queensland (Australia)

Abstract

Irregular patches of Archaean hypersthene-bearing granulites, widely ranging in size, are common throughout the Wheat Belt of southwestern Australia (60000 km^2) which is now composed dominantly of biotite or hornblende gneisses intruded by large coarse 2600 MA grey granitic domes. Some of the granulites were Archaean sediments, volcanic rocks (both mafic and silicic) and highly magnesian ultramafic rocks, whereas some large areas appear to be older Archaean basement now converted to dark greasy rocks like the silicic charnockites of India. Although some mafic granulites are strongly lineated and some silicic granulites show well-developed platy quartz grains, most granulites exhibit ill-defined linear elements, and present the appearance of having suffered thermal metamorphism rather than strong deformation. Cordierite, sillimanite, spinel and olivine are common in rocks of appropriate composition, but kyanite and high pressure clinopyroxene-garnet assemblages are unknown.

Linear belts of granulites, which yield Proterozoic ages by Rb-Sr methods, engirdle on three sides the Archaean block of the Wheat Belt and adjoining greenstones to the east. The Fraser Block (8000 km^2), separated from the Archaean by an extensive mylonite zone and intense Bouguer gravity anomaly, is dominated by anhydrous mafic granulites of higher pressure than the Archaean granulites of the Wheat Belt. The mafic granulites were basic lavas and small intrusives of continental tholeiite type set within an intracratonic linear zone, and metamorphosed at 1338 MA (Rb-Sr).

In the Musgrave Ranges silicic and mafic granulites were formed at about 1380 MA, and are of both anhydrous and hydrous types, and range from low to intermediate granulites. These granulites are intruded by large bodies of "magmatic" charnockitic adamellite at about 1130 MA.

Older silicic and mafic granulites (1860 MA) appear in the Strangways Range within the Arunta Block, and in places have recorded a second granulite event at c. 1470 MA. Archaean basement has not yet been recognised within the Arunta Block, and the granulites form imbricate thrust slices of deep-seated igneous and sedimentary rocks.

Major and trace element and REE studies reveal a close correspondence of the mafic granulites of the Fraser Range to continental tholeiites, whereas those of both the Musgrave and Strangways Ranges mostly show island arc or oceanic tholeiitic characters. Element mobility during granulite facies metamorphism was strongly controlled by thermal and fluid gradients. Large increase in total pressure but at approximately constant temperature produced no significant element mobility in dry mafic granulites along 100 km traverse in the Fraser Range, and U and Th values are not correlated with metamorphic grade.

In the Musgrave Ranges, where granulite hornblende and biotite are common, Th, U, Rb, Ni and Co were depleted with increase of both total pressure and temperature, whereas Ba, Sr, REE, Na, Ca and P were enriched. Elements showing less than 20% change are Si, Al, Mn, Ca, K, Zn, Cr, Y, Nb, Fe^3, Fe^2, Mg, Ti, Zr, V and Pb.

In the aureole of a large charnockitic intrusive pluton where F was active there is a marked increase in U, Th and Rb, but a depletion of Zr and Ti, two elements normally considered immobile elements.

Silicic granulites in all three terrains show depletion, with increase of grade, of U, Th and Rb and increase with K/Rb, as noted in several Archaean shields.

In view of the varying influence of partial pressures of the fluids, and temperature, complications of polymetamorphism, and differing tectonic setting of many terrains, significant chemical differences cannot yet be recognised between Archaean and Proterozoic granulites.

INTRODUCTION

Most of geochemical research on Australian granulites has been done on those yielding Proterozoic Rb-Sr ages. However, it has been found that techniques successfully applied and principles established in studying the Proterozoic granulite terrains are helping to unravel very complex Archaean terrains.

In this paper the tectonic setting of some of the Australian granulites is outlined prior to a discussion of their geochemical features. Some general observations likely to help in further studies are then made.

TECTONIC SETTING OF AUSTRALIAN ARCHAEAN GRANULITES

The largest outcropping expanse of Archaean rocks in Australia is situated in the southwestern portion of the continent (Fig. 1). Most of this region is composed of rocks common to all of the great shields: siliceous gneisses, intrusive and metasomatic granites, and many linear belts of Archaean clastic and chemical sediments, mafic and ultramafic volcanic rocks and associated shallow intrusives and tuffs.

In Australia, as in India, Canada and other shields, much gold (as at Kalgoorlie) is associated with these mafic sequences. The gold mineralisation is restricted to amphibolite and greenschist metamorphic terrains.

In this paper, however, we are concerned with the granulite facies terrains of the shield (Fig. 1). The Archaean granulites are restricted to the central and southern portions of the agricultural section of the shield known as "The

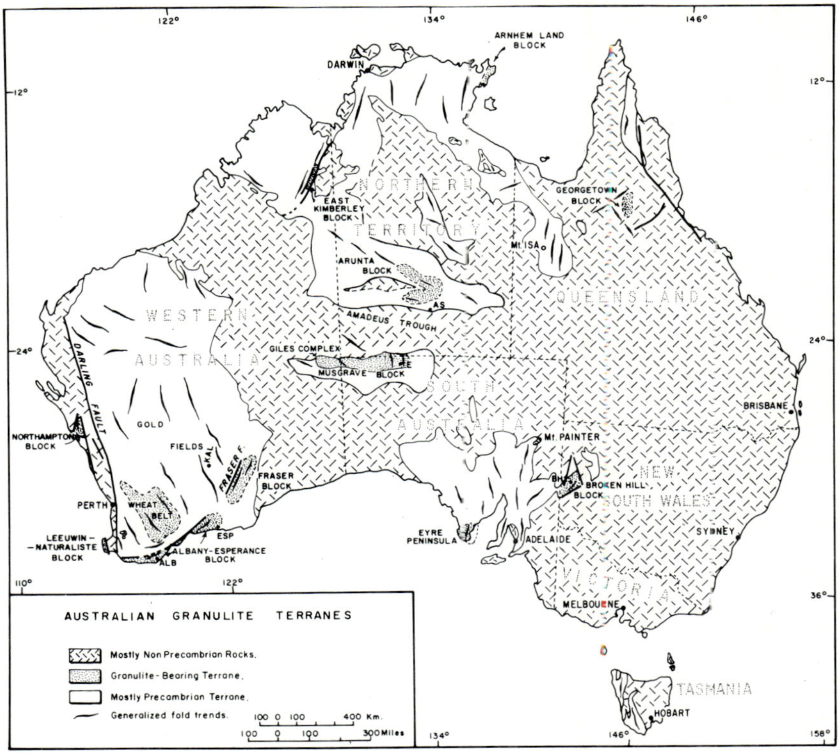

Fig. 1. Granulite terrains of Australia. The Wheat Belt granulites in Western Australia are Archaean, whereas all other granulite belts yield Proterozoic ages by Rb-Sr. KAL, Kalgoorlie; ALB, Albany; ESP, Esperance; E, Ernabella; AS, Alice Springs; BH, Broken Hill.

Wheat Belt". This 60000 km^2 region is composed dominantly of biotite or hornblende gneisses which have been intruded at 2600-2700 MA by large plutons of coarse grey granites. Throughout this region there are irregular patches of hypersthene-bearing granulites ranging from a few square metres to several square kilometres in area. However, even within these patches of granulite facies rocks there are normally linear or anastomosing zones of amphibolite or greenschist retrogression.

Within the Wheat Belt granulite terrain there are numerous relics (now granulite facies) of deformed troughs of post-basement volcanic and sedimentary rocks, or post-basement igneous intrusions (e.g. Wilson, 1969a,

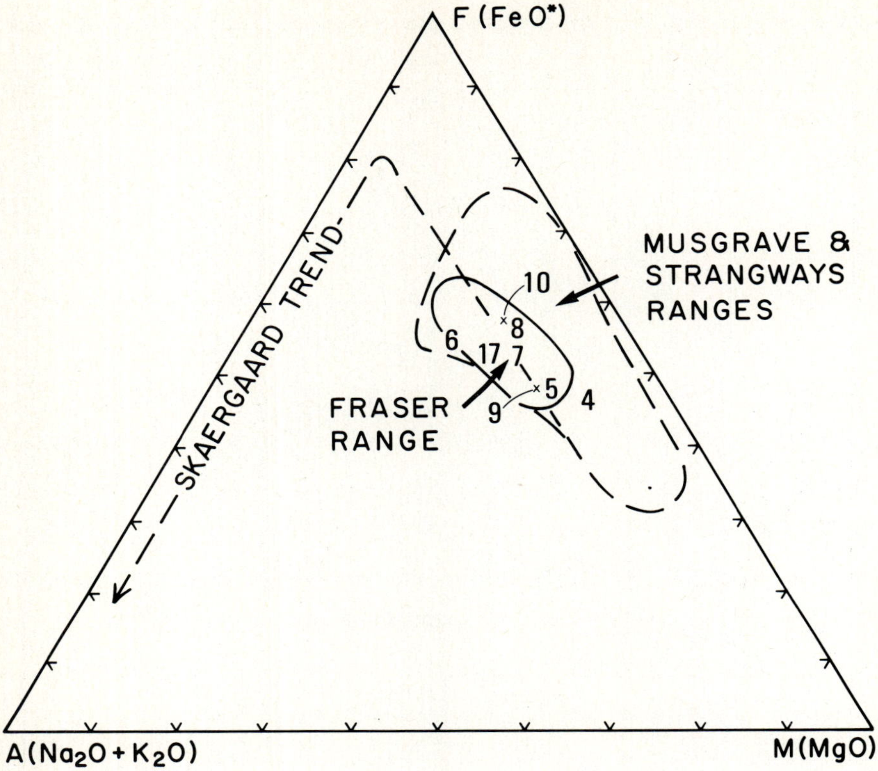

Fig. 2. AFM diagram showing range in composition of mafic granulites. The smaller outlined field represents the Fraser Range mafic granulites; the larger field represents the wider variety of mafic granulites from the Musgrave and Strangways Ranges. Skaergaard trend is after Wager & Deer, 1939. The numbers refer to basic rocks of Table 1.

p. 243). In several places (e.g. Dingo Rock, ibid p. 246) the basement gneisses themselves have suffered a superimposed granulite facies metamorphism, and the greasy dark bluish-grey or greenish-grey colour is indistinguishable from the Indian charnockites.

In earlier papers (Wilson, 1969a, 1971) I have maintained that the post-basement volcanic and sedimentary rocks (including the jaspilites) were of an age comparable with that of less metamorphosed volcanic and sedimentary rocks of the greenstone belts of the Goldfields, especially as the granitic bodies intruding both regions fall within the tight age bracket of 2600-2700 MA.

Arriens (1971) has shown, however, that at least some of the gneisses and granulites of the Wheat Belt are considerably older (c. 3300 MA). Although he does not give a detailed tectonic setting for his samples, sufficient is known to say that at least some of the metamorphosed basic rocks, banded iron formations and associated felsic rocks suffered a regional metamorphism of c.3300 MA, an age that has not yet been recognised in the Goldfields region. It is for this and other reasons that some geologists (e.g. Gee, 1975) favour the view that the banded iron formations and related layered rocks of the granulite terrains of the Wheat Belt probably belong to an older system than those of the Goldfields region. Moreover, some envisage that the most ancient nucleus comprises only the extreme south-western triangle of the continent, and that the layered rocks of the Wheat Belt granulite terrain were deposited to the east of that nucleus on oceanic crust.

Although these views have some attraction, they do not explain some important facts. The most obvious is the presence of relics of jaspilites in many parts of that proposed older nucleus (e.g. at Wundowie, several places in the Manjinup region, etc.). Moreover, the close association of highly metamorphosed relics of Mg-rich mafic and ultramafic rocks with many of the relics of banded iron formations in the nucleus as well as in the Wheat Belt granulite terrain, points to a close similarity between the oldest of the Goldfields sequences and the layered remnants within the whole region south-west of the Goldfields.

The fact that the granulites are restricted to a fairly narrow linear belt separating the extreme south-west gneissic block and the greenstone-rich Goldfields region supports the concept that the granulites were developed within a "mobile belt" between two cratons, somewhat as applies to the Limpopo Belt of Africa and the Eastern Ghats of India. Although the Wheat Belt granulite terrain has not yet been well mapped, it is clear that it contains much larger volumes of 2700 MA silicic batholiths (largely adamellite) than appear in the African or Indian mobile belts. Thus, if the granulites were developed through high heat flow in this linear zone of low P_{H_2O}, the zone must have remained mobile as a region in which Archaean acid magmatism and K-metasomatism and (later) amphibolite facies retrogression were very active.

The Archaean granulite terrain has not yielded high pressure assemblages characterised by kyanite or clinopyroxene-garnet; rather, the granulites of the terrain display high temperature-low pressure assemblages. Thus, cordierite (with or without garnet) is common in the quartzose gneisses and granulites (e.g. near Dangin (Prider, 1945b) and in the Dumbleyung district). Sapphirine has been recorded at Dangin (Prider, 1945a), and south of Quairading among various assemblages rich in one or more of cordierite, corundum, hypersthene aand spinel (Wilson, 1962, 1971b). Forsterite and spinel have been recorded near Dangin (Prider, 1945b), and in the Dumbleyoung region (Wilson, in prep.), and fayalite is known from a metagabbro at Burges Find near Burracoppin, and from metajaspilites near Dumbleyoung and elsewhere.

TECTONIC SETTING OF AUSTRALIAN PROTEROZOIC GRANULITES

All Australian granulites, other than those of the Wheat Belt, yield Proterozoic ages by Rb-Sr techniques. These should be considered minimum ages, for the degree of mobility of either Rb or Sr during granulite facies metamorphism is still not known. Although initial ratios of Sr preclude Archaean ages from some of the Proterozoic belts (e.g. Fraser Range), other techniques may yet uncover an Archaean history in some of these belts.

1. Tectonic setting of the Fraser Range

An earlier publication (Wilson, 1969a) discusses the range of tectonic setting and petrological characteristics of the large number of linear belts of Proterozoic granulites shown in Fig. 1. A striking feature of the granulite distribution is that linear belts of Proterozoic granulites engirdle on three sides the Archaean block containing the Wheat Belt granulites. The best studied of these is the Fraser Block (8000 km^2) which is separated from the Archaean by a complex linear suture containing an extensive mylonite zone and an intense Bouguer gravity anomaly.

"The anomaly has a gradient of 12.5 mgal/km and consists of a positive gravity ridge over the western portion of the Fraser Block (+ 30 mgal) and a gravity trough (down to -100 mgal) immediately to the west of the fault zone. There appear to have been several periods of movement in the Fraser Fault Zone. Minor structures in the pyroxene granulites and the gravity profile suggest a strong reversed movement which has raised the eastern block. Other structures suggest a strong sinistral transcurrent component moving the east block northward" (Wilson 1979a).

The Fraser Block is essentially anhydrous and composed largely of mafic pyroxene granulites, many of which are derived from basalts (many seem to have been vesicular and a few were pillowed - Wilson, 1975) and related mafic intrusions. Felsic and mafic garnet-bearing granulites are common, particularly in the deformed zones in the west near the Fraser Fault Zone.

The age of the granulite metamorphism has been recorded by Rb-Sr as 1328 MA. with an initial $^{87}Sr/^{86}Sr$ ratio of 0.7049 ± 0.0012 (Arriens and Lambert, 1969). but somewhat younger ages record thermal activity down to about 1210 MA (Wilson, 1966).

The block appears to be composed largely of continental tholeiitic basalts and related intrusives, and some shallow water sediments deposited in an intracratonic linear zone. Metamorphism has been intense. However, cordierite is unknown in the sedimentary rocks, and the pyroxenes of the mafic granulites show higher pressure characteristics than those of the Archaean granulites of the Wheat Belt (Wilson, 1976).

Notwithstanding the high-grade of metamorphism, a remarkable feature is that in structurally favourable sites amygdales have been well preserved even in pyroxene granulites. However, in highly transposed granulites nearer the Fraser Fault Zone primary volcanic features have been obliterated (Wilson, 1969b, 1975).

Other petrological data on the Fraser Block appear in Wilson (1969a and c, 1971, 1975, 1976a) and Wilson and Middleton (1968). Geochemical features of the Fraser Block granulites are discussed in a later section.

Some notes on two other belts of Proterozoic granulites need to be given before the geochemical features of these rocks are described. These are in the Musgrave Ranges and in the Strangways Range of central Australia where extensive tracts of granulites are found.

2. Tectonic setting of the Musgrave Ranges

Fig. 1 shows a very large region of granulites within the Musgrave Block. Most of the geochemical data discussed below pertain to the Musgrave Ranges which occupy only the eastern half of that granulite region. Summaries dealing with the tectonic setting and aspects of the petrology and geochemistry of parts of the Musgrave Ranges have been published (Wilson,1960, 1969a; Hudson,1968; Collerson, 1975 and others).

About 75% of the granulites are orthopyroxene-bearing felsic granulites of adamellite, granodiorite or tonalite composition. In colour they are a characteristic mauve-grey, and not the greasy dark bluish or greenish grey of the intrusive ferrohypersthene "granites" of the same region, or of the pyroxene gneisses and "charnockites" of India and elsewhere. Cordierite or garnet-sillimanite felsic rocks of pelitic composition comprise about 10% of the granulites. Pyroxene and pyroxene-hornblende basic granulites, probably representing basalt, and basic feeder dykes and sills, make up about 10% of the granulites.

Layered basic and ultrabasic intrusions form the large Giles Complex (Nesbitt and Talbot, 1966), and are best developed in the western Musgrave Block. These are somewhat later than the granulite metamorphism (1390 MA; Arriens and Lambert, 1969), and are represented in the eastern portion of the block by the Woodroffe Norite.

One of the most distinctive features of the granulite terrain of the eastern Musgrave Block is the intrusion of several greasy dark bluish-grey ferrohypersthene "granites" (actually, mostly adamellites or granodiorites). "One well exposed mass, the Ernabella Adamellite, is about 40 km long and 18 km wide (Wilson,1960, Fig. 7). Contacts with the granulites are normally sharp and locally discordant. Moreover, the contact phenomena together with the attitude and nature of xenoliths show that the granite has not formed in place, but has been intruded to higher levels. The high temperature thermal metamorphism of the granulites in the narrow aureole confirms its mobility." (Wilson, 1969). These bodies were intruded 1120 ± 100 MA ago. This age and the initial $^{87}Sr/^{86}Sr$ of 0.710 ± 0.001, when compared with the 1380 ± 120 MA isochron for the intruded granulites with their initial $^{87}Sr/^{86}Sr$ of 0.707 ± 0.003 support the regenerative origin of these ferrohypersthene granites.

The unexpected movement of certain trace elements within the aureole of these granites is discussed below.

Unlike the Fraser Range region, Archaean basement has not been recognised in the Musgrave Ranges. Collerson et al. (1972) have made a structural analysis of part of the western Musgrave Ranges, and they can see no evidence of the complex tectonic history that would be expected of reworked basement. Indeed, they would interpret the structural sequence in terms of a single orogenic cycle, and suggest, notwithstanding the granulite metamorphism, that the granulites represent a series of "cover rocks" (i.e. Middle Proterozoic Carpentarian sediments) rather than reworked basement.

The Rb-Sr studies of Arriens and Lambert (1969) show that the granulite metamorphism took place 1380 MA ago, yet the granulites have a low initial $^{87}Sr/^{86}Sr$ of 0.707 ± 0.003. This is interpreted as precluding the possibility that Sr isotopic homogenisation was sustained for longer than about 300-500 MA by any process of regional diffusive exchange. The structural studies of Collerson et al. would appear to support this view. Alternatively, the accumulated radiogenic ^{87}Sr would need to have been almost completely lost from the granulite facies rocks over the same period. New data on the geochemistry of these granulites are examined below.

3. Tectonic setting of the Strangways Range

North of the Amadeus Trough in central Australia lies the Arunta Block within which large expanses of granulites are being recorded (Fig. 1). Structural mapping is not well advanced, but sufficient is known to show that the granulites represent imbricate thrust slices of deep-seated "cover rocks" of igneous and sedimentary types.

The metasedimentary rocks are much more varied in composition than in the Musgrave Ranges. In addition to many felsic rocks, a wide range of carbonate rocks, highly magnesian and aluminous pelites, and many types of basic and ultrabasic rocks have been converted to granulite facies rocks.

The Strangways Range has recorded more metamorphism than the Musgrave Ranges. The oldest yet established, however, is the first granulite facies metamorphism which took place at 1860 ± 80 MA (initial $^{87}Sr/^{86}Sr$ = 0.704 ± 0.003 for the felsic granulites - Iyer et al., 1976). Other granulite metamorphic episodes which are being recorded fall in the range 1700 - 1600 MA, and may represent phases of the earlier 1860 event. At 1470 MA another granulitic event is recorded (Iyer et al., ibid; Woodford et al., 1975). Several later non-granulite igneous and metamorphic events and retrogressions have been recorded (Black, 1975; Allen, 1976) but are not pertinent to this study. Some of the mineralogical and geochemical phenomena associated with these metamorphisms have been recorded (Iyer, 1974; Allen, 1976; Woodford and Wilson, 1976a, b, and in prep.; Wilson, 1977; Wilson and Baksi, in press).

GEOCHEMICAL COMPARISONS

One of the aims of this portion of the paper is to see if there are significant geochemical differences between granulites of Archaean and Proterozoic age. Another aim is to survey some of the factors affecting fractionation elements in rocks in the deep crust.

It should be evident from the earlier portion of this paper that an adequate geochemical survey of the granulites cannot be undertaken without knowing the tectonic setting and, preferably, the nature and age of the metamorphism(s) suffered by the rocks in question. Indeed, the understanding of the degree of mobility, under deep crustal conditions, of Rb, Sr, U, Th, K, Pb is basic to valid estimations of age of metamorphism and length of time rocks were in the crust prior to intense metamorphism.

As Archaean rocks commonly show several deformations and metamorphic reconstitutions, I shall first look at some Proterozoic granulites to try to see the effects of metamorphism which are more or less restricted to a single orogenic cycle. An example from the Musgrave Ranges, where the effects of a second granulite metamorphism may be seen, are then considered, after which some of the data from the literature on Archaean granulites are discussed. Most of the discussion in this paper will centre on mafic rocks, but some of the enunciated principles are known to apply to silicic rocks as well.

It must be emphasised that, in the following discussion of the chemical features of the granulites, only synopses of extensive unpublished data (1974) from Wilson, Mateen, Woodford and Iyer are presented. Details of individual rock analyses and correlations with other shields are being prepared for publication in several papers elsewhere by Wilson, and Mateen and Wilson, and others.

1. Geochemical features of the Fraser Range

The tectonic setting of the granulites of the Fraser Range in Western Australia, and several pertinent references have already been given.

Significant features about the sampling of this terrain are as follows: (1) Rocks of basaltic composition are abundant, and enable sampling of closely comparable chemical units throughout the terrain. (2) The mafic granulites are unusually "dry", for they are two-pyroxene granulites. Hornblende-bearing

TABLE 1: Average compositions of mafic granulites and igneous rocks from Precambrian and Recent environments

	GRANULITES						ARCHAEAN THOLEIITES		MODERN THOLEIITES		
	AUSTRALIA			U.K.		BRAZIL	W.A.	CANADA	OCEAN	IS. ARC	CONT.
	1	2	3	4	5	6	7	8	9	10	11
SiO_2	50.27	50.28	49.48	48.00	52.00	51.39	51.30	49.70	49.92	51.10	51.50
TiO_2	1.71	1.06	1.05	0.93	0.83	1.32	0.96	1.00	1.53	0.83	1.20
Al_2O_3	15.62	15.80	14.96	12.80	13.90	16.67	14.80	14.90	15.63	16.10	16.30
Fe_2O_3	2.08	4.33	2.91	-	-	2.84	1.70	2.60	1.65	3.00	2.80
FeO	9.85	7.41	9.91	12.33	10.17	8.68	9.00	8.80	8.19	7.30	7.90
MnO	0.26	0.23	0.23	0.23	0.20	0.19	0.21	-	0.17	0.18	0.17
MgO	6.69	7.10	8.24	11.90	8.6	5.03	6.70	6.31	7.65	5.10	5.90
CaO	9.43	10.54	10.61	9.9	10.1	8.40	10.8	9.39	11.17	10.80	9.80
Na_2O	2.43	2.15	1.39	2.00	2.40	3.14	2.70	2.11	2.75	2.00	2.50
K_2O	0.92	0.59	0.33	0.50	0.50	1.23	0.18	0.32	0.16	0.30	0.86
P_2O_5	0.36	0.17	0.10	-	-	-	0.12	-	0.13	-	0.21
H_2O^+	0.23	0.35	0.53	-	-	0.90	1.20	2.76	0.95		0.81
H_2O^-	0.09	0.16	0.12	-	-						
Rb	27.7	12.5	3.3	4.8	4.7	21.5	9	8	2	5	22
Sr	214	228	92	118	227	451	105	164	121	225	328
Ba	366	200	91	250	328	793	-	105	15	60	246
V	188	220	259	-	-	-	320	-	303	270	251
Cr	193	195	317	620	690	116	367	187	298	50	168
Co	52	55	55	-	-	-	-	35	37	20	48
Ni	80	65	109	332	267	-	170	103	119	25	138
Cu	34	53	58	69	69	61	107	110	66	80	90
Zn	112	111	146	112	105	118	107	100	84	80	100
Pb	9.6	8	8	-	-	-	-	-	1	-	4
Zr	175	99	71	53	82	114	61	105	98	60	137
Nb	10	6	3	3	4	-	-	-	5	-	13
Y	40	34	35	23	22		20	20	35	20	28
La	24	11	8	27	39		4	7	4	4	15
Ce	54	29	18				8	15	12	7	33
Nd	26	16	11				-	-	11	9	19
U	0.61	0.48	0.47								0.5
Th	1.75	1.76	1.25								1.8
K/Rb	331	633	927	870	880	696	430	332	664	700	324
Th/U	3.1	3.6	3.2								3.6

1 = Fraser Range granulites (27); 2 = Musgrave Ranges granulites (19); 3 = Strangways Range granulites (26); 4 & 5 = Scourie granulites, UK (Holland & Lambert, 1975); 6 = Brazilian granulites (Sighinolfi, 1971b); 7 = Archaean tholeiite, W. Australia (Hallberg & Williams, 1972); 8 = Archaean tholeiite, Canada (Condie & Harrison, 1976); 9 & 10 = Ocean and Island Arc tholeiites, resp. (Hart et al., 1972; Wedepohl, 1975; Condie, 1976; Jakeš & White, 1971, 1972); 11 = Continental tholeiite (Manson, 1968; Prinz, 1968; Wedepohl, 1975).

granulites are exceedingly rare. (3) A regular metamorphic gradient over at least 100 km has been established by mineral chemistry (Wilson, 1976), and samples were collected to investigate element variation with gradient.

In the eastern Fraser Range, about 20 km from the Fraser Fault Zone, many zones in the pyroxene granulites contain quartz and plagioclase ellipsoids which represent amygdales in basalt. The major element analyses show that the metamorphosed basalts are quartz-poor tholeiites (Wilson, 1975). Many other analyses of pyroxene granulites from the central and western portions of the range reveal almost identical tholeiitic composition, both in major and trace element composition. Thus, although in these granulites no amygdales can be recognised, there are numerous felsic streaks and other features that could be interpreted as highly deformed and transposed vesicular tops of basalts.

Table 1 sets out the average composition of the mafic granulites from the Fraser, Musgrave and Strangways Ranges, together with data on some Archaean granulites and comparable modern volcanic rocks.

If the element distribution in these granulites is considered as a group (irrespective of their geographical location on metamorphic gradients, presence or otherwise of biotite, etc.) some important conclusions may be drawn:

1. The FeO-MgO-Alkalies plot (Fig. 2) shows the relationship of the Fraser Range mafic granulites to those of the Musgrave and Strangways Ranges, and to some unmetamorphosed basalts, both ancient and modern. It is clear that the Fraser Range granulites form a very tight grouping and reveal normal tholeiitic characteristics. Their somewhat more alkaline character, however, is merely a reflection of their continental tholeiitic nature, features which are much more clearly shown by other discriminants (TiO_2-K_2O-P_2O_5; Fig. 3).
2. The rare earths measured by XRF are La, Ce and Nd, and Y is plotted at the Ho position in the normalised REE patterns, representing the REE portion. When the normalised patterns are plotted against chrondritic values it is clear that the Fraser Range granulites conform with continental tholeiites (Fig. 4). Granulites from the Musgrave and Strangways Ranges, on the other hand, show varying degrees of correspondence to island arc tholeiites.
3. The distributions of TiO_2, MgO, K_2O, P_2O_5, Ba, Zn, Zr, Nb, Y and REE are closely correlated in this group of rocks, and such variations as do occur would appear to indicate the pre-metamorphic character of the mafic rocks and not metamorphic effects. The relationships of MgO with Cr, Ni, and Zn, and of total Fe (as FeO) with Zn reflect the fractionation

trend similar to the middle stage differentiation of a tholeiitic magma. The element ratios of Fe/Mg, Ti/Zr, Zr/Nb, Ti/Y, La/Y and Th/U also support the view that the Fraser Range mafic granulites are the metamorphic derivatives of continental mafic igneous rocks of tholeiitic composition.

4. The small variations in the distribution of SiO_2, Na_2O, Rb, Cu and U, and their lack of correlation with immobile elements, may be attributed to minor alteration prior to and during metamorphism of the granulites. However, when samples are not bulked but are considered in relation to one another on metamorphic gradients, other conclusions may be drawn concerning these and other elements (see Table 2).

TABLE 2.

Element variations in Fraser Range anhydrous mafic granulites along 100 km traverse showing increase in pressure at constant temperature, and variations from the Fraser Range norm in a chemically anomalous zone unrelated to metamorphic grade.

Percentage variation with increase grade of metamorphism		Variations from norm in anomalous zone	
No change (± 10%)	Increase (10%)	+ (%)	− (%)
Si, Ti, Al, Fe^3, Fe^2, Mn, K, Na, P, V, Co, Zn, Rb, Zr, Ba, La, Ce, Nd, Pb, Y	U, Cu(80); Th(70); Ni(60); Cr(40)	U(150); P, Ba(90). La(80); Ce, Nd, Th. Zr(70); Rb, Ti(50); Y(40); K(30); Mn, Zn(15)	Cr(35); V(20); Mg(15)

TABLE 3

Element variations in hydrous mafic granites along 25 km traverse showing increase in both pressure and temperature in the Musgrave Ranges.

Percentage variation with increase in metamorphism		
No change (± 10%)	Increase (%)	Decrease (%)
Si, Al, $Fe^2/\Sigma Fe^2$, Mn, Ca, K, Zn, Cr, Y, Nb	Ba(220); Sr, K/Rb(120); Nd(80); Na(70); Cu(60); Ce(50); P(40); La(30);	Th(75); U(70); Rb(60); Ni(40); Co(30); Fe^3, Fe^2, Mg, Ti, Zr, V, Pb(20);

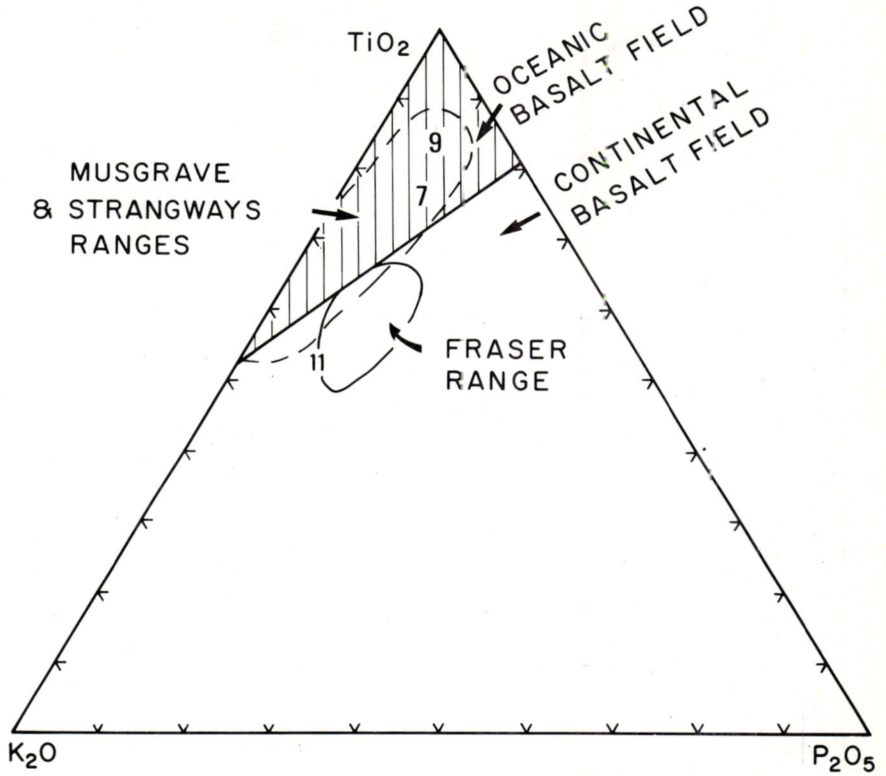

Fig. 3. K_2O-TiO_2-P_2O_5 diagram (after Pearce et al., 1975) showing the Fraser Range mafic granulites in the field of continental basalts, whereas the bulk of the mafic granulites from the Musgrave and Strangways Ranges fall in the field of oceanic basalts. The numbers refer to basic rocks of Table 1.

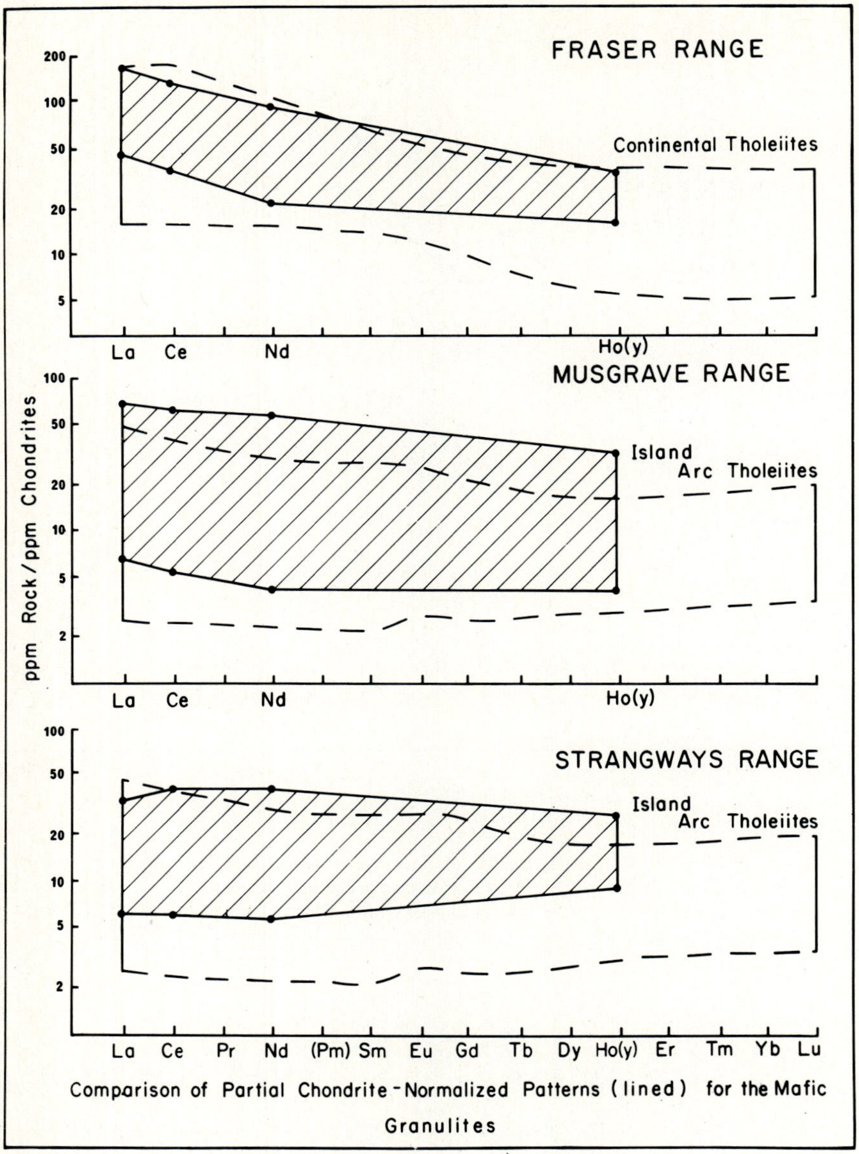

Fig. 4. The partial rare earths pattern of the mafic granulites of the Fraser Range, Musgrave Ranges and Strangways Range compared with normalised chondrite. The fields for continental tholeiites and oceanic tholeiites are shown for comparison.

Table 2 shows the variations in element distribution in a group of eleven mafic granulites located according to their position on a 100 km traverse. The Al_2O_3 content of coexisting pyroxenes and Na in clinopyroxenes indicate a constant decrease in pressure along this traverse, but calculations of temperature (c. 860°C) show no significant differences along the same traverse (Wilson, 1976). Only "dry" pyroxene granulites, i.e. those containing no significant biotite or hornblende, were selected for the Table.

The plotted data (details will be presented elsewhere) show only very small variations (< ± 10%), or no variations with increase in metamorphic grade for 20 elements. The small increase in MgO (only 15%), and increases in the related traces Cu, Ni and Cr and Sr, may indicate, rather, a slight shift in composition of the original magmatic parent rocks, for more likely mobile components such as K, Na, Si and (especially) Rb show no significant changes.

The most striking apparent increases with metamorphic grade are in U (80%) and Th (70%). Even if the most radioactive sample (the sample originally under highest pressure) is omitted from the averages, there are still large increases of 35 and 50%, respectively. These results are contrary to those obtained from the Musgrave Ranges, and may be explicable if one envisages post-metamorphic fluids penetrating rocks that have responded to off-loading on rapid uplift. On the other hand, the possibility that the parent rocks in the higher grade zone were enriched before metamorphism should also be considered, especially as primary enrichments in U and Th appear within the "anomalous" zone mentioned in the next paragraph.

There is a prominent anomalous zone on the traverse beginning about 11 km NNE of the Microwave Station (loc. M, Wilson, 1976) and continuing for about 26 km NNE. The granulites of this zone are presumed to represent continental tholeiites which show significant enrichments above the average values for the rest of the assumed parent rocks of the 100 km traverse. The enriched elements (Table 2) are: U (150%), P and Ba (90%), the rare earths La (80%) and Ce and Nd (70%), Th and Zr (70%), Rb and Ti (50%), Y (40%), K (30%), and possibly Mn and Zn (15%).

The fact that K and P have increased, whereas these elements did not change with increase in metamorphic grade, shows that the K-feldspar and fluor-apatite seen in all of these "anomalous" granulites represents a relict primary igneous composition. The high content of apatite and zircon probably hosts the rare earths, Y, Th and U, and the K-feldspar probably holds the Ba and Rb. Elements depleted in this zone are the commonly correlated elements Cr (35%), V (20%) and Mg (15%). These depletions, as well as the long list of all the other unlisted elements which show no significant change from the Fraser Range norm, make untenable a metasomatic-influx origin for this zone.

2. Geochemical features of the Musgrave Ranges

The mafic granulites from the Musgrave Ranges exhibit wide variations in the major and trace element abundances, and there are some anomalies. The variations and anomalies may be due to one or more of three main factors: (1) some of the sampled mafic layers are narrow bodies only 1 or 2 m in width, and some reaction with their silic host rocks is probable. (2) There is a real range in chemical composition of original igneous material. Some of the more magnesian granulites appear to represent oceanic and arc tholeiites, whereas the more ferriferous rocks resemble the more highly fractionated derivatives of original igneous material. (3) An intense metamorphism, related to a series of intrusive pyroxene granites, has been superimposed on the normal regional granulite metamorphic products, and has caused movement of certain elements which otherwise would be stable.

However, notwithstanding these localised anomalies, the close correlation of TiO_2, MgO, P_2O_5, Zr and REE, and the relationships of MgO-Cr-Ni, the Th-Y-REE and U-Y-REE, show that many characteristics of the original mafic rocks have survived in most of the granulites. Fig. 2 shows the relationship of FeO-MgO-Alkalies in the Fraser Range mafic granulites to those of the Musgrave and Strangways Ranges. However, Figs. 3 and 4 show that the rocks from the Musgrave Range are similar to the oceanic and arc tholeiites.

In Fig. 3 the granulites of the Musgrave Ranges and Strangways Range are pooled, and are shown to fall largely in the field of oceanic and arc tholeiites. Only a few samples transgress into the field of the continental tholeiites into which the Fraser Range granulites fall.

In Fig. 4 the rare earth pattern of the Musgrave Range group is compared with those of the Strangways and Fraser Ranges. These patterns show the general similarity between the mafic granulites of the Musgrave and Strangways Ranges, and the distinctive pattern of the Fraser Range granulites.

As was noticed in the Fraser Range, some of the geochemical features are lost when a large pool of granulites is not subdivided according to possible original magma types, or when either their structural setting or geographic position with respect to metamorphic gradients, or mineralogy are ignored.

In an attempt to see the effect of increasing metamorphism on the granulites (based on Al_2O_3 and Na in pyroxenes, etc.), a selection was made of 17 rocks with similar gross chemistry, size of rock unit and mineralogy. These were plotted on a traverse line which passes from the higher grade granulites in the west (beginning about 7 km NW of Ernabella, Fig. 1) through the intrusive "charnockite" (the ferrohypersthene-bearing Ernabella Adamellite - Wilson, 1960; Arriens and Lambert,1969) at Alalka to a point about 8 km east of the intrusion. Details are not presented here, but Fig. 5 is a sketch to illustrate element distribution along the traverse. The sample is not large, but is sufficient to point out some unexpected geochemical features.

The first feature to notice is some of the chemical characteristics of the "charnockitic intrusion". It is rich in Ba (1670 ppm), Zr (856), Ce (144), Y (81), Rb (77), Sr (347), U (0.40), Th (1.20), TiO_2 (1.60%), P_2O_5 (0.86%) K_2O (3.47%), and very depleted in Cr (1) Ni (8) and Cu (12). Its effect on the enclosing granulites may be noticed for upwards of 1 km, and banded granulites of the first metamorphism (1380 MA; Arriens and Lambert, 1969) were commonly recrystallised at 1120 MA, and now display textures dominated by polygonal grains. Temperatures reached were high, and the rare assemblage orthopyroxene-sillimanite-quartz is developed in an aluminous granulite (14795Q) about 350 m east of the intrusion. Sapphirine is also developed within the aureole (Wilson and Hudson, 1976).

In Fig. 5 the chemical changes found within the aureole are shown. Calculation of percentage increases are meaningless owing to the sharp increase (or decrease) near the contact, and because of the small sample.

The striking increases in U and Th are drawn to scale. The intrusion is more radioactive (0.40 (U) and 1.20 (Th) ppm) than normal granulites, although lower grade granulites in the eastern part of the traverse have similar values (0.44 (U) and 1.74 (Th) ppm). The small increase in Rb is to be expected, as also is the slight increase in $Fe^2/\Sigma Fe^2$. The slight increase (?25%) in Cr was unexpected and may disappear on further sampling.

The most surprising depletions within the aureole are of the two elements Zr and Ti which are generally considered to be immobile. It is known, however, that Ti can be mobile in the presence of the halogens and, as Zr readily forms a fluoride, an influx of fluorine into the aureole may have caused these depletions. The fall in K/Rb is interesting in view of the high K_2O of the intrusion (3.47%) and K/Rb ratio of 387 and the small increase in Rb. The depletion of Cu (c. 100%) is more than would be expected from a very hot "dry" intrusion, and the halogens would appear to be responsible. The economic implications of the movement of Cu in the deep crust may well profit from this observation.

The other features shown in Fig. 5 refer to overall changes in element proportions due to increasing grade of metamorphism. These are shown in Table 3. The following elements shown no significant change: Si, Al, $Fe^2/\Sigma Fe^2$, Mn, Ca, K, Zn, Cr, Y and Nb. Another group show only minor losses of about 20%, and further sampling may show that these may not be significant: Fe^3, Fe^2, Mg, Ti, Zr, V and Pb. There are few surprises here, for most of the elements normally considered to be immobile fall into one or other of these two lists. A very interesting element which shows no significant change is K. This element and Rb are normally considered together. In this traverse, as Rb is heavily depleted (60%), the K/Rb ratio shows a large increase of about 130% from 460 in the east to 1040 in the west.

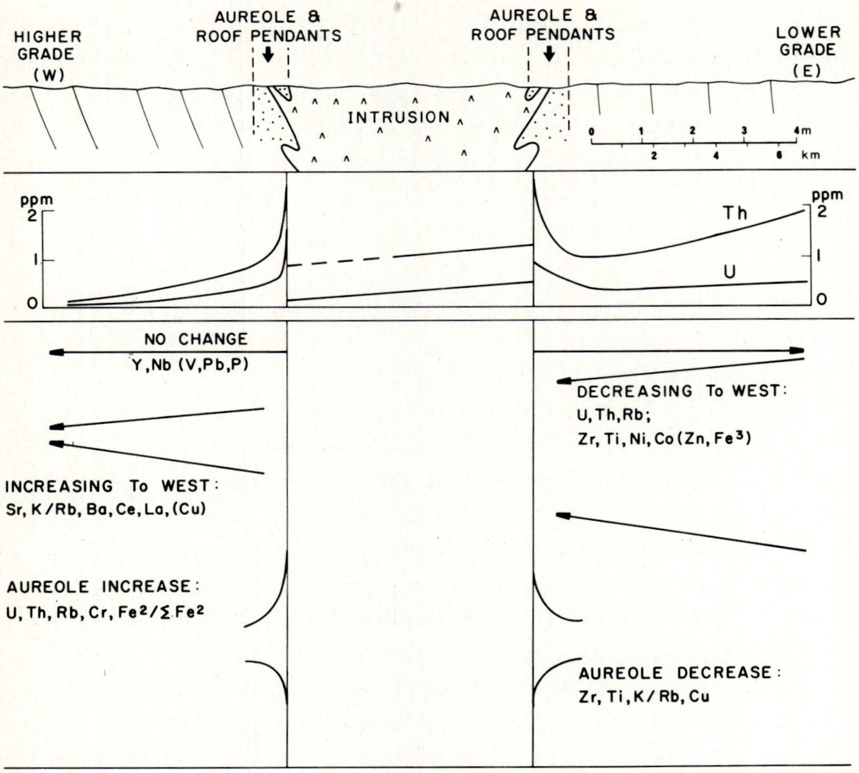

Fig. 5. Sketch to illustrate element distribution in mafic granulites on an E-W 25 km traverse through portion of the eastern Musgrave Ranges. U and Th show a constant fall off toward the higher-grade western regions (average trends are drawn to scale), except in the aureole of the ferrohypersthene adamellite intrusion where both U and Th show a sharp increase. Some other elements show variable but small decreases towards higher grade, and others show small increases. In the aureole several elements are lost and others gained. The only elements which show no significant change across the regional gradient or in the aureole are Y, Nb, and possibly V, Pb and P.

The elements which show a depletion of 30% or more are Th (75%), U (70%), Rb (60%), Ni (40%) and Co (30%). As Mg shows a depletion of about 20% the Ni and Co depletions may be related to slight shifts in parental composition, and may not be metamorphic effects. The largest depletions are in Th (75%) and

U (70%), and many others have noted similar depletions in other deep crustal rocks with increasing metamorphism. These depletions, however, only became apparent when samples were removed which showed obvious chemical differences, or which were affected by the major influx of Th and U in the aureole, or those in which F-rich biotite is an important component. This shows the danger of basing element distribution studies on a large group of somewhat diverse samples.

The elements which are concentrated in the higher grade granulites are: Ba (220%), Sr (120%), the rare earths Nd (80%), Ce (50%) and La (30%), Na (70%), Ca (60%) and P (40%). The increase in P is an indication of an influx of F, for the apatite of all of these rocks is a fluor-apatite (see discussion of role of F and P in the charnockitic granites of the region in Wilson (1960); Hudson (1968) found F = 2.46% in apatite from the intrusive body shown in Fig. 5). The degree to which these element variations are directly related to increase in metamorphic grade is unknown, for variations in parental magmas cannot be ruled out. Indeed, the large number of immobile or near immobile elements on the traverse would support the view that some of the variations in elements such as REE, P, Mg, Ni and Co may be original.

3. Geochemical features of the Strangways Range

The Strangways Ranges mafic granulites appear to be related in composition to the Musgrave Ranges in several respects, and they are bulked together in the AFM diagram (Fig. 2). When the more magnesian granulites are separated out (details are being prepared for publication elsewhere), they are found to be characterised by extremely low abundances of Ti, K, P, Rb, Sr, Ba, Zr, Nb, Y and REE, and relatively high Cr and Ni. This suggests that these granulites could be metamorphosed oceanic tholeiites of low-Zr type (Lambert and Holland, 1977). The more ferriferous granulites, on the other hand, form a loose group of mafic rocks from several localities and structural settings, and some appear to represent more differentiated magmas, exhibiting Skaergaard-like trends.

In Fig. 3 the TiO_2-K_2O-P_2O_5 plot bulks the Strangways and Musgrave Ranges rocks, and shows that most of them fall within the field of the oceanic tholeiites. The partial rare earth pattern of the bulked Strangways Range samples suggests the same general conclusion (Fig. 4).

The use of Ti, Zr, Y, P, and the REE to discriminate magma types and tectonic setting has been attempted. Although several tight groupings of mafic granulites may be made in this way, interpretation of magma type of the groups is rarely consistent with the known geology. Details are not recorded here.

In view of the mobility of some of the so-called immobile elements, as shown by the Musgrave Ranges data, these elements should be used with caution. There it was noted that fluorine-rich emanations combined with very high temperatures would appear to have depleted both Zr and Ti from the aureole of the intrusive (Fig. 5). The Strangways Range is a very complex polymetamorphic terrain in which the effects of at least two granulite facies metamorphisms are commonly seen. Moreover, there are silicic igneous bodies and many fluorine-rich metasomatic zones of several ages. Rocks need to be carefully sampled and their elemental composition considered in relation to their present tectonic and geographic position. Only then is the relation of element migration to metamorphic grade likely to be understood in the Strangways Range.

CHEMICAL FEATURES OF SOME ARCHAEAN MAFIC GRANULITES

Concerning Australian Archaean granulites, there is no body of date large enough for comparison with the Proterozoic granulites. However, several studies of other Archaean granulites have recently appeared and chemical data from two of these regions are shown in Table 1. These are from the Lewisian of Scotland (Holland and Lambert, 1975) and Brazil (Sighinolfi, 1971a and b).

1. Some Lewisian granulites

Mean analyses of the Scourie assemblages in the Lewisian are grouped according to silica content and appear as 4 and 5 of Table 1. The first group (4) is both more magnesian and more ferriferous than the Australian granulites. These rocks are olivine-bearing ultramafic rocks, and are not strictly comparable with the Australian rocks which are of basaltic composition. The second group (5), however, is more similar to our rocks, although the SiO_2 content (52.2%) is notably higher. Rb is much depleted (cf. Strangways Ranges), but both Ni and Cr are high and probably reflect their pre-metamorphic olivine content.

The very low levels of some incompatible elements, especially K, Rb, Nb, Y, Th and U, are thought to result from magmatism and metamorphism related to a process called "slow sub-crustal accretion" (Holland and Lambert, ibid).

It is difficult to assess the geochemical mechanisms involved in bringing about the bulked sample (column 5) without knowing details of sample location. As polymetamorphism was probably involved, the relative roles of total pressure temperature and "volatile" components (especially fluorine) cannot be assessed.

2. Some Brazilian granulites

The Brazilian data for the mafic granulites are averaged from 17 samples with $SiO_2 < 55\%$. The K/Rb ratio of 696 compares well with 633 from the Musgrave Ranges, and is similar to that from other granulite terrains. However, it is much higher than that of the Fraser Range. The Brazilian samples are generally much richer in K than any of the other granulites and basic rocks listed in Table 1. This raises the question of the extent of K-metasomation that these rocks may have suffered.

Sighinolfi (1971), in discussing fractionating effects in deep crustal rocks of Brazil, considers the whole range of granulites rather than only the basic rocks on which the present paper is concentrating. He states: "Assuming that granulites represent residuals of partial melting processes, partial melting is seen as both encouraging the attainment of trends of a magmatic type in the residuals, and causing fractionation between elements'. These conclusions are pertinent when attempts are made (as in this paper) to recognise pre-metamorphic magma types from analyses of granulites.

It remains to be seen, however, whether the data from the Brazilian rocks can be re-assessed in the light of some of the conclusions of this paper. The tectonic and geographic setting of the samples, their mineralogical composition, position of presumed metamorphic gradients, and possible polymetamorphic character, are likely to influence some of the conclusions about factors affecting element fractionation in the deep crust.

CHEMICAL FEATURES OF THE SILICIC GRANULITES

In many granulite terrains quartzose granulites, with SiO_2 ranging from about 60% and upwards, predominate over the mafic granulites which have been selected for discussion in this paper. If, as is obvious, deep crustal rocks suffer partial melting to varying extent, the recognition of the composition of pre-metamorphic silicic rocks will be much more difficult.

Many quartzose granulites from the Fraser, Musgrave and Strangways Ranges and from the Archaean Wheat Belt region have been analysed. A detailed assessment of the data is not proposed for this paper. In these rocks it is found, however, that the metamorphic mineralogy is more important than in basic rocks for studying the effects of metamorphism. For example, the development of garnet (for whatever cause) appears to affect the retention in the rock of Y and the rare earths.

In the Fraser Range, the small sample of silicic granulites shows a general depletion in both U and Th with the increase in total pressure (at constant temperature). It will be recalled that, in the mafic granulites, the distribution of U and Th is markedly different. It was concluded that either the mafic granulites retained originally high values of U and Th during metamorphism, or there was a significant <u>increase</u> in both U and Th with the increasing total pressure. The silicic rocks are garnetiferous and commonly contain biotite, whereas the mafic granulites contain no hydrous phases.

In the Musgrave Ranges the sample of silicic granulites is also small. However, the regional distribution pattern of U and Th is similar to that of the mafic granulites, that is, there is a marked <u>depletion</u> of both U and Th with increasing metamorphic grade. The difference in metamorphic conditions in the Musgrave Ranges is that there the metamorphic gradient from E to W involves an increase in both temperature and pressure. Moreover, the rocks normally contain some hydrous minerals.

In the Strangways Range U and Th were depleted during the first granulite metamorphism, but are commonly found to be enriched during the second granulite metamorphism.

In all three granulite terrains there has been a general depletion of Rb with increasing grade of metamorphism, and, when considered with the K distribution, the ratio K/Rb is found to increase with grade.

Thus, the silicic granulites of the three terrains show distribution patterns for Rb, K, U and Th that are similar to those noted in silicic granulites from several shield regions (Lambert and Heir, 1968; Sighinolfi, 1971a and b; Holland and Lambert, 1975; and others).

CONCLUSIONS

Throughout this paper the importance has been emphasised of the need to try to look at geochemical data of granulites in the context of their tectonic and geographic setting, mineralogical composition, and position on assumed metamorphic gradients, and to try to discriminate in polymetamorphic terrains the effects of individual metamorphisms.

In the Fraser Range many samples can be located on a traverse along which total pressure has changed systematically whereas the temperature of metamorphism appears to have remained constant. In this situation very few changes in whole rock composition have taken place in those mafic rocks which contain no significant biotite or hornblende. In the associated silicic rocks, however, there is a regular increase with grade of the ratio K/Rb. Nevertheless, anomalous element distributions appear where volatile-rich mafic rocks are considered.

In the Musgrave Ranges, however, similar studies related to a metamorphic gradient (where there were increases in both total pressure and temperature) show a depletion in U, Th and Rb in both mafic and silicic granulites. Enrichments appear in Ba, Sr, REE, Na, Ca, and P, but most other measured elements show no significant change. These differences from the Fraser Range are interpreted as due to the higher fluid pressures and the presence of a significant thermal gradient only in the Musgrave Ranges.

The profound changes wrought by the intrusive ferrohypersthene adamellite pose an important question for deep crustal studies. Although the intrusion has chemical features so different from the granulites, the rocks in the aureole have been influenced far more by the heat of the intrusion than by its bulk chemistry. The depletion of Zr and Ti within the aureole was unexpected. These elements are normally considered to be immobile during metamorphism. It is suspected that an influx of fluorine into the aureole has enabled movement of these elements.

The large increases in the aureole of both U and Th are important in view of their general depletion on a regional scale with increasing grade of metamorphism. Although the intrusion was a fairly "dry" body, it appears that its high heat content has caused partial melting of the felsic granulites at depth, and has then mobilised these elements enabling them to become fixed in the mafic granulites.

Suggestions for future studies

In view of these observations the following suggestions are made for element distribution studies of granulites.

1. Random sampling, or statistical sampling on a grid basis, is of limited value in a shield, for rocks showing one or more metamorphisms, and of different parentages, may easily be bulked together. Moreover, owing to the migration, during metamorphism, of some elements from limbs to noses of folds, and during transposition of layers, the tectonic setting of samples should be recorded.

2. Comparisons ideally should be made only between rocks of similar gross chemistry and mineralogy. Thus, hornblende-bearing granulites should not normally be grouped with "anhydrous" granulites where gradient variations are being sought.

3. Silicic and mafic granulites may be pooled to estimate the gross composition of a terrain. However, in view of the marked differences found in the mobility of elements in silicic and mafic rock groups in the Fraser Range, element distributions should be worked out on closely defined compositional groups.

4. Aureoles related to largely hidden very hot igneous bodies may readily be overlooked. Some intrusions have caused retrogression, but others, especially the "magmatic" charnockites, mangerites, anorthosites, fasundites and gabbros, were hot enough and large enough to cause non-deformational granulite metamorphic effects (in reality, a second granulite facies metamorphism) on "regional" granulites in which deformational fabrics are normal.

5. The nature of the fluids or volatiles other than H_2O are critical in assessing element distributions. Fluorine is particularly important in deep crustal rocks, and appears to be responsible for the movement of Ti, Zr and P, some of the elements generally relied upon as remaining immobile during metamorphism.

6. An influx of CO_2 with a corresponding lowering of the activity of H_2O, may cause (at little or no change in temperature) hypersthene-bearing rocks to form at the expense of hydrous (amphibolite facies) rocks. Similarly, hornblende-bearing granulites (where the hornblende is the brownish "granulite" facies hornblende and not an "amphibolite facies" hornblende) are commonly found interlayered with "dry" pyroxene granulites of closely comparable

composition. This relationship would appear to represent juxtaposition of zones composed of differing proportions of CO_2, F, H_2O, SO_3, or their dissociation products. Oxygen isotope studies (Wilson and Baksi, in prep.) are showing that inhomogeneities in the proportions of fluid phases may be very persistent, and have an important bearing on the distribution of elements and their isotopes.

7. Large changes in total pressure without change in temperature, appear to cause very little chemical change to mafic rocks, as was noted in the Fraser Range. Moreover, in that region the whole-rock oxygen isotopes show no significant changes across the total pressure gradients.

ARCHAEAN - PROTEROZOIC DIFFERENCES

In view of the evident lack of correspondence in sampling and assessment procedures adopted by petrologists and geochemists working in this field, it is difficult at this stage to identify significant chemical differences between granulites of Archaean and subsequent ages. Sufficient is known, however, to show that granulites and the related mobilised or magmatic pyroxene-bearing silicic rocks (some so-called charnockites, etc.) may be formed under a range of total pressure, partial pressure of fluids, temperature, and ages. These variable conditions have resulted in significant differences in element mobility from rock to rock. Thus, before subtle chemical differences between Archaean and Proterozoic granulites can be established (I see nothing significant as yet), chemical differences must first be established for their unmetamorphosed equivalents.

There may be some tectonic differences. The impression gained from the literature and Australian studies is that more higher pressure granulites are found in the linear fault-controlled Proterozoic belts than in the broader shield regions where high thermal effects are more evident.

ACKNOWLEDGEMENTS

Abdul Mateen was particularly helpful in chemical analysis and in collating much of the data. A. S. Bagley did many XRF analyses, and G. Langthaler made many determinations of FeO and Na_2O. This research has been supported by research grants from the University of Queensland and by A.R.G.C. grant No. E71/17830, and E76/15913.

REFERENCES

Allen, A.R., 1976. The metamorphic evolution of the Arunta Block in the Harry Creek area, southeastern Strangways Range, Northern Territory. Unpubl. Ph.D. thesis, Univ. Queensland.

Arriens, P.A., 1971. The Archaean geochronology of Australia. Spec. Publs. Geol. Soc. Aust., 3: 11-23.

Arriens, P.A., and Lambert, I.B., 1969. On the age and strontium isotope geochemistry of granulite facies rocks from the Fraser Range, Western Australia, and Musgrave Ranges, central Australia. Spec. Publs. Geol. Soc.

Austr., 2: 377-388.

Black, L.P., 1975. Present status of geochronological research in the Arunta Block, N.T. Geol. Soc. Aust. Geol. Convention Adelaide, Abstr. 37.

Collerson, K.D., 1975. Contrasting patterns of K/Rb distribution in Precambrian high grade metamorphic rocks from central Australia. J. Geol. Soc. Aust., 22: 145-158.

Collerson, K.D., Oliver, R.L. and Rutland, R.W.R., 1972. An example of structural and metamorphic relationships in the Musgrave Orogenic Belt, central Australia. J. Geol. Soc. Aust., 18: 379-393.

Condie, K.C., 1976. Trace element geochemistry of Archaean greenstone belts. Earth-Sci. Rev., 12: 393-417.

Condie, K.C. and Harrison, N.M., 1976. Geochemistry of the Archaean Bulawayan Group, Midlands Greenstone Belt, Rhodesia. Precamb. Res., 3: 253-271.

Gee, R.D., 1975. Regional geology of the Archaean nuclei of the Western Australian shield. In: C. Knight (Editor), Economic Geology of Australia and Papua-New Guinea. Austr. Inst. Mining Metall., Melbourne, Vol. 1, pp. 43-55.

Hallberg, J.A., and Williams, D.A.C., 1972. Archaean mafic and ultramafic rock associations in the Eastern Goldfields region, Western Australia. Earth Planet. Sci. Lett., 15: 191-200.

Hart, S.R., Glassley, W.E. and Karig, D.E., 1972. Basalts and sea floor spreading behind the Mariana Island Arc. Earth Planet. Sci. Lett., 15: 12-18.

Holland, J.G. and Lambert, R. St. J., 1975. The chemistry and origin of the Lewisian gneisses of the Scottish mainland: the Scourie and Inver assemblages and sub-crustal accretion. Precamb. Res., 2: 161-188.

Hudson, D.R., 1968. Some mafic minerals from the granulites and charnockitic granites of the Musgrave Ranges, central Australia. Unpubl. Ph.D. thesis, Univ. Queensland.

Iyer, S.S., 1974. Granulites and associated schists from northern Strangways Range, central Australia: geochemical and isotopic studies. Unpubl. Ph.D. thesis, Univ. Queensland.

Iyer, S.S., Woodford, P.J. and Wilson, A.F., 1976. Rb-Sr isotope studies of a polymetamorphic granulite terrane, Strangways Range, central Australia. Lithos, 9: 211-224.

Jakes, P. and White, A.J.R., 1972. Major and trace element abundances in volcanic rocks of orogenic areas. Geol. Soc. Amer. Bull., 83: 29-40.

Lambert, I.B. and Heir, K.S., 1968. Geochemical investigations of deep-seated rocks in the Australian shield. Lithos, 1: 30-53.

Lambert, R. St. J. and Holland, J.G., 1977. Trace elements and petrogenesis of DSDP 37 basalts. Can. J. Earth Sci., 14: 809-836.

Manson, V., 1968. Geochemistry of basaltic rocks: major elements. In: H.H. Hess and A. Poldervaart (Editors), Treatise on rocks of basaltic composition. Interscience, New York, pp. 215-270.

Nesbitt, R.W., and Talbot, J.L., 1966. The layered basic and ultrabasic intrusives of the Giles Complex, central Australia. Contr. Miner. Petrol., 13: 1-11.

Pearce, T.H., Gorman, B.E. and Birkett, T.C., 1975. The TiO_2-K_2O-P_2O_5 diagram: a method of discriminating between oceanic and non-oceanic basalts. Earth Planet. Sci. Lett., 24: 419-426.

Prider, T.H., 1945a. Sapphirine from Dangin, Western Australia. Geol. Mag.,

82: 49-54.

Prider, T.H., 1945b. Charnockitic and related cordierite-bearing rocks from Dangin, Western Australia. Geol. Mag., 82: 145-172.

Prinz, M., 1968. Geochemistry of basaltic rocks: trace elements. In: H.H. Hess and A. Poldervaart (Editors), Treatise on rocks of basaltic composition. Interscience, New York, pp. 271-324.

Sighinolfi, G.P., 1971a. Investigations into the deep levels of the continental crust: petrology and geochemistry of the granulite facies terrains of Bahia (Brazil). Atti. Soc. Tosc. Sc. Nat. Mem., 77: 327-341.

Sighinolfi, G.P., 1971b. Investigations into deep crustal levels: fractionating effects and geochemical trends related to high-grade metamorphism. Geochim. Cosmochim. Acta, 35: 1005-1021.

Tyrwhitt, D.S. and Orridge, D.H., 1975. Regional geology and mineralization of the Fraser Range Orogenic Belt, Western Australia. In: C. Knight (Editor), Economic Geology of Australia and Papua-New Guinea. Austr. Inst. Mining Metall., Melbourne, Vol. 1, pp. 405-408.

Wager, L.R. and Deer, W.A., 1939. The petrology of the Skaergaard intrusion, Kangerdlugssuaq, east Greenland. Meddr. Grønland, 105, No. 4.

Wedepohl, K.H., 1975. The contribution of chemical data to assumptions about the origin of magmas from the mantle. Fortschr. Miner., 52: 141-172.

Wilson, A.F., 1960. The charnockitic granites and associated granites of central Australia. Trans. Roy. Soc. S. Aust., 83: 37-76.

Wilson, A.F., 1962. A new occurrence of sapphirine and the significance of sapphirine in the granulite terrains of Western Australia. Amer. Mineral., 47: 207-208.

Wilson, A.F., 1966. Metamict allanite from pegmatites cutting basic charnockitic granulites in the Fraser Range, Western Australia. J. Roy. Soc. W. Aust., 49: 85-87.

Wilson, A.F., 1969a. Granulite terrains and their tectonic setting and relationship to associated metamorphic rocks in Australia. Spec. Publs. Geol. Soc. Aust., 2: 243-258.

Wilson, A.F., 1969b. The pyroxene granulites and associated gabbros of the Fraser Range, Western Australia, and their economic significance. Proc. Aust. Inst. Mining Metall., 231: 47-57.

Wilson, A.F., 1969c. The mineral potential of granulite terranes. Proc. Aust. Inst. Mining Metall., 231: 41-46.

Wilson, A.F., 1971a. Some geochemical aspects of the sapphirine-bearing rocks and related highly metamorphosed rocks from the Archaean metamorphic belt of South Qairading, Western Australia. Spec. Publs. Geol. Soc. Aust., 3: 401-411.

Wilson, A.F., 1971b. Some structural, geochemical and economic aspects of the metamorphosed East Fraser Gabbro and associated pyroxene granulites of the Fraser Range, Western Australia. Ind. Mineral., 10: 1-21.

Wilson, A.F., 1975. Pyroxene granulites developed by metamorphism of amygdaloidal tholeiitic basalts, Fraser Range, Western Australia. In: C. Naganna (Editor), Studies in Precambrians. Bangalore Univ., pp. 32-49.

Wilson, A.F., 1976. Aluminium in coexisting pyroxenes as a sensitive indicator of changes in metamorphic grade within the mafic granulite terrane of the Fraser Range, Western Australia. Contrib. Mineral. Petrol., 56: 255-277.

Wilson, A.F., 1977. A zincian högbomite and some other högbomites from the Strangways Range, central Australia. Mineral. Mag., 41: 337-344.

Wilson, A.F. and Hudson, D.R., 1967. The discovery of beryllium-bearing sapphirine in the granulites of the Musgrave Ranges, central Australia. Chem. Geol., 2: 209-215.

Wilson, A.F. and Baksi, A.K. (in press). Unusually light oxygen in granulites from the Strangways Range, central Australia. In: B.W. Robinson (Editor), Stable isotopes in the Earth Sciences. D.S.I.R. Spec. Bull., Govt. Printer, Wellington.

Wilson, A.F. and Middleton, D.D., 1968. Some petrological features of a spinel-bearing metagabbro in the pyroxene granulites of the Fraser Range, Western Australia. J. Roy. Soc. W. Aust., 51: 89-96.

Woodford, P.J., 1974. Polymetamorphism in granulites from the north-eastern Strangways Range, central Australia. Unpubl. Ph. D. thesis, Univ., Queensland.

Woodford, P.J. and Wilson, A.F., 1976a. Sapphirine, högbomite, kornerupine and surinamite from aluminous granulites, north-eastern Strangways Range, central Australia. N. Jb. Miner. Mh., 15-35.

Woodford, P.J. and Wilson, A.F., 1976b. Chemistry of coexisting pyroxenes, hornblendes and plagioclases in mafic granulites, Strangways Range, central Australia. N. Jb. Miner. Abh., 128: 1-40.

Woodford, P.J., Mateen, A., Green, D.C. and Wilson, A.F., 1975. $^{40}Ar/^{39}Ar$ geochronology of a high grade polymetamorphic terrain, north-eastern Strangways Range, central Australia. Precamb. Res., 2: 375-396.

GEOCHEMICAL COMPARISON OF ARCHAEAN GRANULITES IN INDIA WITH PROTEROZOIC GRANULITES IN CANADA

G.V. Anantha Iyer and T.R. Narayanan Kutty

Department of Inorganic and Physical Chemistry, Indian Institute of Science, Bangalore-560012, India

Abstract

The charnockitic gneisses of North Arcot, Tamil Nadu, have a bimodal compositional distribution. The basic granulites show a tholeiitic Fe-rich trend, while the intermediate and acid charnockites show a calc-alkaline one. A similar compositional variation is apparent in amphibolite facies gneisses and associated mafic rocks and in the low-grade schist belts of Kolar and Holenarasipur of southern Karnataka. However there are notable differences in their respective trace element compositions. The charnockitic rocks are low in large-ion lithophile (LIL) elements (i.e. K, Rb, Th, U, Pb) but not in other normally incompatible elements (i.e. Ba, Sr, Zr), and have higher K/Rb and lower Rb/Sr ratios compared with the lower grade gneisses. The geochemical data suggest that the amphibolite facies gneisses of South Karnataka are not retrogressed charnockites; on the other hand, the charnockites may be derived from the high-grade gneisses. There is no major chemical distinction between Archaean and Proterozoic granulites. As such, the hypothesis that granulites are derived from an igneous parent magma that contained low levels of LIL trace elements does not seem to be valid. The basic members of the charnockitic suite compare better with the mafic enclaves within the high-grade gneisses than with the mafic components of the schist belts.

INTRODUCTION

One aim of this paper is to compare the geochemistry of the charnockites of Tamil Nadu with those of the amphibolite facies gneiss terrain of southern Karnataka, in order to provide geochemical constraints on the prevailing ideas on the relation between these two terrains. Field, petrographic and geochemical investigations have so far mainly centred around magmatic (Holland, 1900; Smeeth, 1916; Subramaniam, 1967) versus metamorphic (Pichamuthu, 1953, 1965; Rama Rao, 1962) origin of charnockites. The field evidence is controversial. The transgressive nature of charnockites with the associated gneisses has been used by magmatists in their favour, while the conformable nature of the rock units, including the metasediments, lent support to the metamorphic theory. The transformation of high-grade gneiss to anhydrous charnockites is reported in many localities of southern Karnataka (Pichamuthu, 1960; Ziauddin and Yadav, 1975), while the same field evidence is used to explain the retrogressive migmatisation of charnockites (Devaraju and Sadashivaiah, 1969; Ray, 1972). The regional tectonic evidence indicates the truncation of the N-S trend of the high-grade gneisses of Karnataka by the ENE-trending belt of charnockitic gneisses (Narayana Swamy, 1966; Nautiyal, 1966). It is hoped, therefore, that the geochemical evidence may constrain the possible alternative models.

Diverse opinions prevail on the geochemistry of granulites. A problem often discussed is whether chemical distinctions can be preserved under granulite-grade metamorphism (Lambert and Heier, 1968; Heier, 1973; Holland and Lambert, 1973). Depletion of many of the large-ion lithophile elements in the lower Archaean crust takes place during granulite facies conditions. Decrease of LIL elements with depth and increasing grade from amphibolite to granulite facies has been demonstrated in several regions (Ramberg, 1951; Eade and Fahrig, 1971; Lambert, 1971; Glikson and Lambert, 1973). Partial melting, depletion and migration of the elements into higher crustal levels are said to be the accompanying processes (Sheraton, 1970; Lambert, 1971; Tarney et al., 1972). The alternative prevailing view is that the present composition and the trace element levels of granulites are inherited from the igneous parent rocks and, as such, there is no 'depletion' of any elements whatsoever. Spooner and Fairbairn (1970) and Holland and Lambert (1975) suggested that calc-alkaline magmas were intruded directly into deeper levels of the continental crust and recrystallised as granulite and amphibolite facies gneisses. The findings of McGregor (1973) and Davies (1976) imply that amphibolite and granulite facies gneisses were derived from calc-alkaline tonalites containing low values of LIL elements. Thus, a comparative study of the geochemistry of Archaean and Proterozoic granulites is deemed worthwhile in order to provide a constraint on the above arguments.

The charnockites of North Arcot district, Tamil Nadu, are taken as an example of Archaean granulites. Their chemistry is compared with that of amphibolite facies gneisses and of two of the typical, but lithologically distinct, schist belts of South Karnataka - Kolar and Holenarasipur. The granulites of the Ford and North River areas of the Churchill-Nain boundary of the Canadian Shield are our Proterozoic example, investigated by the first author.

CHARNOCKITES OF NORTH ARCOT, TAMIL NADU

Charnockites constitute the predominant rock type in the North Arcot district where metasedimentary and meta-igneous rocks of acid, basic and ultramafic affinities occur as closely interbanded sequences. The complex tectonic history of this area was studied by the Geological Survey of India (Sugavanam et al., 1976). The Pb/Pb age of syngenetic sulphide mineralisation associated with the pyroxene granulites of this area is about 2600 my (Venkatasubramanian et al., 1977).

TABLE 1: Chemical composition of granulites, gneisses and granites of North Arcot District

	1	2	3	4	5	6	7	8
SiO_2	76.18	61.49	56.50	47.18	47.13	47.94	71.82	68.45
TiO_2	0.23	0.32	0.76	1.64	1.47	0.85	0.26	0.27
Al_2O_3	13.40	14.30	13.84	15.47	16.10	12.15	14.42	15.21
Fe_2O_3	0.34	1.24	1.93	1.21	1.92	2.20	0.59	0.10
FeO	1.61	6.08	10.40	11.98	15.70	15.78	3.04	2.71
MnO	0.04	0.04	0.28	0.06	0.34	0.13	0.07	0.03
MgO	1.02	3.21	5.55	8.28	6.21	11.18	1.43	0.89
CaO	2.82	6.09	5.88	12.25	9.23	8.27	2.73	4.34
Na_2O	3.52	5.82	3.83	1.52	1.42	0.93	3.09	3.40
K_2O	0.93	1.27	0.77	0.32	0.31	0.37	2.50	4.56
P_2O_5	0.09	0.16	0.12	0.09	0.17	0.16	0.07	0.03
Cr (ppm)	75	142	123	499	109	803	94	37
Ni	68	125	116	368	180	495	78	31
Rb	16	19	13	9	9	12	45	52
Sr	380	475	579	187	175	163	440	412
Ba	463	524	830	740	577	616	380	415
Zr	154	193	245	181	152	224	270	178
Y	4	-	13	7	-	6	8	-
Ce	42	71	82	41	-	54	59	28
La	-	-	30	29	-	-	24	-
Pb	12	7	14	19	13	27	23	18
Th	0.9	1.3	0.9	1.5	1.2	2.7	2.1	3.8
U	0.3	0.7	1.5	0.9	1.0	1.1	1.7	2.5

1. Hypersthene-biotite charnockite (av. of 6 samples)
2. Hypersthene-biotite-hornblende charnockite (av. of 4 samples)
3. Hypersthene-diopside-hornblende charnockite (av. of 4 samples)
4. Pyroxene granulite (av. of 3 samples)
5. Norite (av. of 2 samples)
6. Mafic enclaves (mostly amphibolites) in migmatites (av. of 4 samples)
7. Migmatitic charnockites (av. of 6 samples)
8. Granite-Gingee group (av. of 4 samples)

All analyses recalculated to H_2O- and CO_2-free basis.

The charnockitic rocks range from acid to basic in composition; they include gabbroic, noritic and pyroxene granulites, and high-grade gneisses and granites are constant associates.

The average compositions of the charnockites, pyroxene granulites, norites, mafic enclaves, migmatitic charnockites and the Gingee granites are given in Table 1. In the Alk-F-M diagram (Fig. 1) there is a bimodal compositional distribution, with the acid groups indicating a calc-alkaline trend while the basic members show an iron-rich tholeiitic trend. The scattering of points in this diagram indicates that the rocks cannot have a simple igneous history. The compositional variation may be largely original, but with some modification due to allochemical recrystallisation during tectonic activity. The triangular

Na_2O-CaO-K_2O plot (Fig. 1) depicts the calcic nature of these rocks. The increased K_2O content in the migmatitic charnockites and the associated granitic gneisses is evident. The tonalitic ($K_2O/Na_2O < 0.6$) and granodioritic ($K_2O/Na_2O < 1.0$) compositions of the charnockites grade into adamellitic ($K_2O/Na_2O = 1.0$-1.5) compositions of the Gingee group. The mafic-tholeiitic, felsic/calc-alkaline trends are also shown in the FeO/MgO vs SiO_2 and FeO/MgO vs FeO diagrams (Fig. 2). The basic members plot in the field of abyssal and island arc tholeiites, while the acid group plot in the calc-alkaline field of Miyashiro (1975). The bimodal compositional variation is also indicated by the charnockites of the type area near Madras, the compositions of which are also plotted in Figs. 1 and 2 (compilation of T.V. Viswanathan, 1969). The tholeiitic Fe-enriched fractionation pattern of the basic group does not suggest komatiitic chemistry, contrary to the report of S. Viswanathan (1974).

The bimodal compositional variation of these rocks may suggest a dual origin, but Yoder's (1973) experimental work demonstrated that rhyolitic and basaltic liquids can be formed contemporaneously from the partial melting of the same parent material as a result of adiabatic decompression.

GREENSTONE/HIGH-GRADE GNEISS COMPLEX OF SOUTH KARNATAKA

It is often argued that basic charnockites are derived from the mafic components of the greenstone-granite complex through granulite-grade metamorphism. A comparative study of their geochemistry may yield better insight into the origin of charnockites.

A complicated suite of orthogneisses and granitic rocks (Peninsular Gneiss) occupies the southern parts of Karnataka (south of 13°N). They carry abundant vestiges of supracrustal rocks metamorphosed in the amphibolite and, locally, in the granulite grade. The supracrustals have been termed the Sargur high-grade schists (Viswanatha and Ramakrishnan, 1975). There are several other schist belts, mainly composed of mafic rocks, but with ultramafites, gabbros and anorthosites as well as metasediments. These schist belts are surrounded by tonalitic, granodioritic and granitic rocks. The available Rb/Sr isochron data for the high-grade gneisses range from 3000-2600 my (Crawford, 1969; Venkatasubramanian, 1974). Two typical greenstone belts (Kolar and Holenarasipur) are chosen for comparison with the chemistry of the basic granulites.

The mafic rocks of the Kolar schist belt (Table 2) show an iron-rich tholeiite trend (Fig. 3), with FeO/FeO + MgO varying from 0.5-0.83. They have low TiO_2, and high Cr and Ni, the Na/K ratio ranges from 4.7-11.3, K/Rb from 287-332, and CaO/Al_2O_3 from 0.62-0.98. The Kolar mafic rocks are low-alumina tholeiites, as indicated by the Al_2O_3/SiO_2 ratio which is less than 0.30. The associated granodiorites, granites and gneisses show a calc-alkaline trend. On the FeO/MgO vs FeO or SiO_2 diagrams (Fig. 4) the mafic rocks mostly plot in the field of

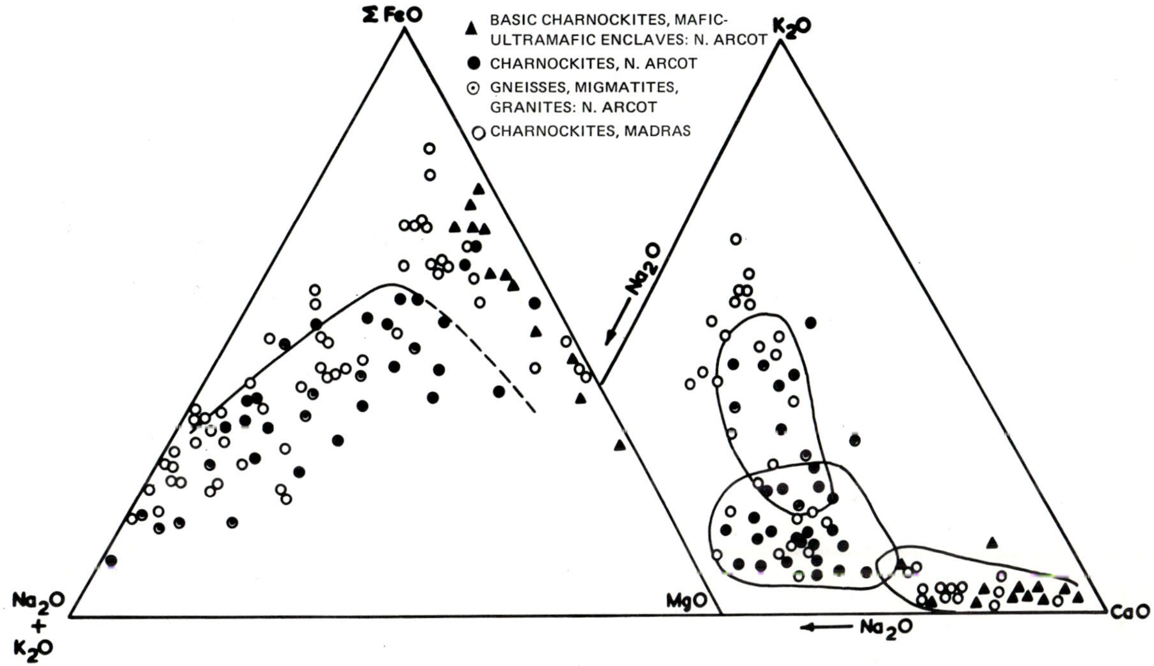

Fig. 1. Alk-F-M and CaO-Na_2O-K_2O (wt %) diagrams for charnockites of North Arcot district and Madras. The line separating the calc-alkaline and the tholeiitic field is after Irvine and Baragar (1971). Separate fields shown in CaO-Na_2O-K_2O diagram for basic and acid groups.

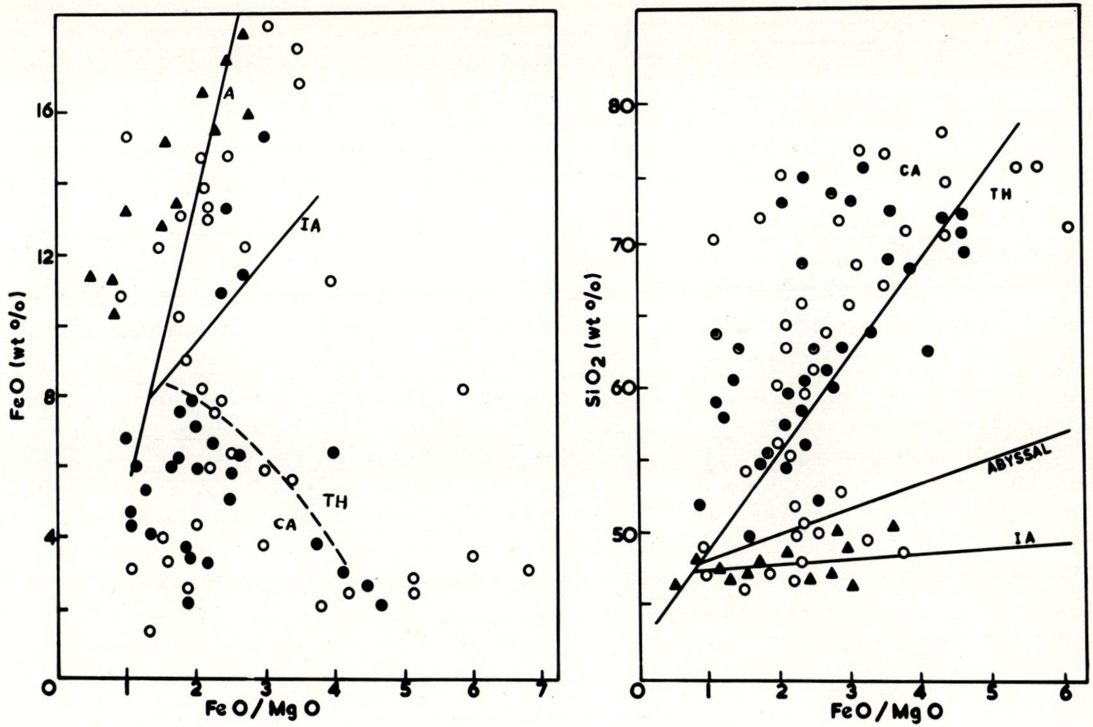

Fig. 2. FeO/MgO vs FeO and SiO_2 diagrams for the charnockites of North Arcot and Madras. The fields corresponding to abyssal, island-arc (IA), continental tholeiites (TH) and calc-alkaline basalts (CA) are indicated (Miyashiro, 1975). Symbols as in Fig. 1.

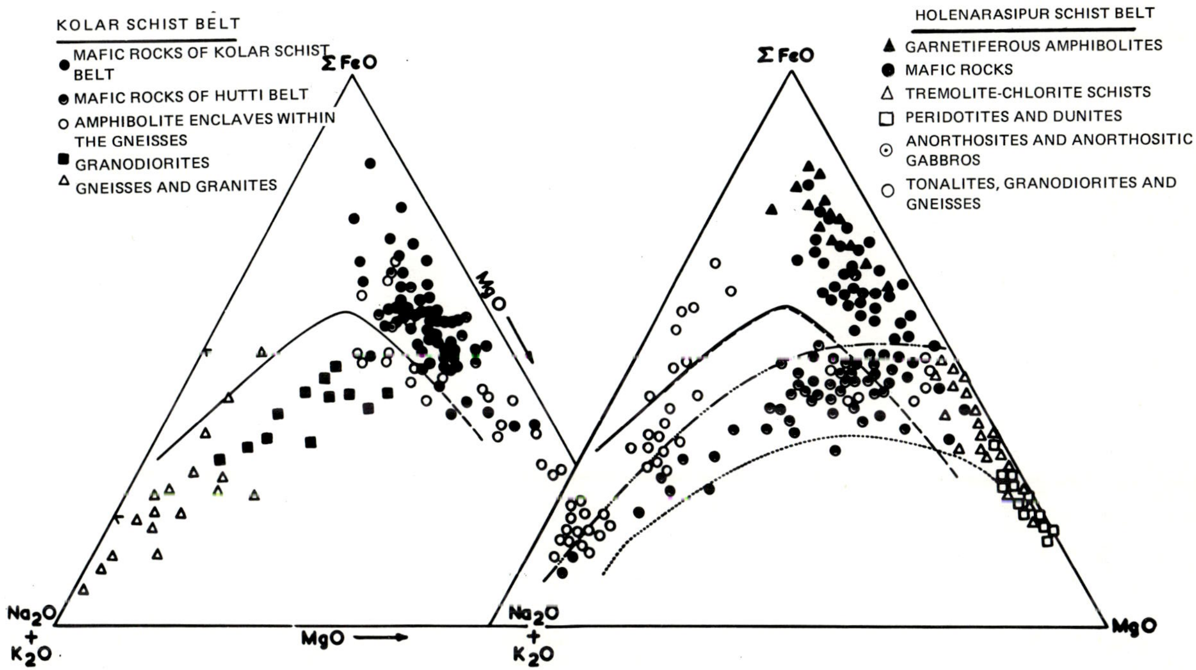

Fig. 3. Alk-F-M diagrams (wt %) for the rocks of Kolar and Holenarasipur schist belts. Dotted lines indicate the compositional limits for the anorthosite suite from Holenarasipur schist belt.

TABLE 2: Chemical composition of mafic rocks of Kolar, high-grade gneisses and mafic enclaves

	1	2	3	4	5	6	7	8
SiO_2	50.67	51.05	52.31	50.83	65.50	64.96	55.51	50.57
TiO_2	0.68	0.61	1.02	0.81	0.31	0.48	0.38	0.65
Al_2O_3	14.96	13.53	12.59	13.24	18.86	16.30	6.94	13.20
Fe_2O_3	1.31	0.97	1.17	0.83	0.31	0.80	1.60	1.75
FeO	12.02	10.96	16.80	11.19	3.32	4.57	7.08	10.13
MnO	0.21	0.14	0.03	0.17	0.02	0.09	0.12	0.21
MgO	6.53	11.44	4.23	8.02	1.40	2.78	15.59	6.45
CaO	10.36	8.39	9.10	13.00	2.65	5.32	11.47	11.48
Na_2O	2.50	3.33	1.86	1.59	4.07	2.72	0.79	4.69
K_2O	0.45	0.27	0.40	0.19	3.36	1.89	0.40	0.45
P_2O_5	0.32	0.21	0.50	0.13	0.12	0.09	0.11	0.42
Cr (ppm)	195	214	232	349	37	58	552	247
Ni	118	123	175	197	16	42	304	198
Rb	17	7.8	10	5	119	132	19	17
Sr	190	125	98	104	285	297	105	96
Ba	85	61	47	49	307	324	122	130
Zr	74	50	60	65	102	98	80	144
Th	5.5	4.0	5.1	4.5	13	7	3.2	2.8
U	2.4	1.2	1.5	1.3	21	14	2.5	1.7

1. Amphibolite, west of Golconda shaft, K.G.F. (av. of 4)
2. Amphibolite, east of Golconda shaft, K.G.F. (av. of 5)
3. Amphibolite, 1850 level, Nandidurg Mine (av. of 4)
4. Amphibolite, 4800 level, Nandidurg Mine (av. of 6)
5. Gneiss, south of Bangalore (av. of 3)
6. Granodiorite, southwest of Kolar schist belt (av. of 7)
7. Mafic enclave within gneisses, southwest of Kolar schist belt (av. of 2)
8. Mafic enclave within gneisses, southwest of Bangalore (av. of 3)

abyssal tholeiites. The ultramafic-gabbro-anorthosite suite of rocks from the Holenarasipur belt follows an Fe-deficient calc-alkaline trend, while the metavolcanics, represented by the amphibolites, show an Fe-rich tholeiitic trend. Tonalites, granodiorites and gneisses surrounding this schist belt follow both the tholeiitic and calc-alkaline trends; seemingly they could be differentiates of both the above series.

The mafic enclaves within the high-grade gneisses of South Karnataka also show a bimodal compositional distribution with FeO/FeO + MgO ranging from 0.23-0.75. The Mg-rich enclaves have lower Cr (700-1200 ppm) and Ni (500-1000 ppm) contents than the ultramafites of Holenarasipur. A few of these enclaves, coexisting with metasedimentary rocks, show modified spinifex texture and may correspond to the high-Mg basalts of the supracrustal sequence. In the K_2O-Na_2O-CaO diagram, the different quartzo-feldspathic components of the high-grade gneisses of South Karnataka plot in the same field as the charnockites, with the average analysis corresponding to granodiorite though tonalite and granite components are not uncommon.

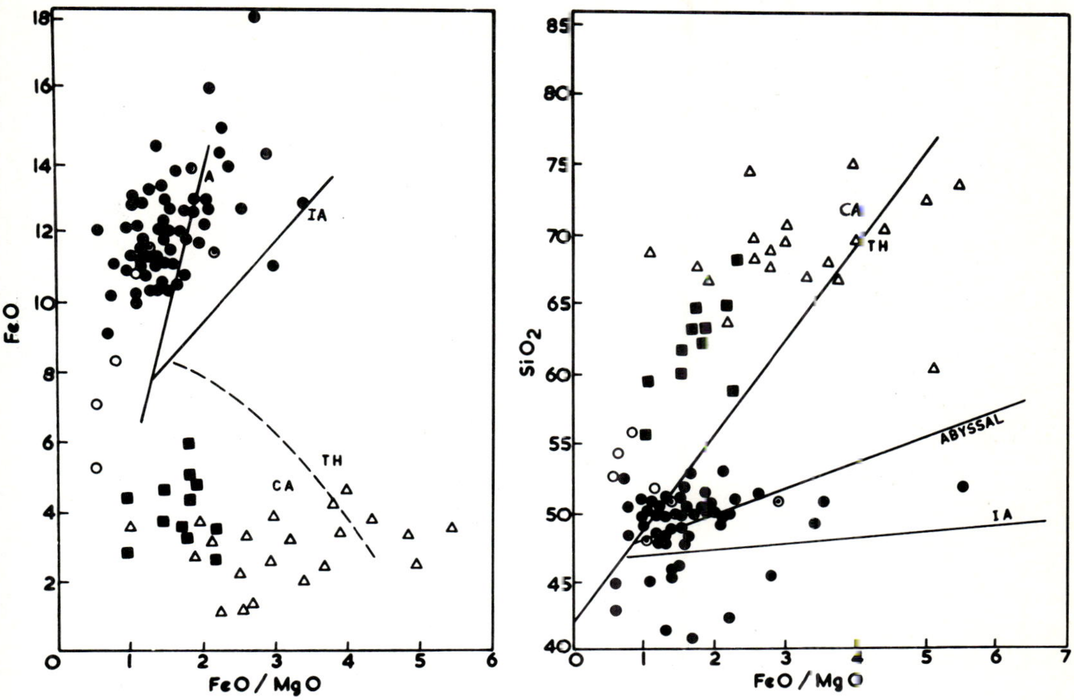

Fig. 4. FeO/MgO vs FeO and SiO_2 diagrams for the mafic rocks of Kolar schist belt and associated granitoids. Symbols as in Fig. 3.

TRACE ELEMENT GEOCHEMISTRY

Distinct differences in trace element geochemistry are observed between the charnockites of North Arcot and the high-grade gneiss complex of South Karnataka. The petrographically identified charnockites of North Arcot are generally low in potassium, although many of the reported analyses of the charnockitic rocks of

Madras indicate higher potassium values, similar in magnitude to the gneissic charnockites of North Arcot. This wide variation in potassium values possibly indicates the migmatitic character of the charnockites. The other LIL elements, i.e. Rb (9-20 ppm), Th (0.2-1.5 ppm), U (0.1-1.0 ppm) and Pb (5-25 ppm), are also low in the charnockites of North Arcot. The gneissic charnockites and quartz monzonite of the Gingee group have higher Rb (30-60 ppm) and other LIL element contents. On the other hand, other normally incompatible elements such as Ba (350-850 ppm), Sr (160-600 ppm) and Zr (80-250 ppm) show enrichment. In the Ti-Zr diagram (Fig. 5a) the charnockite suite falls out of the delineated fields for low-K tholeiites, ocean floor and calc-alkaline basalts (Pearce and

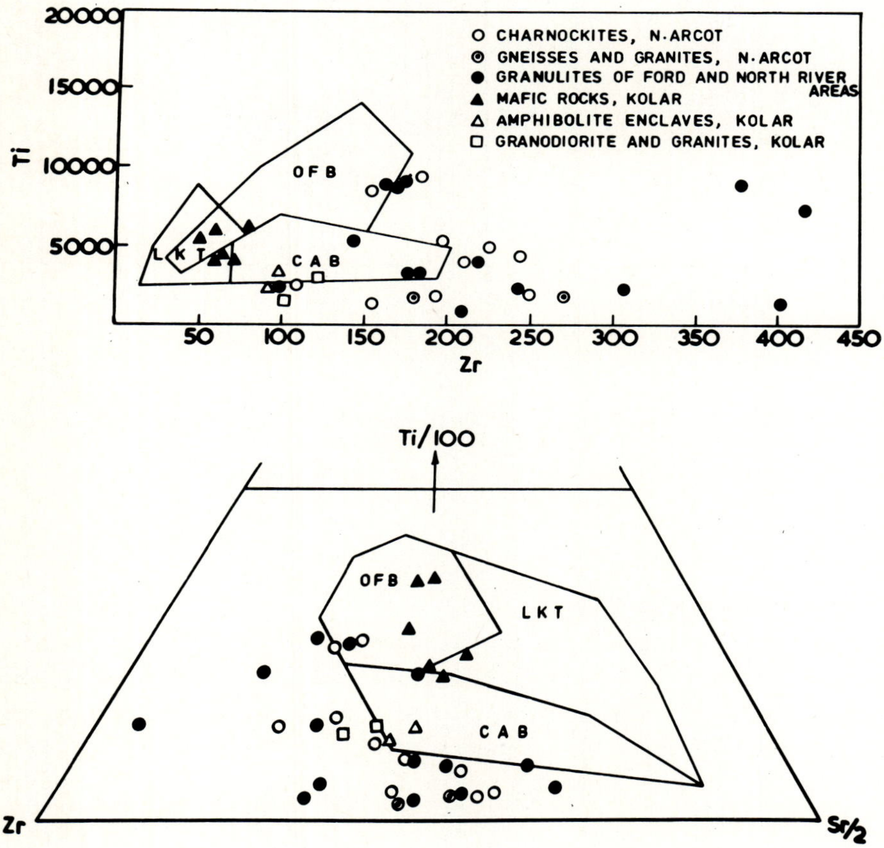

Fig. 5. Discrimination diagrams according to Pearce and Cann (1973). LKT = low potassium tholeiites, OFB = ocean floor basalts, CAB = calc-alkaline basalts.

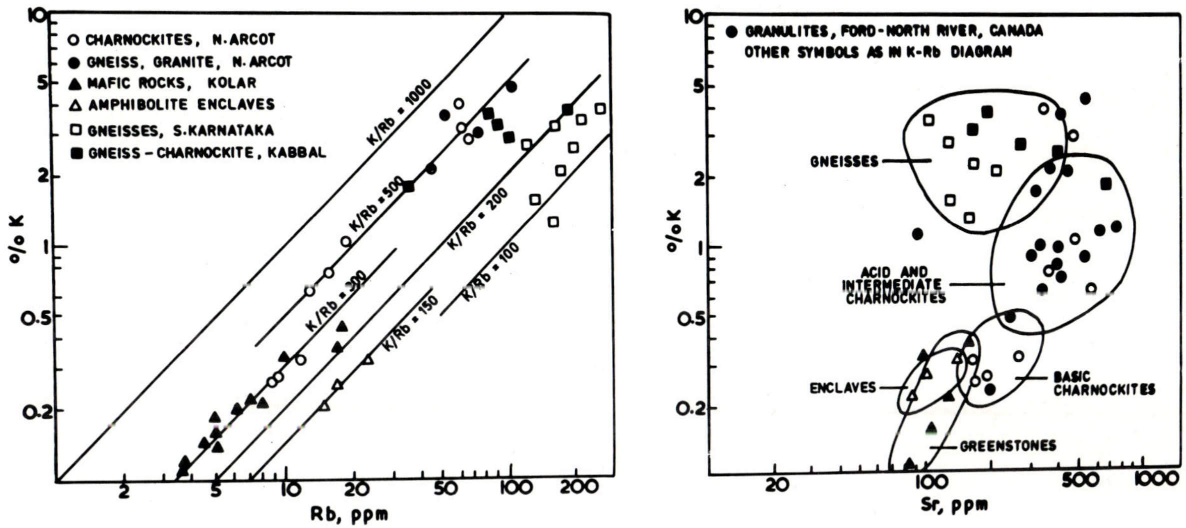

Fig. 6. K/Rb and K/Sr relations in charnockites, high-grade gneisses and mafic rocks.

Cann, 1973. Similar behaviour is observed in the Ti-Sr-Zr triangular diagram (Fig. 5b). Lighter rare earth elements represented by La and Ce show higher concentrations than heavier REE. In contrast, amphibolites from the Kolar schist belt have low K (0.12-0.45% K_2O) and Rb (3.4-17 ppm), while U (1.2-2.4 ppm) and Th (3-5 ppm) are higher than those of the basic charnockites. The granitoids of the Kolar area have high K (1.9-3.5% K_2O), Rb (139-209 ppm), U (13-31 ppm) and Th (7-13 ppm). Rb/Sr ratio is about 0.04 for the charnockites of North Arcot, while for the Kolar gneisses it is about 0.40. Higher LIL element contents are noted for the high-grade gneisses coexisting with charnockites of the Kabbal area, SE of Channapatna, where transformation of gneiss to charnockite has been reported by Pichamuthu (1960). Coexisting granites, gneisses and charnockites of Kabbal yield a Rb/Sr isochron age of 2670 ± 60 my (Venkatasubramanian and Jayaram, 1976). Remarkable differences are observed in the K/Rb ratios (Fig. 6a) which are around 500, comparable with those of acid and intermediate charnockites of North Arcot. Basic members of North Arcot plot close to 300. The high-grade gneisses of Karnataka have low K/Rb ratios around 150 and the mafic rocks of the Kolar schist belt have low-K and Rb concentrations, with K/Rb around 300. Sheraton et al. (1972), Heier (1973) and Tarney (1976) have pointed out that the retrogressed granulites often retain their high K/Rb ratios. Using this criterion, it is evident that the high-grade gneisses of South Karnataka were never in the granulite grade and are not retrogressed charnockites. On the other hand, the K/Rb ratios in the Kabbal area show a systematic increase in going from gneisses (K/Rb = 200), through greasy gneisses (K/Rb = 300) to charnockites (K/Rb = 460-500). Had all these units been derived from the retrogression of charnockites, all of them should show uniformly higher K/Rb ratios. The lower K/Rb ratios of the mafic enclaves (around 150) compared with those of Kolar amphibolites indicate the genetic differences between these two groups and reflects the more continental nature of the mafic enclaves.

Corresponding lower K/Sr and K/Ba ratios are evident in the charnockites compared with the amphibolite-grade gneisses, with mafic rocks as well as charnockites falling in different areas of the K/Sr diagram (Fig. 6b). Ba/Sr ratios for the charnockites are around 1-1.5 for acid to intermediate charnockites of North Arcot, whereas the basic members have much higher Ba/Sr ratios, ranging to 4.0. Oceanic basalts and island arc tholeiites and andesites have low Ba/Sr values and comparatively low Ba content. The rock suites with higher Ba/Sr ratios are found in the calc-alkaline complexes of Andean-type continental margins (Jakes and White, 1972; Tarney, 1976). Similar differences are noticed between the Ba/Sr ratios of Kolar amphibolites (0.4-0.6) and the mafic enclaves (0.8-1.3); this indicates that the basic charnockites and mafic enclaves within the high-grade gneisses, though tholeiitic have continental margin affinities.

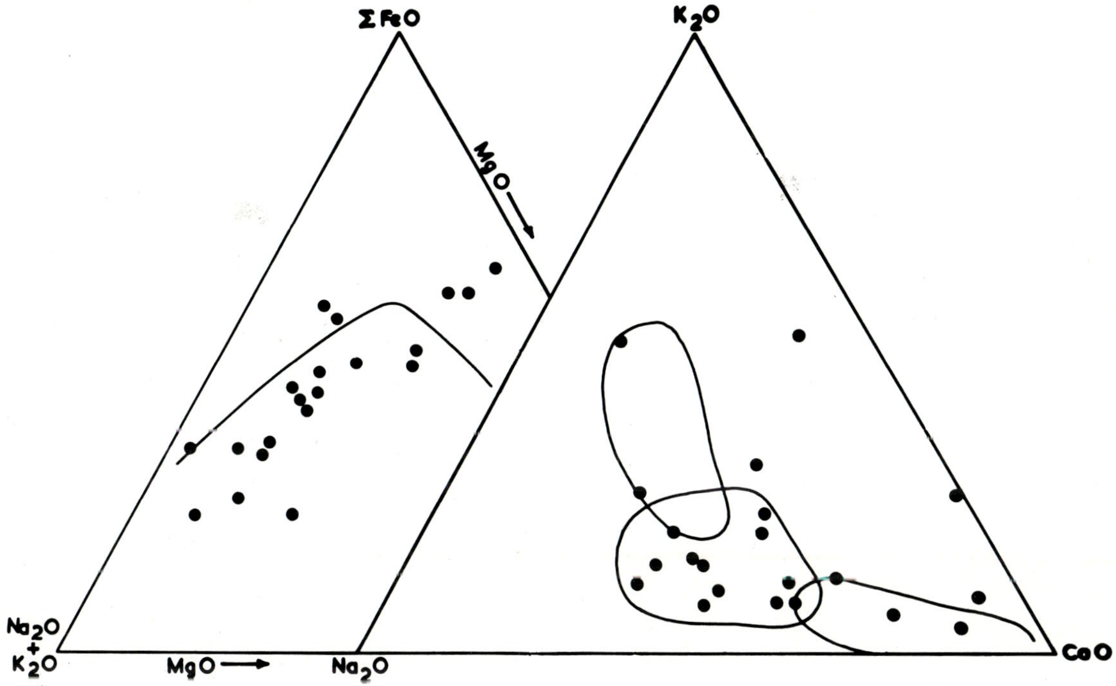

Fig. 7. Alk-F-M and CaO-Na_2O-K_2O (wt %) diagrams for the granulites of Ford and North River areas (Canada). Fields delineated in Ca-Na_2O-K_2O diagram as in Fig. 1.

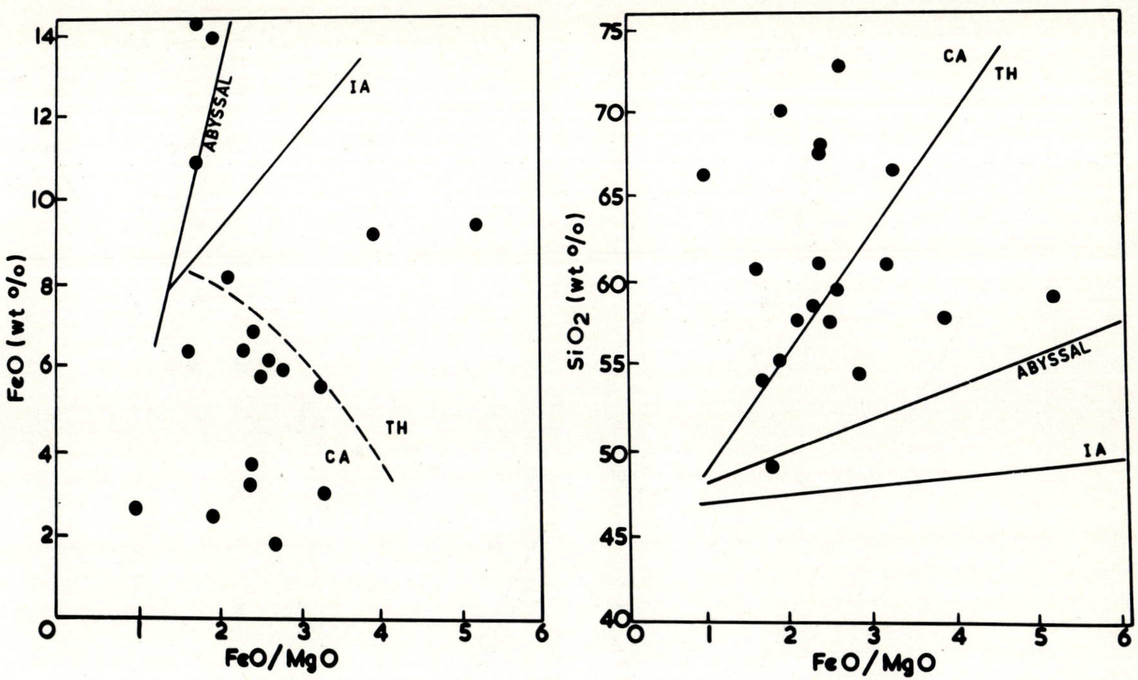

Fig. 8. FeO/MgO vs FeO and SiO_2 (wt %) diagrams for the granulites of the Ford and North River areas (Canada)

TABLE 3: Chemical composition of Proterozoic granulites (Ford and North River area)

	1	2	3	4	5	6	7	8
SiO_2	49.35	53.87	61.20	67.86	59.53	60.28	72.80	61.73
TiO_2	1.49	1.28	0.57	0.40	1.46	1.33	0.26	1.48
Al_2O_3	14.34	22.05	16.42	17.55	16.06	16.93	15.40	16.45
Fe_2O_3	1.71	0.31	1.53	0.10	1.10	1.31	0.80	0.10
FeO	13.44	5.93	5.61	2.96	8.38	4.94	1.10	6.40
MnO	0.17	0.30	0.11	0.05	0.16	0.04	0.02	0.08
MgO	8.12	2.18	2.96	0.92	1.79	2.32	0.70	4.16
CaO	8.02	7.90	6.53	4.59	5.48	6.05	3.20	5.99
Na_2O	2.21	4.47	3.98	4.59	3.59	3.83	4.40	2.34
K_2O	0.60	1.04	0.92	0.82	2.09	2.72	1.20	1.12
CO_2	0.22	0.00	0.08	0.03	0.01	0.00	0.01	0.01
P_2O_5	0.22	0.66	0.09	0.12	0.35	0.25	0.11	0.14
Cr (ppm)	230	-	-	-	-	-	-	-
Ni	134	-	-	-	-	-	-	-
Zr	377	419	238	307	169	786	405	164
Ba	331	628	479	250	868	1173	270	683
Sr	236	500	559	332	549	383	402	302
Ce	131	-	-	-	-	-	-	-

1. Hornblende-hypersthene granulite
2. Biotite-hypersthene granulite
3. Two-pyroxene granulite
4. Biotite-hypersthene-garnet granulite
5. Garnet-hornblende-hypersthene granulite
6. Hornblende-hypersthene granulite
7. Biotite-hypersthene granulite
8. Biotite-hypersthene granulite

GEOCHEMISTRY OF PROTEROZOIC GRANULITES

Eastern Labrador is part of the disrupted portion of the North Atlantic craton. The Archaean block is bounded by linear mobile zones which were active during the Proterozoic. Rocks of the Churchill province west of the Ford river area are metamorphosed in the granulite facies, with whole-rock isotope ages ranging from 1175-1680 my (Wanless, 1969). Only one period of folding and metamorphism has been recognised in this area and according to Taylor (1971) they are truly Proterozoic and not reworked Archaean gneisses.

Relevant geochemical data for the granulites of this area are presented in Table 3. In the Alk-F-M diagram they show a calc-alkaline trend (Fig. 7), though in the plots of FeO/MgO vs FeO or SiO_2 (Fig. 8) they fall both in the calc-alkaline and tholeiitic fields. The basic granulites have fairly high K_2O (> 0.6%), Ba (250-628 ppm), Sr (236-559 ppm) and Zr (238-419 ppm). They have trace element characteristics comparable to the charnockites of North Arcot, as shown in Ti-Zr, Ti-Zr-Sr (Fig. 5) and K-Sr diagrams (Fig. 6). The geochemistry of the granulites of the North River area of Nain compares with those of the adjacent Ford River area of the Churchill province (Table 3). If the distinctive geochemistry of granulites is attributed only to the chemistry of the parent rocks, one might expect marked differences in the geochemistry of Archaean and Proterozoic granulites. Since they exhibit similar geochemical trends, their distinct geochemistry can only be attributed to the nature of the granulite-grade metamorphism.

THE EVOLUTION OF A CHARNOCKITE TERRAIN

Apparently uniform igneous trends are observed in the Niggli variation diagrams for the charnockites, as in mg vs c, c vs (al-alk) and 100 mg-c-(al-alk) plots. However, the metamorphic nature of these rocks is evident from the scatter of points in the Alk-F-M diagram and FeO/MgO vs FeO or SiO_2 diagrams as well as from field and texture evidence. Particularly in the K_2O-SiO_2 diagram (Fig. 9), charnockites lack positive correlation between these two constituents, which is unusual for calc-alkaline igneous rocks. The Niggli al-fm-c diagram also shows

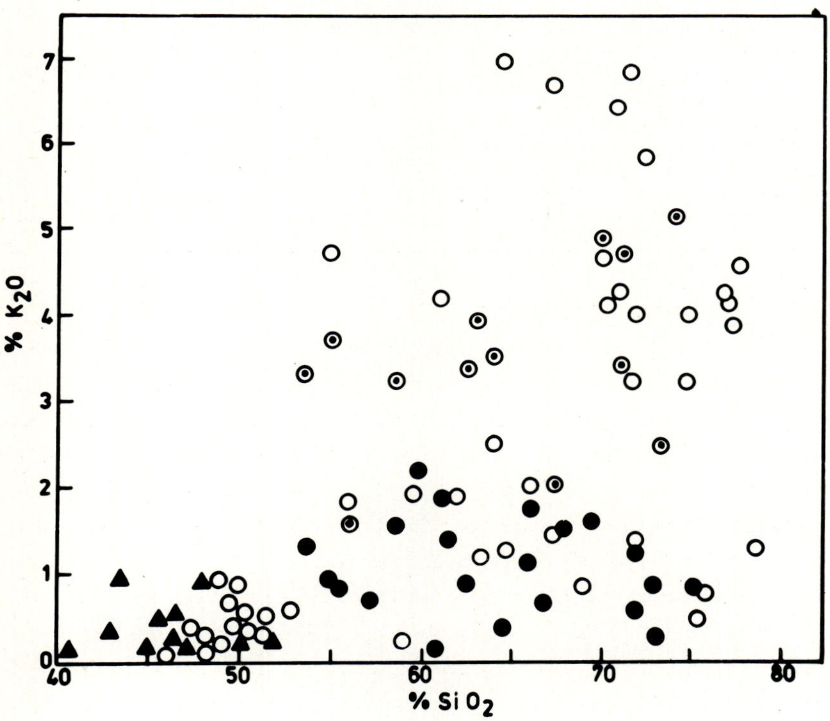

Fig. 9. SiO_2-K_2O (wt %) diagram for charnockites and associated rocks of North Arcot and Madras. Symbols as in Fig. 1.

an irregular pattern for the charnockites, which has been used by Sen (1974) to illustrate their migmatitic nature. Charnockites thus represent a hybrid suite, the major part of which is a meta-igneous quartzo-feldspathic rock of tonalite-granodiorite composition, compositionally modified through contamination from the associated supracrustals during the granulite grade metamorphism, and accompanied by progressive dehydration and partial separation of a granitophile melt.

Mineralogical investigations of the charnockites and associated rocks of North Arcot have demonstrated that orthopyroxene coexists with high-grade hornblende and biotite (Iyer et al., 1976). It has been shown that the hornblende and biotite within the charnockite form equilibrium assemblages along with the pyroxenes and that their interconversion is a two-way process, generally called sliding equilibrium. Moreover, the presence of opx-bearing and opx-free metapelites containing cordierite, garnet, biotite and sillimanite are unmistakable indicators of prograde metamorphism. It is obvious from experimental petrology that granulite-grade metamorphism takes place under anatectic conditions, which marks the broad boundary between magmatic and metamorphic processes. The amphibolite-to-granulite facies boundary is controlled by $P_{load}/P_{load} + P_{fluid}$ ratios which may approach unity but are not equal to unity. The fluid phase composition can be variable and may not represent the condition of $P_{fluid} = P_{H_2O}$. The presence of crystalline limestones and co-metamorphosed calcsilicates, of hornblende and biotites containing chlorides and fluorides in charnockites, and of graphite in the associated metapelites indicates the presence of CO_2, HCl, HF and even hydrocarbons in the fluid phase. Open system conditions during granulite-grade metamorphism facilitates the removal of granitophile melts and migration of lithophile elements along with the fluids, causing the depletion of selected incompatible elements of larger ionic radii. The geochemical differences between the amphibolite facies gneisses of southern Karnataka and the associated charnockites support the above arguments.

The general geochemistry of the charnockites does not lend support to the concept of continental growth during the Archaean by lateral accretion of oceanic island arc or calc-alkaline volcanics, which implies uniform depletion of all lithophiles from the lower to upper crust. However, it would appear from the trace element studies that the amphibolite enclaves within the high-grade gneisses and the basic granulites within the charnockites have continental rather than oceanic affinities, and are different geochemically from the mafic rocks of the Kolar schist belt. A natural corollary is that charnockites may be derived by the granulite facies metamorphism of the high-grade gneiss complex with its mafic enclaves and overall granodiorite composition. The geothermal gradient necessary for the granulite facies metamorphism could be facilitated by horizontal tectonic movements rather than by down-sagging, and hence the rocks need not be taken to great depths in the earth's crust to attain granulite facies conditions. The structural discordance between the charnockite and high-grade gneiss may be of later origin, possibly during the reactivation of the granulite terrain during a Proterozoic orogeny.

CONCLUSIONS

1. Geochemical studies reveal that the amphibolite facies gneisses of South Karnataka are not retrogressed charnockites but that, conversely, the charnockites could be derived by prograde metamorphism of the gneisses.

2. Both the charnockite and high-grade gneiss terrains incorporate identical lithological units, including shelf-type metasediments; greywackes and turbidites are conspicuously absent from both.

3. The major element compositions of both intermediate and acid charnockites compare with the quartzofeldspathic components of the high-grade gneisses. The relation between charnockites and the high-grade gneiss does not support the origin of granulites from an igneous parent magma that contained low levels of LIL trace elements.

4. There is no apparent chemical distinction between Proterozoic and Archaean granulites.

5. Geochemically the basic charnockites compare better with the mafic enclaves of the high-grade gneisses than with the mafic components of the schist belts. They do not have true komatiite chemistry. Both basic charnockites and mafic enclaves have the characteristics of continental tholeiites. Mafic rocks of the schist belts have primitive oceanic basalt character. Indeed, metamorphosed greenstone belts are absent in the charnockite terrains.

ACKNOWLEDGEMENT

The authors thank Professor A.R. Vasudeva Murthy for his keen interest in this work. Thanks are due to U.N.D.P. for the award of a Fellowship to the first author to work on the high-grade rocks in the Geological Survey of Canada, Ottawa, the Department of Geology, Syracuse University, U.S.A., the University of Glasgow, Scotland, and the Mineralogisk-Geologisk Museum in Oslo, Norway. The first author thanks Dr F.C. Taylor of the Geological Survey of Canada for discussions on the granulites of Churchill and Nain provinces of the Canadian shield.

REFERENCES

Crawford, A.R., 1969. Reconnaissance Rb-Sr dating of the Precambrian rocks of southern Peninsular India. J. geol. Soc. India, 10: 117-166.

Davies, F.B., 1976. Early Scourian structures in the Scourie-Laxford region and their bearing on the evaluation of the Laxford Front. J. geol. Soc. Lond., 132: 543-554.

Devaraju, T.C. and Sadashivaiah, M.S., 1969. The charnockites of the Satanur-Halaguru area, Mysore State. Indian Mineral., 10: 67-88.

Glikson, A.Y. and Lambert, I.B., 1973. Relations in space and time between major Precambrian shield units. Earth Planet. Sci. Lett., 20: 395-403.

Heier, K.S., 1973. Geochemistry of granulite facies rocks and problems of their origin. Phil. Trans. R. Soc. Lond., A273: 429-442.

Holland, T.H., 1900. The charnockite series, a group of Archaean hypersthenic rocks in Peninsular India. Mem. geol. Surv. India, 28: 119-249.

Holland, J.G. and Lambert, R.St.J., 1973. Comparative major element geochemistry of Lewisian of the mainland of Scotland. In: R.G. Park and J. Tarney (Editors), The Early Precambrian of Scotland and Related Rocks of Greenland. Univ. Keele, pp. 51-62.

Irvine, T.N. and Baragar, W.R.A., 1971. A guide to the chemical classification of the common volcanic rocks of orogenic areas. Bull. geol. Soc. Am., 83: 29-40.

Iyer Anantha, G.V., Narayanan Kutty, T.R., Vasudeva Murthy, A.R., Sugavanam, E.B. and Venkata Rao, V., 1976. Mineral chemistry investigations of charnockites and associated rocks of North Arcot District, Tamil Nadu, South India. J. geol. Soc. India, 17, p.162 (Abstr.).

Jakes, P. and White, A.J.R., 1972. Major and trace element abundances in volcanic rocks of orogenic areas. Bull. geol. Soc. Am., 83: 29-40.

Lambert, I.B., 1971. The composition and evolution of deep continental crust. In: J.E. Glover (Editor), Symposium on Archaean Rocks. Geol. Soc. Austr. Sp. Publ., 3: 419-428.

Lambert, I.B. and Heier, K.S., 1968. Geochemical investigations of deep-seated rocks in the Australian shield. Lithos, 1: 30-53.

McGregor, V.R., 1973. The early Precambrian gneisses of the Godthaab District, West Greenland. Phil. Trans. R. Soc. Lond., A273: 343-358.

Miyashiro, A., 1975. Classification, characteristics and origin of ophiolites. J. Geol., 83: 249-281.

Narayana Swami, S., 1966. Tectonic problems of the Precambrian rocks of Peninsular India. Symposium on Tectonics, Nat. geophys. Res. Inst., Hyderabad, pp. 77-94.

Nautiyal, S.P., 1966. Precambrians of Mysore Plateau. Presid. Address, Indian Sci. Congr. 53rd Sess., Sect. 3, pp. 1-14.

Pearce, J.A. and Cann, J.R., 1973. Tectonic setting of basic volcanic rocks determined using trace element analyses. Earth Planet. Sci. Lett., 19: 290-300.

Pichamuthu, C.S., 1953. The charnockite problem. Mysore Geol. Assoc. Sp. Publ., 2.

Pichamuthu, C.S., 1960. Charnockite in the making. Nature, 188: 135-136.

Pichamuthu, C.S., 1965. Regional metamorphism and charnockitisation in Mysore State, India. Ind. Mineral., 6: 119-126.

Rama Rao, B., 1962. Handbook of the Geology of Mysore State. Bangalore Publ. Co., 190 pp.

Ramberg, H., 1951. Remarks on the average chemical composition of granulite facies and amphibolite-to-epidote amphibolite facies gneisses in West Greenland. Meddr. dansk geol. Foren., 12: 27-34.

Ray, S., 1972. Charnockite of Kabbal, Mysore - a brief study. Quart. J. Geol. Min. Met. Soc. India, 44: 163-166.

Sen, S.K., 1974. A review of some geochemical characters of the type area charnockites. J. geol. Soc. India, 15: 413-420.

Sheraton, J.W., 1970. The origin of the Lewisian gneisses of North West Scotland with particular reference to the Drumbeg area, Sutherland. Earth Planet. Sci. Lett., 8: 301-310.

Sheraton, J.W., Skinner, A.C. and Tarney, J., 1973. The geochemistry of the Scourian gneisses of the Assynt District. In: R.G. Park and J. Tarney (Editors), The Early Precambrian of Scotland and Related Rocks of Greenland. Univ. Keele, pp. 13-30.

Smeeth, W.F., 1916. Outline of the geological history of Mysore. Bull. Dept. Mines Geol. Mysore State, 6: 15-16.

Spooner, C.M. and Fairbirn, H.W., 1970. Strontium 87/strontium 86 initial ratios in pyroxene granulite terrains. J. geophys. Res., 75: 6706-6713.

Subramaniam, A.P., 1967. Charnockites and granulites of India: a review. Meddr. dansk geol. Foren., 17: 473-493.

Sugavanam, E.B., Venkata Rao, V., Anantha Iyer, G.V. and Narayanan Kutty, T.R., 1976. Geology and whole-rock chemistry of granulites and associated rocks of North Arcot District, Tamil Nadu. J. geol. Soc. India, 17, p. 160 (Abstr.).

Tarney, J., 1976. Geochemistry of Archaean high-grade gneisses, with implications as to the origin and evolution of the Precambrian crust. In: B.F. Windley (Editor), The Early History of the Earth. Wiley, Lond., pp. 405-417.

Tarney, J., Skinner, A.C. and Sheraton, J.W., 1972. A geochemical comparison of major Archaean gneiss units from North-West Scotland and East Greenland. Rep. 24th Int. geol. Congr., Montreal, 1: 162-174.

Taylor, F.C., 1971. A revision of Precambrian structural provinces in north-eastern Quebec and northern Labrador. Can. J. Earth Sci., 8: 579-584.

Venkatasubramanian, V.S., 1974. Geochronology of the Dharwar craton - a review. J. geol. Soc. India, 15: 463-468.

Venkatasubramanian, V.S. and Jayaram, S., 1976. Geochemical trends of high-grade metamorphism as reflected in the charnockites of Karnataka State. J. geol. Soc. India, 17, p. 150 (Abstr.).

Venkatasubramianian, V.S., Radhakrishna, B.P. and Jayaram, S., 1977. Model galena ages from Karnataka and surrounding areas. J. geol. Soc. India, 18: 49-52.

Viswanathan, M.N. and Ramakrishnan, M., 1975. The Pre-Dharwar supracrustal rocks of the Sargur schist complex in southern Karnataka and their tectono-metamorphic significance. Indian Mineral., 16: 48-65.

Viswanathan, S., 1974. Contemporary trends in geochemical studies of the early Precambrian greenstone-granite complexes. J. geol. Soc. India, 15: 347-379.

Viswanathan, T.V., 1969. The granulitic rocks of the Indian Precambrian shield. Mem. geol. Surv. India, 100: 37-66.

Wanless, R.K., 1969. Isotopic age map of Canada. Geol. Surv. Can. Map 8-1970.

Yoder, H.S., Jr., 1973. Contemporaneous basaltic and rhyolitic magmas. Am. Mineral., 58: 153-171.

Ziauddin, M. and Yadav, P.K., 1975. Acid charnockite (metasomatic) near Sivasamudram, Karnataka State. J. geol. Soc. India, 16: 215-219.

SIGNIFICANCE OF OXYGEN ISOTOPE STUDIES ON GRANULITE FACIES ROCKS

Allan F. Wilson and Ajoy K. Baksi*
Department of Geology and Mineralogy, University of Queensland, St. Lucia (Brisbane), Queensland, Australia. 4067

Abstract

Oxygen isotope geothermometry, which has been used to estimate the temperature of metamorphism of low-grade and hydrous metamorphic rocks, may also be used for rocks of granulite metamorphic facies, provided they contain suitable pairs of minerals which still retain the oxygen isotopic fractionation developed at the time of initial metamorphism. Coarse-grained orthopyroxene and quartz appear to be very suitable and yield "reasonable" temperatures of about 700°C. Magnetite and ilmenite, however, are generally unsuitable for they show oxygen isotope exchange with intergranular fluids during the slow cooling of the granulites. There is also evidence that under some conditions oxygen isotopic temperatures may be more meaningful than temperatures deduced from $K_{D.Fe-Mg}$ in coexisting pyroxenes. Whole-rock oxygen values of some large bodies of mafic granulites show little change during granulite facies metamorphism. However, some narrow mafic layers set within a thick silicic metasedimentary sequence appear to have become equilibrated with the pore fluids of the sedimentary rocks. In one extensive metasedimentary terrane in central Australia abnormally light oxygen values of $\delta^{18}O \sim 3\%$ for some quartzose granulites, and $\delta^{18}O \sim 2\%$ for some interlayered mafic granulites, are common and may reflect climatic or depositional conditions of the Precambrian sedimentation.

INTRODUCTION

For many years the partitioning of cations between coexisting minerals has provided an important basis for estimation of temperature, oxygen fugacity and other parameters of igneous activity and metamorphism. Anomalous results, however, have commonly been reported, and are normally interpreted as disequilibrium phenomena.

In recent years the partitioning of the oxygen isotopes (^{18}O and ^{16}O) in coexisting minerals has been used successfully for geothermometry

* Present address: National Geophysical Research Institute, Hyderabad - 500007, India.

(Taylor and Epstein, 1962a, 1962b; Taylor, 1968; Clayton et al., 1972; and others). Anderson et al. (1971) have shown, however, that oxygen isotopes, like the cations, commonly reveal disequilibrium. Slow cooling of large igneous bodies such as the Bushveld Complex, granite batholiths, etc., has resulted in many mineral phases of an intrusion participating in oxygen isotope exchange for a long time after the igneous body has essentially solidified. Thus, partition of oxygen isotopes may indicate temperatures some hundreds of degrees below the temperature of initial formation of crystalline phases in an igneous body (Anderson, 1966).

Because of the problems of late or post-magmatic isotope exchange, most attention has been given to the study of the minerals in rapidly quenched igneous rocks such as basalts (Onuma et al., 1970), or in low-grade metamorphic rocks (Garlick and Epstein, 1967). In low-grade metamorphic rocks P_{H_2O} is normally equal to P_{load}, and equilibrium assemblages are commonly registered by oxygen isotope and cation partition.

Garlick and Epstein (1967) were among the first to show that the minerals of high-grade metamorphic rocks were difficult to use for geothermometry. Thus, it is not surprising to find that little work has been put into the study of the isotopes of very high-grade rocks such as those of the granulite facies. One of us (A. F. W.) for many years has considered that it should be possible to get meaningful temperatures from granulites <u>provided suitable minerals are studied</u>. The basic reason is that, as most of the minerals are anhydrous, it is assumed that these rocks were formed under extreme conditions where P_{H_2O} was much less than P_{total}. Thus it is assumed that there would have been insufficient fluids available to enable significant post-crystallization isotopic adjustments to take place.

The oxygen isotope studies in granulites have not turned out as was first hoped. Many "anomalies" have appeared in the data and there have been unexpected experimental problems. However, attempts to explain the anomalies and to solve the technical problems have combined to throw much light on the metamorphic and post-metamorphic processes operative in the deep crust. In this paper we shall discuss some of these problems, and point out the significance of the oxygen isotope studies on granulite facies rocks.

SELECTION OF APPROPRIATE MINERALS FOR GEOTHERMOMETRY

The basis for oxygen isotope geothermometry is well stated by Taylor (1968), thus: "In all known cases the oxygen isotope fractionations between any two minerals decrease with increasing temperature of formation, in a simple systematic manner (generally falling off linearly on a $1/T^2$ plot, where T is the absolute temperature in $^{\circ}K$). These relations allow us to make quantitative inferences about existence of equilibrium and temperature of crystallization in a mineral assemblage (e.g., see O'Neil and Clayton, 1964; and O'Neil and Taylor, 1967)".

Laboratory calibrations have been done for oxygen isotope fractionation in the systems feldspar-water (O'Neil and Taylor, 1967), magnetite-water (O'Neil and Clayton, 1964) and muscovite-water (O'Neil and Taylor, 1969; Blattner and Bird, 1974). However, by a series of concordance studies of appropriate mineral assemblages containing a non-calibrated mineral, expressions have been deduced for the temperature dependence of the fractionation of oxygen isotopes between several minerals whose fractionation with water has not yet been experimentally determined. Thus it is possible to use (with varying reliability) the following minerals for geothermometry: quartz, plagioclase, K-feldspar, pyroxene, olivine, garnet, amphibole, biotite, muscovite, ilmenite and magnetite (Bottinga and Javoy, 1976). All of these minerals may be found in granulites, with the exception of muscovite. Their compositions, however, may be very different from those for which the temperature expressions have been calculated. For example, fluorine and titanium commonly are high in granulite facies amphiboles and biotites, and as these elements affect the (OH) and Si sites, respectively, some anomalies can be expected.

Petrographic studies show that the slow cooling of the granulites facilitates the development of exsolution phenomena. For this reason a granulite showing perthitic and/or antiperthitic feldspars must be treated with caution, for oxygen isotope exchange at sub-solidus temperatures is likely in such a rock. The problem arises, however, as to whether <u>all</u> mineral phases in a granulite are affected equally by slow cooling or later metamorphism.

Anderson (1966) has shown that in the Labrieville anorthosite recrystallization during cooling has affected all of the measured minerals to the extent that he postulates that the rock must be considered to be a closed system with respect to oxygen. He states that "as temperature decreases the fractionation of oxygen isotopes between oxides and silicates increases, and material balance within a locally closed system requires that the resultant oxygen isotopic composition of all minerals participating in the retrograde recrystallization reflects the modal composition of the rock" (ibid.,

p. 1690). He showed that plagioclase (both bulk feldspar and megacrysts), apatite, ilmenite and magnetite have partaken of this mutual oxygen isotope exchange. Although ortho- and clinopyroxene were not measured, it has been assumed that they have also shared in this oxygen exchange. As Anderson records an abundance of exsolution lamellae in the pyroxenes of this igneous body this assumption is reasonable.

Our work shows, however, that the pyroxenes of metamorphic rocks (rather than plutonic igneous rocks) may not be as susceptible to slow cooling effects as the opaque oxides and feldspar. Under equilibrium conditions both ortho- and clinopyroxenes should display virtually identical oxygen isotope fractionation (Broecker and Oversby, 1971, p. 162). We have found that the two pyroxenes of our granulites normally exhibit identical fractionation only in "dry" granulite rocks. However, in the presence of granulite facies hornblende we find that the $^{18}O/^{16}O$ ratios of these three minerals (which are in textural equilibrium) are commonly opx > cpx > hbe (Wilson and Baksi, in prep.). We assume that of the three chain silicates, orthopyroxene is the most likely to retain its initial (i.e. its pre-retrogression) oxygen isotopic composition.

Quartz is another mineral which seems to retain its initial oxygen isotopic composition. Quartz is especially good when it occurs as large grains. Petrographic study of a rock can reveal suspect grains such as those showing granulation and local recrystallization. Care must be taken not to include in a mineral fraction for analysis quartz particles of more than one generation. Thus we believe that the two minerals most likely to be useful for geothermometry of normal granulites are coarse grains of orthopyroxene and quartz. An example of the use of this mineral pair has been published (Wilson and Baksi, 1977), where analysis of a hypersthene-cordierite-quartz granulite indicated a temperature of 760°C. Other minerals which could be expected to retain their pre-cooling oxygen character are some very refractory minerals such as spinel, corundum, sapphirine, rutile, garnet, coarse sillimanite and zircon. Notwithstanding their general refractory character, petrographic studies need to be done to ensure that second generation spinel, corundum, etc., are not present (e.g. Woodford and Wilson, 1976a). There are technical problems, however, with most of these minerals. For example, an earlier paper reported that only partial extraction of oxygen could be obtained from sapphirine from Western Australia (Wheat Belt - Fig. 1) (Wilson and

Green, 1971). Very fine grinding (under anaerobic conditions) and longer reaction at higher temperatures than is normal for the more common minerals will be needed to extract oxygen quantitatively from these minerals on a routine basis. Experimentation is in progress.

Laboratory temperature calibration of these minerals will be difficult, but an empirical approach using carefully selected mineral assemblages containing calibrated phases should be possible.

Studies of the Australian granulites show that the opaque oxides are particularly susceptible to oxygen isotope exchange during cooling. The magnetite and ilmenite of both silicic and mafic granulites are much depleted in ^{18}O. It has already been pointed out for the Musgrave Ranges (Fig. 1) that concordant temperatures in excess of

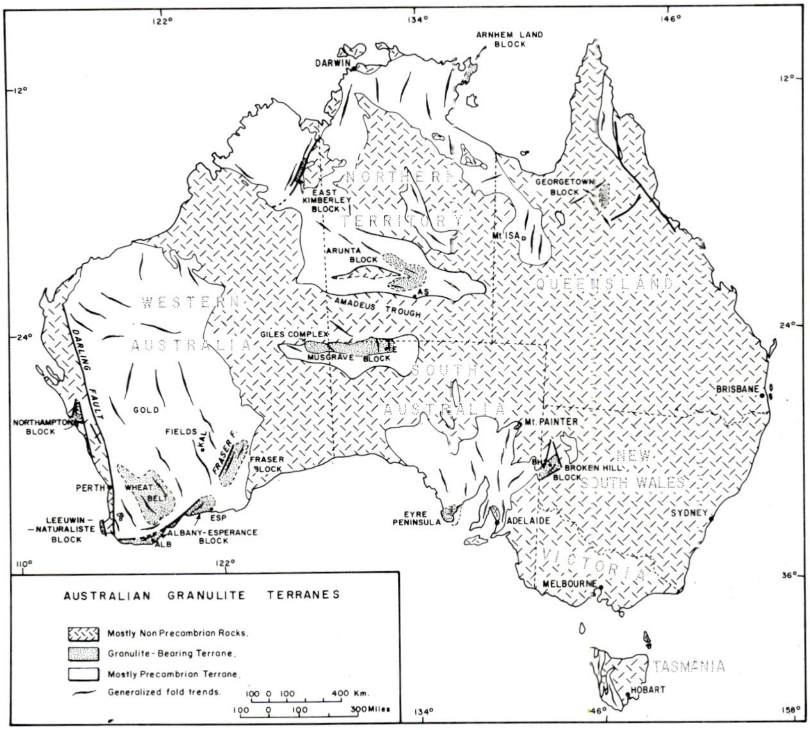

Fig. 1. - Granulite terranes of Australia. The Wheat Belt granulites are Archaean, whereas all other granulites yield Proterozoic ages by Rb/Sr. KAL, Kalgoorlie; ALB, Albany; ESP, Esperance; E, Ernabella; AS, Alice Springs; BH, Broken Hill. The Fraser Range occupies most of the Fraser Block; the Musgrave Ranges form the eastern third of the granulite portion of the Musgrave Block; the Strangways Range forms the eastern portion of the granulite portion of the Arunta Block.

700°C may be obtained if the measured differences between the major minerals and the magnetite are reduced by 2½% . (Wilson et al., 1970,

p. 17; Wilson and Green, 1971, p. 302). This assumes that the silicates (plagioclase and the two pyroxenes) were unaffected by the cooling, and that the oxygen of the opaque oxides exchanged with fluids from outside of the rock. This may have been possible in view of the special tectonic setting of the samples: they form part of a large xenolith or roof pendant within a large body of intrusive (magmatic) pyroxene adamellite (Wilson and Green, 1971, Fig. 2).

The temperature expressions as refined by Bottinga and Javoy (1975) confirm the earlier conclusions of Wilson et al. (1970), that magnetite and ilmenite have suffered oxygen isotope exchange down to threshold temperatures a little higher than 500°C. Quartz and pyroxene, on the other hand, appear to conserve a state of oxygen isotopic equilibrium and indicate a temperature of at least 730°C in a quartz-rich pyroxene granulite.

If these concepts are accepted, a reinterpretation of the isotope data of Wilson and Green (1971) from the Razor Hill locality in the Musgrave Ranges is possible. The total isotopic, petrological and field evidence would seem to suggest the following sequence:

(a) Intrusion of the pluton at a temperature well in excess of 730°C (Bottinga and Javoy, 1975, Table 1, No. 174). The quartz-rich portion of the large granulite xenolith retained a temperature of 730°C (ibid. No. 73).

(b) Cooling fissures in the pluton and roof pendant were filled first with pegmatite (ibid. No. 110) and then with aplite (ibid. No. 175), and temperatures of 720°C and 600°C, respectively, were registered.

(c) The quartz and pyroxenes of the pluton (ibid. No. 73) ceased oxygen isotope exchange at about 640°C, whereas the feldspar and pyroxenes of the basic granulites did not cease exchange until about 600°C (ibid. Nos. 6,7 and 8).

(d) As the whole rock mass further cooled the opaque iron oxides continued to react, possibly with late fluids from the pluton, and temperatures of 500°C and less were recorded (ibid. all samples: Nos. 73, 110, 174-178).

This research was completed several years ago. Further samples should now be collected to test more rigorously the conclusions that rocks containing some inequilibrium isotope features may still be used for geothermometry.

RELATIONSHIP OF OXYGEN ISOTOPE TEMPERATURES AND THOSE DEDUCED FROM $K_{D.Fe-Mg}^{opx-cpx}$

Anomalous temperature relationships are not uncommonly found between coexisting pyroxenes in metamorphic rocks, even though the mineral phases appear to be in textural equilibrium (Woodford and Wilson, 1976b). A detailed study of this problem will be reported elsewhere. However, some of the pertinent preliminary conclusions may be stated.

1. Where the ortho- and clinopyroxene of a two-pyroxene granulite have the same $^{18}O/^{16}O$ ratio we may assume that they conserve a state of oxygen isotopic exchange equilibrium. Such a pyroxene pair may be expected to retain cation equilibrium and thus yield a reliable temperature by one of the cation procedures (such as K_D; Wood and Banno, 1973).
2. In hornblende-bearing two-pyroxene granulites, however, anomalous oxygen values are common, even though the hornblende, orthopyroxene and clinopyroxene appear to be in textural equilibrium. (Note: the hornblende is stable in the granulite facies and is not due to retrogression to amphibolite facies.) In some of these rocks the $^{18}O/^{16}O$ ratio is found to be opx > cpx > hbe, where the differences are about 0.2 - 0.3% (see Appendix); in others, the relationship is hbe > opx > cpx. Theoretically, δ for opx and cpx in equilibrium at granulite facies temperatures should show no measurable difference (Broecker and Oversby, 1971, p. 160).
3. We presume that the fluid pressures enabling the granulite hornblende to form have allowed some oxygen isotope exchange to proceed after the two pyroxenes have become stable. The fact that the two pyroxenes in such rocks now show measurable differences in δ suggests that orthopyroxene may be somewhat more stable in this situation than clinopyroxene, to which the hornblende is possibly more commonly adjoined.
4. Oxygen occurs in two "sites" in hornblende: as part of the normal cation linkages, and in OH. The relative stability of the oxygen isotopes in these two sites is unknown. Moreover, F or Cl may replace some of the OH in some granulites. Both of these factors could affect the normal relationship between the oxygen isotopes of hornblende and the coexisting pyroxenes.
5. An example of these conclusions is set out in Table 1.

The two samples are from homogeneous layers about 20m apart in a stable granulite block about 3 km WNW of Kenmore Park Station,

Musgrave Ranges (Fig. 1; and Wilson, in prep.). Sample 17967 is pyroxene-quartz-andesine granulite. The two pyroxenes, which are in perfect textural equilibrium, are also in oxygen isotopic equilibrium, for both opx and cpx have δ = 5.4‰ . The $K_{D.Fe\text{-}Mg}^{opx\text{-}cpx}$ is 1.754, a common figure for a pyroxene granulite.

TABLE 1

Sample	17967	Scap F
opx ‰	5.39	6.57
cpx ‰	5.40	6.37
hbe ‰	-	6.17
$K_{D.Fe\text{-}Mg}^{opx\text{-}cpx}$	1.754	1.949
$T^{o}C$	848	869

Sample Scap F, on the other hand, is a hornblende-rich two-pyroxene mafic granulite. The mafic and felsic minerals of the granulite display perfect textural equilibrium. The mafic minerals, however, are not in perfect oxygen isotope equilibrium, for opx, cpx and hbe have δ = 6.6, 6.4 and 6.2‰ , respectively. The anomalously high K_D for this rock is 1.949, and this would normally be taken to indicate a lower temperature of equilibration than that of sample 17967, with its K_D of 1.754. We suspect that, because the oxygen isotopes of the hornblende-bearing granulite have been disturbed in sample Scap F, the cations should also be disturbed. This would account for the anomalously high K_D, and for a Wood-Banno cation temperature of 869°C that is spuriously higher than the 848°C of the two-pyroxene granulite.

WHOLE-ROCK STUDIES

Whole-rock oxygen isotope ratios are measured on deep crustal rocks for two main reasons. One is to monitor the extent of change of whole rock oxygen during high grade metamorphism. Another is to use oxygen isotopes to help determine the nature of the unmetamorphosed rocks and their depositional or diagenetic environment.

1. Effect of Change in Oxygen During Metamorphism

Detailed studies of the cations of major and trace elements of many mafic granulites from several parts of Australia show that very little (if any) change takes place during high-grade metamorphism in most of these basic rocks, especially if they are in large compact igneous bodies or wide (say, greater than about 10 m) tabular bodies (Wilson, in prep.). Whole-rock analyses of oxygen in some of these metabasaltic rocks yield values of about 6 to 7‰ (Fraser Range - see Fig. 1; Wilson and Baksi, in prep.). These fall within the range of those obtained for modern basalts (Bottinga and Javoy, 1975).

Narrow mafic bodies (a few cm to 2m wide) are commonly found interlayered in metasedimentary rocks. Some appear to be basic tuffaceous layers; others narrow basaltic flows or dolerite sills. Some of these may have been emplaced in the sedimentary pile prior to its metamorphism. The oxygen isotope ratios of these narrow mafic layers are commonly grossly different from normal basaltic rocks, and are much more compatible with those of their sedimentary host rocks. This implies isotopic exchange throughout the sedimentary pile during the metamorphic process. It appears, however, that whereas narrow permeable basaltic layers have been altered, the less permeable larger mafic bodies have retained their "normal" basaltic oxygen isotopic values.

2. Use of Oxygen Isotopes to Study Precambrian Depositional or Diagenetic Environments

Oxygen isotope studies of many granulite facies rocks from Australia show unexpectedly light oxygen in many quartzose granulites ($\delta \sim 3‰$) and in some mafic granulites ($\delta \sim 2‰$), whereas "normal" igneous or gneissic values are common in other quartzose granulites ($\delta \sim 8‰$) and mafic granulites ($\delta \sim 6‰$).

Veizer and Hoefs (1976) point out that the tendency towards lighter $^{18}O/^{16}O$ ratios of both carbonate rocks and cherts with increasing geologic age has been well documented. Many consider these oxygen isotope trends were caused by a continuous post-depositional equilibration with isotopically light meteoric waters, whereas others suggest that the trends reflect lighter $^{18}O/^{16}O$ ocean waters in Precambrian times. As our studies of metamorphosed Precambrian sediments show a wide range of oxygen isotope ratios, other explanations for our large body of Precambrian rocks with unusually light oxygen must be sought.

The explanation given for the presence of unusually light oxygen in rocks usually invokes an interaction between a hot igneous body and light meteoric waters (Taylor and Forester, 1971). Light waters are known to be caused in modern times by precipitation in high latitudes. Moreover, inland regions and high altitudes are also important contributing factors (Epstein and Mayeda, 1953).

Thus the presence of light oxygen in Precambrian metasedimentary rocks raises the question as to whether the variations of meteoric waters were comparable in Precambrian times. The discovery by Taylor (1974) of a low ^{18}O late Precambrian batholith in the Seychelles Islands suggests that similar processes indeed were active in late Precambrian times. The normal model involving reaction of light meteoric waters and hot igneous rocks could not apply among the regional metamorphic rocks of the Strangways Ranges (Fig. 1) where the largest number of light oxygen rocks have been found. However, we have suggested elsewhere (Wilson and Baksi, 1977) that these metasediments may have derived their light oxygen by reaction with light high latitude waters that were heated not by igneous activity but by virtue of being buried in a deep artesian basin. Studies are in hand to test the reaction of the 90°C waters in the Great Artesian Basin of Australia with the minerals of the aquifers.

The measurement of D/H ratios of the water in whole rocks and minerals will greatly extend the usefulness of the $^{18}O/^{16}O$ of deep crustal rocks. This will enable the provenance of the water to be defined much more accurately, as has been done so elegantly, for example, in many recent studies of porphyry copper deposits (Taylor, 1974; Sheppard and Taylor, 1974). The aim of this approach will be to study the fluids found in rocks of the deep crust, and to try to discriminate between juvenile, connate and meteoric fluids.

ACKNOWLEDGEMENTS

In addition to generous support from the University of Queensland, the Australian Research Grants Council (Refce. E71/17830) has strongly supported the oxygen isotope project, and awarded the post-doctoral fellowship (to A.K.B.).

REFERENCES

Anderson, A.T., 1966. Mineralogy of the Labrieville Anorthosite, Quebec. Amer. Mineral., 51:1671-1711.

Anderson, A.T., Clayton, R.N. and Mayeda, T.K., 1971. Oxygen isotope thermometry of mafic igneous rocks. J. Geol., 79:715-729.

Blattner, P. and Bird, G.W., 1974. Oxygen isotope fractionation between quartz and K-feldspar at 600°C. Earth Planet. Sci. Lett., 23:21-27.

Bottinga, Y. and Javoy, M., 1975. Oxygen isotope partitioning among the minerals in igneous and metamorphic rocks. Rev. Geophys. Space Phys., 13:401-418.

Broecker, W.S. and Oversby, V.M., 1971. Chemical Equilibria in the Earth. McGraw Hill, New York.

Clayton, R.N. and Mayeda, T.K., 1963. The use of bromine pentafluoride in the extraction of oxygen from oxides and silicates for isotopic analysis. Geochim. Cosmochim. Acta, 27: 43-52.

Clayton, R.N., O'Neil, J.R. and Mayeda, T.K., 1972. Oxygen isotope exchange between quartz and water. J. geophys. Res., 77:3057-3067.

Craig, H., 1961. Standards for reporting concentrations of deuterium and oxygen in natural waters. Science, 133: 1833-1834.

Epstein, S. and Mayeda, T.K., 1953. Variation of O^{18} content of waters from natural sources. Geochim. Cosmo. Acta., 4:213-224.

Garlick, D. and Epstein, S., 1967. Oxygen isotope ratios in coexisting minerals of regionally metamorphosed rocks. Geochim. Cosmo. Acta, 31:181-214.

Matsuhisa, Y., 1974. $^{18}O/^{16}O$ ratio of NBS-28 and some silicate reference samples. Geochim. J., 8:103-107.

O'Neil, J.R. and Clayton, R.M., 1964. Oxygen isotope geothermometry. In : Isotopic and Cosmic Chemistry (Urey volume), H. Craig et al. (Editors), North Holland, Amsterdam, 157-168.

O'Neil, J.R. and Taylor, H.P., 1967. The oxygen isotope and cation exchange chemistry of feldspars. Amer. Mineral., 52:1414-1437.

O'Neil, J.R. and Taylor, H.P. 1969. Oxygen isotope exchange between muscovite and water. J. geophys. Res., 74:6012-6022.

Onuma, N., Clayton, R.N. and Mayeda, T.K., 1970. Apollo 11 rocks : oxygen isotope fractionation between minerals, and an estimate of the temperature of formation. Geochim. Cosmo. Acta, supp. 1, 1429-1434.

Sheppard, S.M.F. and Taylor, H.P., 1974. Hydrogen and oxygen isotope evidence for the origins of water in the Boulder Batholith and the Butte Ore Deposits, Montana. Econ. Geol., 69:926-946.

Taylor, H.P., 1968. The oxygen isotope geochemistry of igneous rocks. Contrib. Mineral. Petrol., 19:1-71.

Taylor, H.P., 1974a. The application of oxygen and hydrogen isotope studies to problems of hydrothermal alteration and ore deposition. Econ. Geol., 69: 843-883.

Taylor, H.P., 1974b. A low - ^{18}O, late Precambrian granite batholith in the Seychelles Islands, Indian Ocean : evidence for formation of ^{18}O depleted magmas and interaction with ancient meteoric ground waters. Geol. Soc. Amer. Abstracts, 981-982.

Taylor, H.P. and Epstein, S., 1962a. Relationship between O^{18}/O^{16} ratios in coexisting minerals in igneous and metamorphic rocks, pt. 1 ; Principles and experimental results. Bull. geol. Soc. Am., 73:461-480.

Taylor, H.P. and Epstein, S., 1962b. Relation between O^{18}/O^{16} ratios in coexisting minerals in igneous and metamorphic rocks, pt. 2 : Applications to petrologic problems. Bull. geol. Soc. Am., 73: 675-694.

Taylor, H.P. and Forester, R.W., 1971. Low-O^{18} igneous rocks from the ring-dike complexes of Skye, Mull and Ardnamurchan, Western Scotland. J. Petrol., 12 : 465-497.

Veizer, J. and Hoefs, J., 1976. The nature of O^{18}/O^{16} and C^{13}/C^{12} secular trends in sedimentary carbonate rocks. Geochim. Cosmo Acta, 40: 1387-1395.

Wilson, A.F. and Baksi, A.K., 1977. Unusually light oxygen in granulites from the Strangways Range, central Australia: In : B.W. Robinson (Editor), Stable isotopes in the Earth Sciences, D.S.I.R. Spec. Bull., Gov. Printer, Wellington (in press).

Wilson, A.F. and Green, D.C. 1971. The use of oxygen isotopes for geothermometry of Proterozoic and Archaean granulites. Geol. Soc. Austral., Spec. Pub., 3: 389-400.

Wilson, A.F., Green, D.C. and Davidson, L.R., 1970. The use of oxygen isotope thermometry on the granulites and related intrusions, Musgrave Ranges, central Australia. Contrib. Mineral. Petrol., 27: 166-178.

Wood, B.J. and Banno, S., 1973. Garnet-orthopyroxene and orthopyroxene clinopyroxene relationships in simple and complex systems. Contrib. Mineral. Petrol., 42: 109-124.

Woodford, P.J. and Wilson, A.F., 1976a. Sapphirine, hogbomite, kornerupine, and surinamite from aluminous granulites, north-eastern Strangways Range, central Australia. N.Jb. Miner. Mh. Jg. 1976: 15-35.

Woodford, P.J. and Wilson, A.F., 1967b. Chemistry of coexisting pyroxenes, hornblendes and plagioclases in mafic granulites, Strangways Range, central Australia. N.Jb. Miner. Abh., 128:1-40.

APPENDIX 1

EXPERIMENTAL TECHNIQUES

Mineral separations were made from the sample by standard methods. Particular care ensured at least 99.9% purity.

Oxygen was extracted by means of the bromine pentafluoride ($Br\ F_5$) method described by Clayton and Mayeda (1963). Our procedures differ from other laboratories in that no conversion of oxygen to carbon dioxide is made, as has been reported in earlier papers (Wilson et al., 1970; Wilson & Green, 1971).

Results are reported as values, defined as :

$$\delta = \frac{R_{sample}}{R_{SMOW}} - 1 \times 1000$$

where $R_{sample} = {}^{18}O/{}^{16}O$ in the sample, and $R_{SMOW} = {}^{18}O/{}^{16}O$ in Standard Mean Ocean Water standard (Craig, 1961).

Fractionations between coexisting minerals in the same rock are reported as Δ values, defined as $\Delta_{A-B} = 1000 \ln \alpha_{A-B} \simeq \delta_A - \delta_B$ where α_{A-B} is the fractionation factor for the coexisting minerals A and B.

Our data are tied to SMOW by repeated analyses of NBS Quartz 28 which we are currently accepting to be 9.5‰ (Matsuhisa, 1974). Repeated analysis of many samples indicates that the σ precision in our δ values is about $\pm$ 0.05‰.

As was mentioned on p.7, the general difference in δ values between coexisting triplets of opz, cpx and hbe is 0.2 - 0.3‰. This indicates that for some of the mineral pairs the δ values overlap at the 2σ or 3σ confidence levels. However, we are sure that the individual values are meaningfully different, for the same discordant pattern has now been noted for many coexisting triplets.

VOLCANICS AND SEDIMENTS

IMPLICATIONS OF PHASE EQUILIBRIA AND CHEMICAL PARAMETERS FOR THE ORIGIN OF ARCHAEAN ULTRAMAFIC AND MAFIC LAVAS

P. Krishnamurthy
Atomic Minerals Division, Begumpet, Hyderabad 500016, India

Abstract

Polybaric phase equilibria of komatiitic liquids along with possible upper mantle compositions in the pseudo-quarternary CaO-MgO-Al_2O_3-SiO_2 (CMAS) system indicate that the parental rocks from which these liquids were derived were probably richer in CaO and hence in clinopyroxene than comparable source rocks. Available data indicate that the ultramafic and mafic liquids originated at c. 20-40 Kb (c. 60-120 km); a remaining problem relates to the availability of higher temperatures required (c. 1600-1700°C) to produce these liquids at such depths. Melting at deeper levels poses problems of quick transport which is essential for the preservation of the primitive character of these liquids. Consideration of major and minor oxide variation of the komatiitic suite indicates that the ultramafic members may have been the parental liquids to the basaltic and/or the pyroxenitic komatiites, and that fractionation of olivine and to a lesser extent clinopyroxene ± orthopyroxene may have controlled such an evolution.

A comparative study is made of the textural and chemical features of Archaean komatiites, especially basaltic types, with those of non-cumulus and possibly primary or primitive picritic liquids of younger age. (1) Palaeozoic basaltic komatiites of the Rambler area, Newfoundland, are closely comparable in composition with similar rocks of the Archaean (mean of 9: MgO 14.21%, TiO_2 0.16%, CaO/Al_2O_3 = 1, K_2O 0.12%), although they lack the spinifex or quench textures. (2) Picritic rocks of the North Atlantic Province (such as Baffin Bay) and the Karroo Province (Nuanetsi) are characterised by skeletal forms of olivine indicating rapid cooling or quenching, akin to the quench textures of komatiites. Whereas the former have chemical affinity with komatiites, the latter are richer in K and Ti. (3) Deccan picritic basalts characteristically have high CaO/Al_2O_3 ratios (i.e. > 1), but unlike basaltic komatiites they are enriched in incompatible elements (e.g. Ti, K and P) and have evidently undergone much more heat loss before eruption because spinifex textures are not developed and olivine phenocrysts are abundant. (4) Solomon Island picrites lack spinifex texture but are analogous to komatiites in chemistry (MgO > 10%, K_2O < 1%, TiO_2 < 0.6%, CaO/Al_2O_3 = 0.9-1.0). (5) Among minor oxides, TiO_2 appears to be very diagnostic in distinguishing komatiitic from other types, although the enrichment of minor and trace elements in ultramafic and mafic liquids could be attributable to several factors.

Broad chemical similarity between Archaean and younger basaltic komatiites and picritic rocks, especially of major elements, suggests that melting and crystallisation events that produced these rocks were repeated throughout earth history, whereas peridotitic komatiites are unique to the Archaean implying the importance of early steep thermal gradients in their formation.

INTRODUCTION

Several important questions arise when we consider the phase equilibria and chemical features of Archaean ultramafic and mafic lavas: (1) What are the pressure (depth) and temperature conditions of their formation and the extent or degree of melting of the source rock to produce liquids of this type? (2) What were the source rocks from which they might have been derived and what are the possible residual types after their extraction? (3) Are these ultrabasic and basic lavas, presently known by the term komatiitic suite, unique to the Archaean when compared with younger primary or very primitive liquids such as the non-cumulus picritic types from continental and island arc regions? (4) Can any tectonic significance be attached to the presence of such a rock suite?

An attempt is made here to provide possible answers to some of these questions from a consideration of komatiitic compositions from the Archaean terrains of South Africa (Viljoen and Viljoen, 1969), Canada (Arndt et al., 1977) and Western Australia (Anhaeusser, 1971; Nesbitt and Sun, 1976) as plotted in established polybaric phase diagrams and in simple oxide variation diagrams. Non-cumulus picritic rocks of younger age, which have been shown from mineralogical, chemical and phase equilibria to be either primary or very primitive, at least with reference to major oxides, occur (see Table 1) in the Nuanetsi in southern Africa (Cox and Jamieson, 1974), the Deccan Traps, India (Krishnamurthy, 1974; Krishnamurthy and Cox, 1977), the Baffin Bay area of eastern Canada and Western Greenland (Clark, 1970), and the New Georgia area in the British Solomon Islands (Stanton and Bell, 1969; Cox and Bell, 1972). These are compared with the Archaean rocks to find their common features in order to better understand the evolution of the Archaean rocks. This is relevant to the concept of uniformitarianism.

It must be emphasised at the outset that the origin of the Archaean ultramafic and mafic suite cannot be treated in isolation, but as part of the major basic volcanic episode during the Archaean. It is well known that both tholeiitic or sub-alkaline and calc-alkaline types of magmatism predominate in Archaean greenstone belts in different parts of the world, and the proportion among these of komatiitic rocks, especially of the ultrabasic members, is insignificant: they

TABLE 1. Brief summary of the salient features of picritic types

Locality and Age	Tectonic Setting	Field Relations	Mineralogical and Textural Features	Remarks
Solomon Islands	Island arc	Lava flows and tuffs	Phenocrysts of olivine (Fo 89-88) up to 5 mm across and clino-pyroxenes in a groundmass which is glassy to partly crystalline	Possible primary or very primitive ultra-mafic liquids with c. 15-20% MgO (Stanton & Bell, 1969; Cox & Bell, 1972)
Nuanetsi, Karroo Province. Triassic-Lower Jurassic	Continental flood basalt association (CFBA), Karroo volcanic cycle	Lava flows at the base of the succession	Olivine (Fo 78-73) crystals usually show skeletal and embayed form. Minor clinopyroxene in skeletal forms. Orthopyroxene phenocrysts (high pressure relics) mantled by clinopyroxene overgrowths; glassy to finely crystalline groundmass	Primitive or primary liquids with c.15-18% MgO (Cox and Jamieson, 1974)
Baffin Bay, Baffin Island in eastern Canada, and Svartenhuk Peninsula in W. Greenland. Tertiary	CFBA, North Atlantic Province	Lava flows	Skeletal crystals of olivine (Fo 90-88) indicating rapid cooling; clinopyroxenes rare	Possible primary magmas from the upper mantle produced by c.30 Kb with c. 18-20% MgO (Clarke, 1970)
Bore hole flows at Dhandhuka, Wadhwan. Junction and Botad Deccan Traps, western India	CFBA, Deccan Province	Picritic lava flows from bore holes associated with mildly alkaline types, generally at base of succession	Equant olivine (Fo 80) phenocrysts up to 2 cm across along with varying amounts of clinopyroxene; groundmass glassy to finely crystalline	Possible primary or very primitive ultramafic liquid with at least 16% MgO (Krishnamurthy, 1974; Krishnamurthy & Cox, 1977)

constitute less than 1% in the Canadian Superior Province (A.M. Goodwin, pers. comm.) and in Western Australia (Nesbitt and Sunn, 1976) and probably in South Africa.

PHASE EQUILIBRIA

There is general agreement among earth scientists that the upper mantle is largely composed of a peridotitic type of rock. Mineralogically it comprises forsteritic olivine, orthopyroxene, clinopyroxene and an aluminous phase which may be plagioclase, spinel or garnet depending upon the pressure and temperature. Bulk compositions of such source rocks as well as their constituent minerals can be shown to best advantage in the pseudo-quarternary CaO-MgO-Al_2O_3-SiO_2 (CMAS) system of O'Hara (1968). The system can also show empirically the compositions of liquids resulting from partial melting under varying pressures along with their possible course of descent. Sub-projections within the system (especially of compositions or mineral phases which are commonly present either in the solidus or liquidus phases in the melt or in the residue such as olivine, clinopyroxene or orthopyroxene) can be used to look into the nature of possible liquid lines of descent at varying depths, the probable nature of the source rocks and the residual types resulting from varying degrees of partial melting of the source.

Fig. 1 shows the salient features of the CMAS system and the positions of the two sub-projections within this system used here. Fig. 2 shows the perspective view of the two surfaces or loci of liquids, namely the harzburgite and eclogite surface, which are important in ultrabasic and basic magma genesis at higher pressures (c. 30 Kb), as well as the position of the two possible source rock compositions for which the CaO/Al_2O_3 ratio is either > 1 (composition Z, relatively rich in clinopyroxene) or < 1 (composition O, relatively rich in garnet). Also indicated is the possible composition of the initial liquid formed as a result of partial melting, and the subsequent movement of the liquid as a result of increased or advanced stages of melting. Melting starts when the temperature exceeds the solidus and initially the liquid so produced is in equilibrium with all four crystalline phases, viz. olivine, orthopyroxene, clinopyroxene and garnet. This is the composition of the liquid at 'B' in Fig. 2. During subsequent stages as melting continues either garnet or clinopyroxene will be totally consumed. For source rocks whose compositions lie in the critical plane (i.e. the plane containing olivine, orthopyroxene and liquid B), the liquid composition essentially remains at B as melting proceeds. The critical plane also divides source rock compositions which are relatively rich in Al_2O_3 (i.e. in garnets such as in O) from compositions relatively rich in CaO (in clinopyroxenes such as Z). However, the locus of liquids which can be developed for source rocks such as O or Z is as follows.

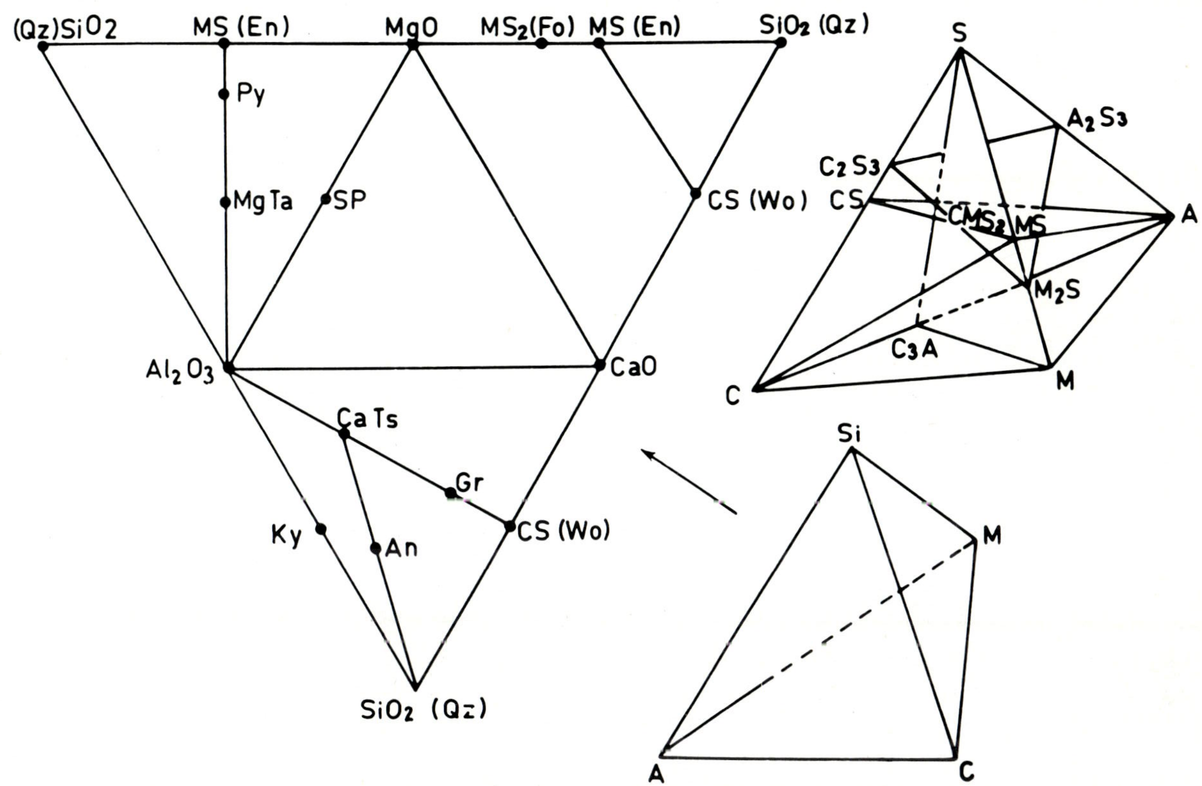

Fig. 1. The CaO-MgO-Al_2O_3-SiO_2 (CMAS) system of O'Hara (1969) showing the different minerals involved, i.e. olivine (Fo), enstatite (En), pyrope (Py), grossular (Gr), spinel (Sp), anorthite (An), wollastonite (Wo), kyanite (Ky), calcium tschermackite (Ca Ts), magnesium tschermackite (Mg Ts) and quartz (Qz), along with important planes of subprojections within the CMAS tetrahedron.

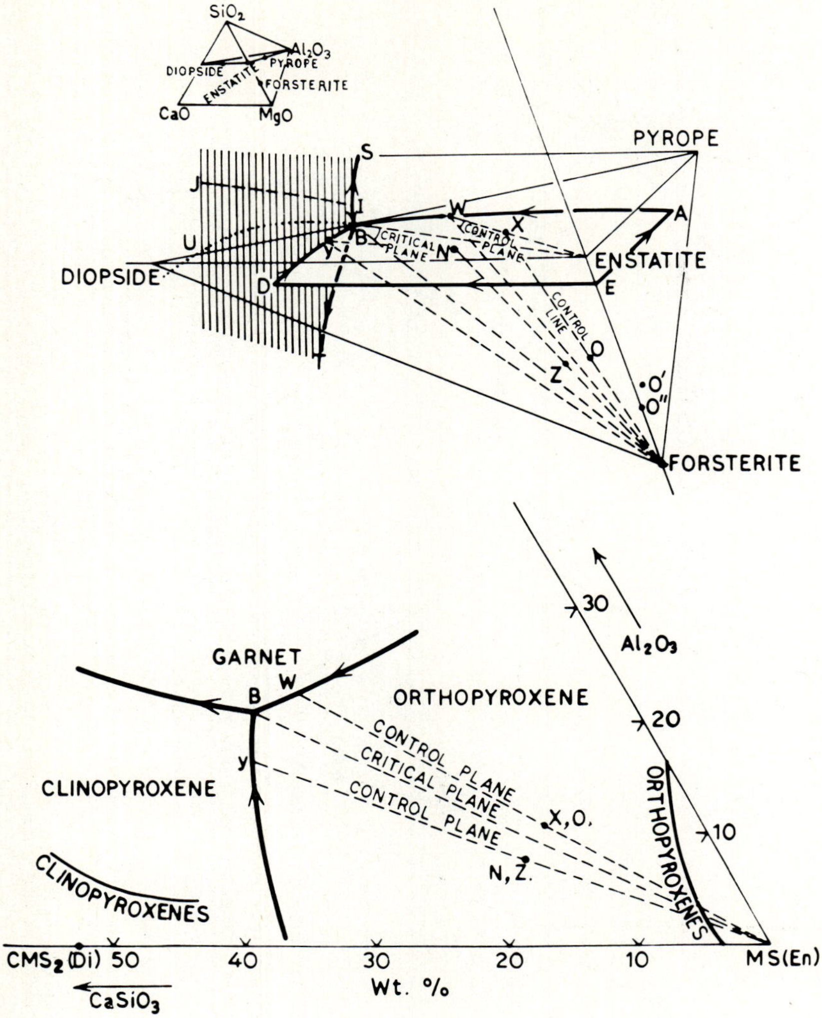

Fig. 2. The upper part gives a perspective view of the harzburgite and eclogite surface. BAED is the locus of liquids in equilibrium with olivine and orthopyroxene (harzburgite) while the shaded surface extending to the left of the lines S-B-T is the locus of liquids in equilibrium with garnet and clinopyroxene (eclogite) (after O'Hara, 1970). The lower diagram is a projection from olivine onto the plane diopside-enstatite-alumina showing the position of postulated upper mantle compositions, i.e. O or Z, and their probable melting paths. Clinopyroxene and orthopyroxene solid solutions are also shown (after O'Hara, 1970; Cawthorn and Strong, 1974).

For O, which has a CaO/Al_2O_3 ratio of < 1 and in which garnet is more than clinopyroxene, clinopyroxene is the first phase to disappear and the locus of the liquid, as melting proceeds, moves along the path BW. During this stage, once diopside is exhausted, garnet starts going into the liquid until the point W, where it has all gone into the liquid. The successive residual phases here are garnet-lherzolite, garnet-harzburgite (gt + opx + cpx), harzburgite (opx + cpx) and various residual phases such as O' and O"; all these (B-O-C' or B-O-O") are co-linear. For a source rock composition Z, which has a CaO/Al_2O_3 ratio of > 1 and in which clinopyroxene is in excess of garnet, garnet is the first phase to disappear and as melting advances the locus of the liquid moves along the path BY; after garnet, clinopyroxene is gradually exhausted until at Y the whole of it goes into the liquid. The successive residual phases when this type of source rock is melted are garnet-lherzolite, lherzolite and harzburgite. For any further increase in temperature leading to still higher stages of melting, enstatite preferentially goes into the liquid and the liquid compositions lie in the control plane WXO or YNZ, the lines joining the residue and liquid invariably passing through O or Z. Thus, wherever a higher degree of melting has taken place (as appears to be the case with the ultrabasic members of the komatiitic suite) the residue invariably may be either harzburgite or between harzburgite and dunite.

Having seen the possible loci of liquid compositions during advanced stages of partial melting from relatively CaO-rich and/or Al_2O_3-rich source compositions, it is interesting to treat the natural compositions, i.e. the komatiitic suite of rocks, in a similar plot. This is done in Fig. 3 which shows that the data from South Africa, Western Australia and Canada (Munro Township) seem to support the postulation that the loci of liquids generally followed the predicted path of BYZ or away from it towards the critical plane. Further support for the probable relative enrichment of the source rock in CaO is provided by clinopyroxene-rich nodules from the Motsoku kimberlite pipe (Cox et al., 1973) (Fig. 3) and recently reported 3.4 by old spinel lherzolite containing as much as 35% of modal diopside (Basu and Murthy, 1977). It is pertinent to point out at this stage that there are wide variations in the modal percentage of postulated source rocks such as garnet lherzolite and spinel lherzolite (Cox et al., 1973; Maaløe and Aoki, 1977); in the latter clinopyroxene predominates over spinel. Thus it is not improbable that the source rocks from which the komatiitic suite was derived were relatively richer in CaO and hence in clinopyroxene (thereby giving a CaO/Al_2O_3 ratio of > 1 or nearly equal to 1) than closely comparable postulated sources. However, the present scheme differs from that of Cawthorn and Strong (1974) in that it does not imply a layered mantle in which there is a clinopyroxene-rich layer at shallow depths. Their postulation seems to arise from the fact that they considered only a limited number of source rocks, all of

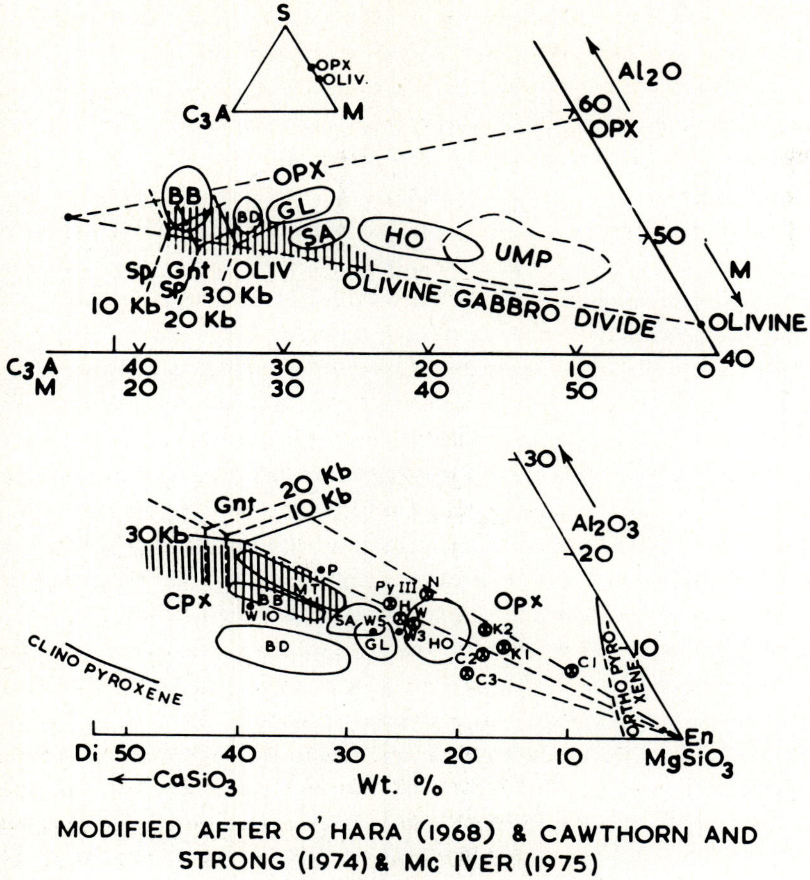

Fig. 3. Projections from olivine into part of the system $CaSiO_3$-$MgSiO_3$-Al_2O_3 indicating the positions of the pseudo-invariant points at various pressures and the location of the komatiitic compositions, picritic liquids and some postulated source rocks. South Africa: BB = Barberton type; BD = Badplass type; GL = Geluk type; HO = Hoogenoeg meta-tholeiites; SA = Sandspruit ultramafic type; MT = Onverwacht meta-tholeiites. W. Australia: W 3, W 5, W 10 (Williams, 1972, table 1, Nos.3, 5, 10). Munro Township: P (Pyke et al., 1973, table 1, No. 3) Phanerozoic picritic types: G = Rambler area, Newfoundland (Gale, 1973). Other samples, such as from Deccan (Krishnamurthy, 1974) and Baffin Bay (Clarke, 1970), plot within the shaded area. Upper mantle source rocks: C1-C4 = nodules from kimberlites (Cox et al., 1973; Nos. LMB 9, 36A, 27, 11); K1-K2 = lherzolite and peridotite nodules from kimberlite (Ito & Kennedy, 1968) (for refs. to others see Fig. 4). The upper part is a projection of the above data from olivine into part of the system C_3A-M-S. UMP encloses the postulated upper mantle compositions (after Cawthorn & Strong, 1974; McIver, 1975).

which had initially low CaO/Al_2O_3 ratios ranging from 0.59 to 0.69, compared with the wider range detailed here in a later section. It may be possible that under relatively shallow depths (say between 20-30 Kb) within the stability field of spinel-lherzolite melting of the source rock took place resulting in the parental liquids of the komatiitic suite. Support for such a possibility comes from the melting studies conducted by Mysen and Kushiro (1976) on a sheared Cretaceous garnet-peridotite nodule from a kimberlite. They produced liquids analogous to peridotitic komatiite at c. 20 Kb at temperatures of 1700°C (anhydrous) and 1555°C at 1600°C, and 1750°C with 1.9% H_2O in the source rock. These liquids correspond to anything over 45% melting leaving generally an olivine or dunite residue (see Figs. 21-23 and Table 16 of Mysen and Kushiro, op. cit.). Their work also strongly supports the contention that partial melting of peridotite in the upper mantle is closely approximated by experiments in some simple systems such as the $CaSiO_3$-$MgSiO_3$-Al_2O_3 (CS-MS-A).

Green (1975) postulated a steeper geothermal gradient during the Archaean with a minimum depth of c. 200 km for mantle diapirism, leading to ultramafic extrusions at the surface at 1650 ± 20°C. However, during their ascent on adiabatic P-T paths from c. 200 km, they pass through a stage of 10-20% melting in which garnet is selectively left behind and olivine and enstatite are entrained to higher levels where higher degrees of partial melting (c. 70%) ensue. Fractionation or depletion in garnet not only yields the high CaO/Al_2O_3 ratios observed in some komatiitic suites, as in South Africa, but also accounts for the depletion in heavy rare earth elements observed in some peridotitic extrusions. From a consideration of rare earth fractionation patterns in different melt fractions obtained after melting a sheared garnet peridotite nodule from a kimberlite, Mysen and Holloway (1976) question the primary nature of some of the peridotitic komatiites as well as some picritic rocks, since both are slightly depleted in lighter rare earths relative to chondrites. Therefore, the possibility that peridotitic komatiites were derived from still more primitive liquids cannot be ruled out. From melting and crystallisation studies at varying pressure conditions (1 atmosphere, 10 kb, 20 kb, 30 kb and 40 kb) Arndt (1976) concludes that either peridotitic komatiitic liquids are minimum melting components at c. 50-60 kb (or 150-180 km) from peridotitic source rocks, or they result from high degrees of melting of normal mantle or lower degrees of melting of mantle material depleted earlier in fusible components at c. 120 km depth.

Clarke (1970) considered the picritic rocks of Baffin Bay as primary magmas from a treatment of their compositions in the CMAS system and of their geochemistry. A depth of origin of c. 100 km and c. 20% melting of a garnet-lherzolite was proposed. From high pressure phase equilibria studies on picritic samples from the Nuanetsi area, Cox and Jamieson (1974) proposed shallow depths of origin and very high local thermal gradients. Initially these picritic liquids were in

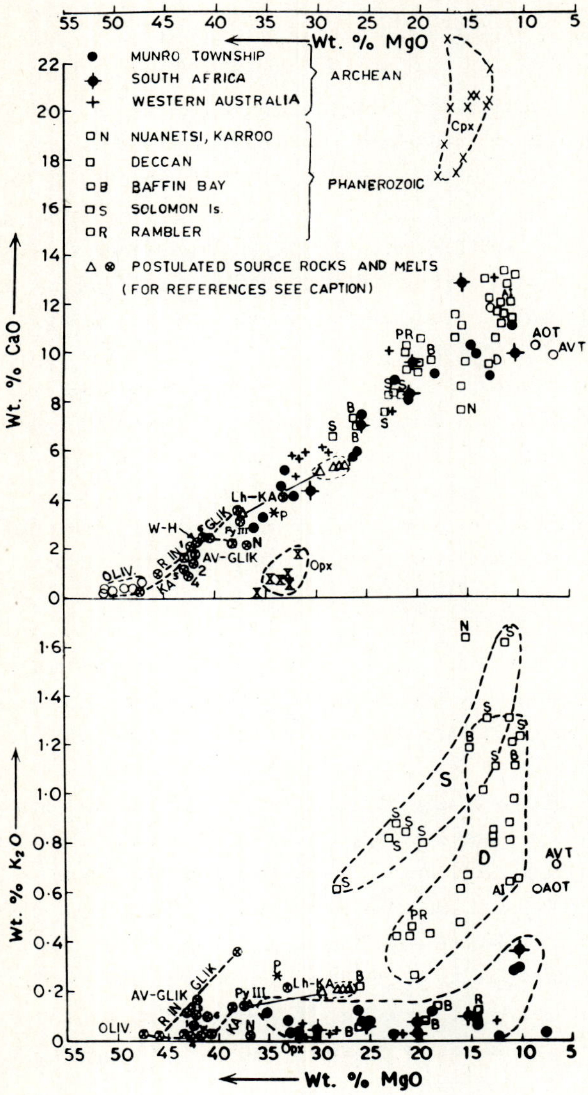

South Africa: Viljoen and Viljoen (1969) in Viswanathan (table III, 1974).
Munro Township, Canada: Arndt et al. (table 7, 1977).
Western Australia: Anhaeusser (1971) in Viswanathan and Sankaran (table II, 1973); Nesbitt and Sun (table I, 1976).
India: solid dots suffixed with I-K or I-Ku, Viswanathan (table II, 1974).
Phanerozoic picritic samples: Baffin Bay (Clarke, table 1, 1970); Deccan (Krishnamurthy, 1974; Krishnamurthy and Cox, table 5, 1977); Nuanetsi (Cox, table 2, 1972).
R = Rambler (Gale, table 1, 1973);
PR = Puerto Rico;
AI = Aleutian Islands (Brooks and Harte, table 1, 1974);
S = Solomon Islands (Stanton and Bell, table 1, 1969).

Upper mantle compositions:
P = peridotite (Nockolds, 1954);
Rin Glik = range of compositions in 15 garnet lherzolite inclusions in kimberlite (Carswell and Dawson, table 3, 1970);
AV-GLIK = average of the above;
KA = lherzolite nodules based on MgO/Σ FeO ratio (Kuno and Aoki, table 16, 1970);
Lh-KA = postulated primitive upper mantle source (Kuno and Aoki, table 17, No PL, 1970);

Fig. 4a. Variation of some major and minor oxides with MgO as an index of evolution.

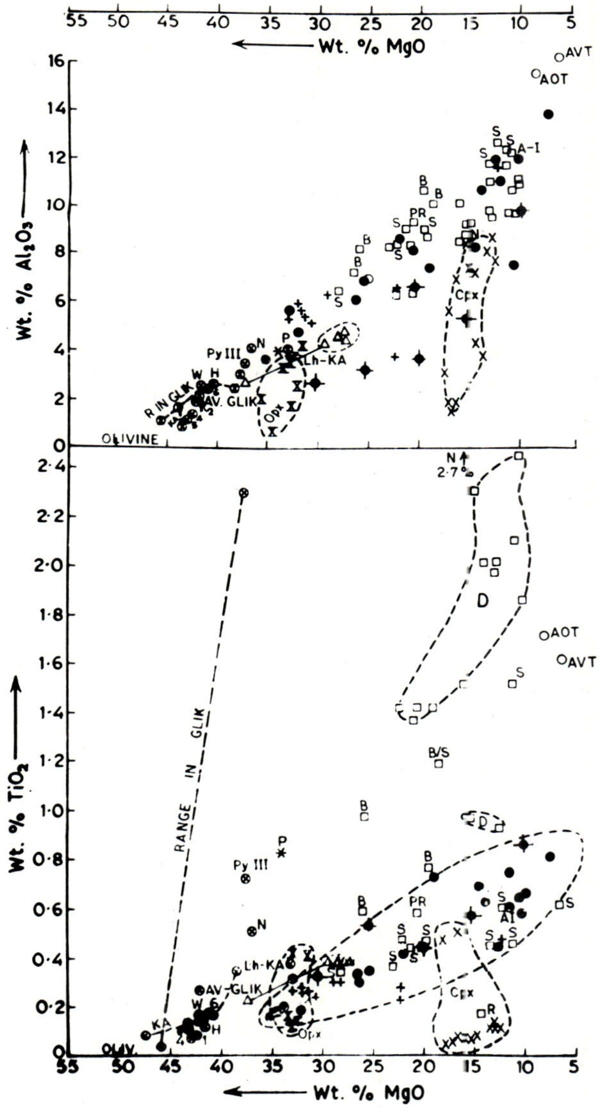

Py III = pyrolite III (Ringwood, 1966);
H = Harris (1967);
W = White (1967);
N = Nicolls (1967);
Nos.1,2,4-6 = compositions of the upper mantle estimated from the major element composition and modal percentage of constituent minerals (Maaloe and Aoki, table 2, 1977);
triangles plot the experimentally melted sheared garnet peridotite and the peridotitic komatiitic compositions of Mysen and Kushiro (1976).

AOT and ACT are average olivine tholeiite and average tholeiite, respectively (Manson, 1967, p. 217).

Olivine, clino- and orthopyroxene compositions from Arndt et al. (tables II & III, 1977); Krishnamurthy and Cox (tables 1 & 2, 1977); Kuno and Aoki (table 13, columns 2,9,12-14, 1970); Deer et al. (table 2, Nos. 3,4,6, 1963).

Fig. 4b. Variation of some major and minor oxides with MgO as an index of evolution.

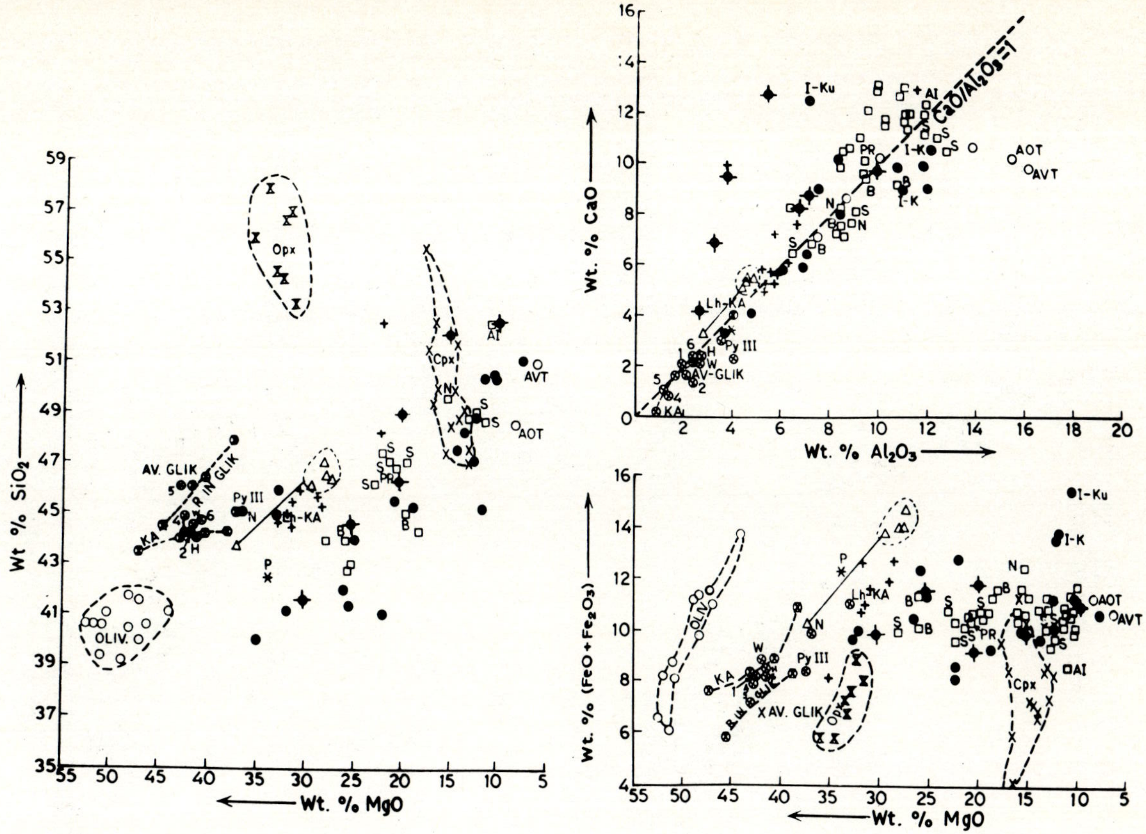

Fig. 4c. Variation of some major and minor oxides with MgO as an index of evolution.

equilibrium with olivine and orthopyroxene at c. 35 km (c. 12 kb), then they ascended to a depth of 18 km (c. 6 kb) fractionating olivine and orthopyroxene, followed by rapid eruption. Initiation of the Karroo volcanic cycle in general and the eruption of the picritic rocks in particular has been attributed to the uprise of mantle material to such shallow depths during the initial stages of the break up of Gondwanaland (Cox, 1972). From phase equilibria and chemical features a depth of c. 100 km was proposed for the parental liquids of the picritic rocks of the Deccan (Krishnamurthy, 1974).

Thus, taking into consideration evidence by different workers for the depth of formation of these ultramafic and mafic liquids, it appears that the depth of equilibration could have varied depending upon the local tectonic and thermal framework from 12 Kb (as at Nuanetsi) to as high as 40 Kb (as suggested by Arndt, 1976). A crucial factor is the temperature required (1600-1750°C) to obtain compositions akin to peridotitic liquids and quick transportation of these melts with very little modification to the surface. Steeper geothermal gradients and shallower depths coupled with triggering by meteoritic impacts (Green, 1972) may provide a comparison for some of the Archaean occurrences.

CHEMICAL CONSIDERATIONS

Major and some minor oxide variations of the Archaean komatiitic suite from Western Australia, Canada and South Africa are plotted in Fig. 4 along with Phanerozoic non-cumulus picritic rocks. Postulated upper mantle source rocks are also plotted along with olivine, orthopyroxene and clinopyroxene, which are important either as liquidus phases during cooling or in the source rock. Several important features emerge from such a treatment.

The broad chemical similarity between the komatiitic suite and the non-cumulus picritic rocks with reference to certain major oxides (i.e. SiO_2, Al_2O_3, CaO, FeO + Fe_2O_3) can clearly be seen. With regard to certain minor oxides (i.e. K_2O and TiO_2), using the revised criteria of Brooks and Harte (1974) and Arndt et al. (1977), there are some similarities as well as differences. The similarities are valid with regard to samples from Baffin Bay, the Solomon Islands and from sundry localities such as the Aleutian Islands and Puerto Rico.

Olivine control for the peridotitic komatiites can clearly be seen in the plots of CaO and Al_2O_3, for the Western Australian and Munro Township occurrences, whereas the South African samples seem to have been influenced by some clinopyroxene and orthopyroxene fractionation. This feature for the South African occurrence was pointed out by McIver (1975) using the various sub-projections within the CMAS system of O'Hara (1968). Clinopyroxene fractionation at c. 15% MgO level can also be seen in the Munro Township samples.

The SiO_2 variation shows a scatter generally away from the olivine-clinopyroxene join which may be attributed to the influence of these phases. Total iron for most of the samples remains at the 10-12% level for the braod range of MgO values from 35-10%, indicating the non-participation of magnetite or ilmenite in their evolution. It is interesting to note that the peridotitic komatiitic compositions obtained from an experimentally melted garnet peridotite at 20 Kb and at temperatures varying from 1600°C to 1750°C under anhydrous and hydrous conditions (Mysen and Kushiro, 1976) are rather rich in iron compared with the natural samples, inspite of about 20% iron loss during runs at 20 Kb. The reported basaltic komatiitic compositions from Kolar and Kulamara (Viswanathan, 1974) are also richer in iron than comparable types.

TiO_2 varies from c. 0.2% (in the peridotitic types) to c. 0.9%, the higher values being characteristic of the basaltic komatiites. Some picritic rocks from the Solomon Islands, Aleutian Islands and Puerto Rico also have low TiO_2 values at comparable stages of evolution. Basaltic komatiite from the Rambler area, Newfoundland, appears to have the lowest TiO_2 content (0.16%). Some of the Deccan picritic rocks from the Dedan area have relatively low TiO_2 contents (c. 1%) when compared with the majority (Krishnamurthy, 1974). K_2O contents in the komatiitic suite as a whole, and particularly in the peridotitic and pyroxenitic types, are low (less than 0.15%), and some of the Baffin Bay picrities have comparable values. However, the picritic rocks from the Deccan, Solomon Islands and Nuanetsi have increasing amounts of K_2O, even at a particular stage of evolution, as at the 10% MgO level. The picritic rocks of the Nuanetsi (average of 47) show maximum values of K_2O at c. 15% MgO level. Enrichment of incompatible minor and trace elements in basic and ultrabasic magmas has been dealt with by a number of workers and may be explained by a variety of possible factors, such as source rock heterogeneity, mantle wall-rock reaction, degree of melting and crustal contamination (Jamieson and Clarke, 1970).

CaO/Al_2O_3 RATIOS

Although high CaO/Al_2O_3 ratios (normally 1 and above) form an important criterion for distinguishing rocks of komatiitic chemistry in South Africa (Viljoen and Viljoen, 1969), subsequent workers from other Archaean areas Nesbitt and Sun, 1976; Arndt et al., 1977) have pointed out that a ratio around 1 or even slightly less than 1 is also equally characteristic of this suite of rocks. It can be seen from Fig. 4 that the South African samples are characterised by higher ratios, whereas those from Canada and Western Australia are closer to 1. Picritic rocks from Baffin Bay, the Solomon Islands and the Deccan have comparable ratios, although the Deccan picrites consistently show ratios of slightly more than 1. The importance of CaO/Al_2O_3 ratios in this suite of rocks is because CaO and Al_2O_3 behave like incompatible elements under high degrees of melting and in a way reflect the CaO/Al_2O_3 ratios of the source rocks, and also provide a measure of the fertile or barren nature of the source. Postulated source rock compositions give a range of CaO/Al_2O_3 ratios varying from c. 0.6 to 1.2 (see Fig. 4), and hence the higher CaO/Al_2O_3 ratios indicate their enrichment or otherwise in these components. Furthermore, the high CaO/Al_2O_3 ratio of the South African rocks has been attributed to selective removal of garnet during the early stages of adiabatic diapiric uprise of mantle material from c. 200 km depths (Green, 1975). Thus fractionation of mineral phases during magma uprise en route to the surface can also cause variations or increases in the CaO/Al_2O_3 ratios. In addition, Nesbitt and Sun (1976) point out the strong possibility of non-isochemical metamorphism for increased CaO/Al_2O_3 ratios in which Al can be lost and Ca may be gained. However, it must be noted that higher CaO/Al_2O_3 ratios in isolated basaltic types could also be due to several factors, such as, firstly, fractionation and accumulation of clinopyroxenes, aluminous phases (i.e. garnet or spinel) being left behind as residues if the basaltic types themselves are partial melts representing lower degrees of melting of the source (less than 20%); and, secondly, eclogite fractionation of the primitive parental liquid from which the basaltic type was possibly derived. Thus the CaO/Al_2O_3 ratios of the whole suite should be taken into consideration.

An important point which emerges from these chemical considerations is that in the Phanerozoic there is a paucity of lavas akin to peridotitic komatiites, or such types have not yet been reported. Most of the primary or very primitive picritic liquids have MgO contents in the range 15-20%, and more basic compositions than this within a particular suite may have been influenced by olivine accumulation from the 15-20% liquids because of the relatively slower rates of cooling in some of the picritic types when compared with the komatiites in this MgO range. These younger picritic liquids apparently resulted through relatively lower degrees of melting. Nesbitt and Sun (1976) proposed 30% melting for the most primitive Baffin Bay picrites containing c. 18-20% MgO, and for the most primitive Deccan picritic rock (c. 16-19% MgO) the estimated degree of melting is around 20% (Krishnamurthy, 1974). Therefore 20-30% melting of the source rock could, in general, account for the major element characters of these younger primitive or primary liquids when compared with the peridotitic komatiites which apparently result from more than 45% melting (cf. Mysen and Kushiro, 1976). Thus it appears that the thermal gradient was slightly lower in the Phanerozoic compared with the Archaean.

An important aspect which seems to arise from both these phase equilibria and chemical parameters is that given a fertile source rock of a particular type, say garnet or spinel lherzolite, it is only the degree of melting which appears to be very important and which in turn depends essentially on the temperature available above the solidus, resulting in mafic or ultramafic liquids which are broadly comparable with respect to major element compositions. Differences in minor and trace element chemistry within and between these suites of rocks at comparable

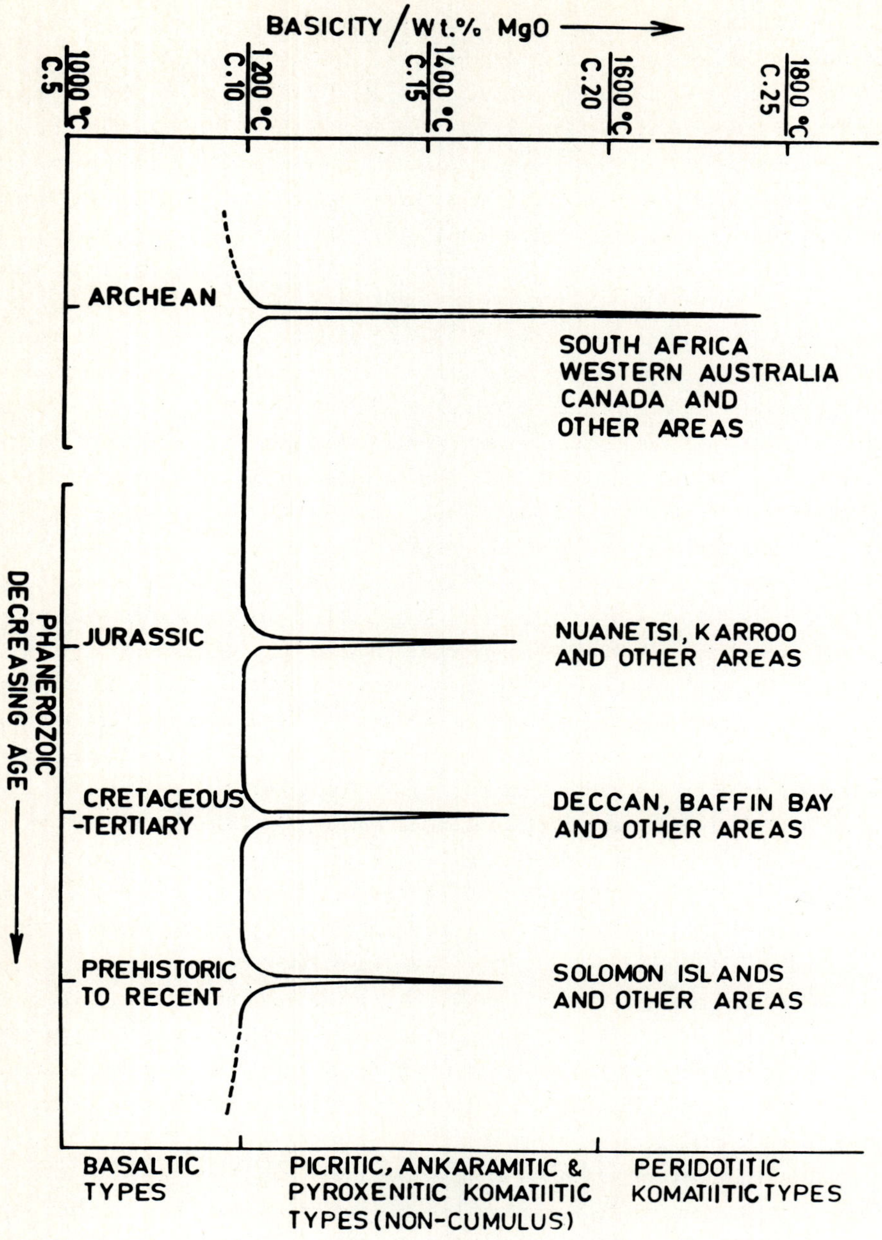

Fig. 5. Schematic diagram showing the broad relationship between basicity of erupted liquids and decreasing age. MgO contents of the liquids and the temperatures needed to produce these are very approximate. A simplified classification of the erupted products is also given.

stages of evolution, say on an MgO index, may be attributable to several factors such as degree of melting, high pressure fractionation, mantle wall rock reaction or crustal contamination (Jamieson and Clarke, 1970; Harris et al., 1972; Sun and Nesbitt, 1977). Additional problems posed in this context are effects of weathering and metamorphism on the chemistry, especially with regard to minor and trace elements such as K, Ba, Sr, Rb, etc. (Condie, 1976).

Considering that peridotitic or picritic types are very minor in erupted sequences as a whole, whether in the Archaean or the Phanerozoic, it appears that certain thermal regimes, apparently cyclical in nature, are of paramount importance in their formation. This is shown schematically in Fig. 5; the available data suggest that peridotitic komatiites are unique to the Archaean, whereas the lower magnesian types were repeatedly erupted in the Phanerozoic in association with major episodes of basaltic volcanism.

Finally, plausible answers to some of the questions raised here are: (1) Peridotitic komatiites, which are most probably parental to the komatiitic suite, may have originated at depths of c. 60-120 km (c. 20-40 kb) at c. 1600-1750°C, as a result of high degrees (greater than 45%) of partial melting. (2) The source rock is most probably akin to sheared garnet peridotite nodules in kimberlite in which the clinopyroxenes are possibly in excess of garnet or spinel. (3) Peridotiitic komatiites appear to be unique to the Archaean, whereas types analogous to basaltic komatiites were repeatedly formed in younger times. (4) Considering the fact that rocks akin to basaltic komatiite occur in younger and different tectonic environments, tectonic models using or based on the presence of the komatiitic suite appear to be of limited value.

ACKNOWLEDGEMENTS

I wish to express my sincere thanks to C.R. Udas, Director, Atomic Minerals Division, for permission to publish this paper. Thanks are also due to T.N. Mahadevan, N. Nagaraja Rao, S. Viswanathan and Bajaj for encouragement and assistance during the preparation of the manuscript. During my stay in the U.K. I benefited immensely from discussions on basalt genesis with Dr K.G. Cox, my Guru at the Department of Geology and Mineralogy, University of Oxford, and Professor M.J. O'Hara and Drs B.G.J. Upton and J.G. Fitton at the Grant Institute of Geology, University of Edinburgh, and to them I owe a deep sense of gratitude.

REFERENCES

Anhaeusser, C.R., 1971. Cyclic volcanicity and sedimentation in the evolutionary development of Archaean greenstone belts of shield areas. Sp. Publ. geol. Soc. Austr., 3: 37-70.

Arndt, N.T., 1976. Melting relations of ultramafic lavas (komatiites) at one atmosphere and high pressure. Carnegie Inst. Washington Yearbk., 75: 555-562.

Arndt, N.T., Naldrett, A.J. and Pyke, D.R., 1977. Komatiitic and iron-rich tholeiitic lavas of Munro Township, Northeast Ontario. J. Petrol., 18(2): 319-369.

Basu, A.R. and Murthy, V.R., 1977. Ancient lithospheric lherzolite xenolith in alkali basalt from Baja California. Earth Planet. Sci. Lett., 35: 239-249.

Brooks, C. and Hart, S.R., 1974. On the significance of komatiite. Geology, 2: 107-110.

Carswell, D.A. and Dawson, J.B., 1970. Garnet peridotite xenoliths in South African kimberlite pipes and their petrogenesis. Contrib. Miner. Petrol., 25: 163-184.

Clarke, D.B., 1970. Tertiary basalts of Baffin Bay: possible primary magma from the mantle. Contrib. Miner. Petrol., 25: 203-225.

Condie, K.C., 1976. Trace-element geochemistry of Archean greenstone belts. Earth Sci. Rev., 12: 393-417.

Cox, K.G., 1972. The Karroo volcanic cycle. J. geol. Soc. Lond., 128: 311-336.

Cox, K.G. and Bell, J.D., 1972. A fractionation model for the basaltic rocks of the New Georgia Group, British Solomon Islands. Contrib. Miner. Petrol., 37: 1-13.

Cox, K.G., Gurney, J.J. and Harte, B., 1973. Xenoliths from the Motsoku pipe. In: P.H. Nixon (Editor), Lesotho Kimberlites. Cape and Transvaal Printers, Cape Town, pp. 76-101.

Cox, K.G. and Jamieson, B.G., 1974. The olivine-rich lavas of Nuanetsi: a study of polybaric magmatic evolution. J. Petrol., 15: 269-301.

Deer, W.A., Howie, R.A. and Zussman, J., 1963. Rock forming minerals. 2: Chain silicates. Longmans, London.

Gale, G.H., 1973. Palaeozoic basaltic komatiite and ocean floor type basalts form northeastern Newfoundland. Earth Planet. Sci. Lett., 18: 22-28.

Green, D.H., 1972. Archean greenstone belts may include terrestrial equivalents of lunar maria. Earth Planet. Lett., 15: 263-270.

Green, D.H., 1975. Genesis of Archean peridotitic magmas and constraints on Archean geothermal gradients and tectonics. Geology, 3: 15-18.

Harris, P.G., 1967. Segregation processes in the upper mantle. In: S.K. Runcorn (Editor), Mantles of the Earth and Terrestrial Planets. Interscience, New York, pp. 305-317.

Harris, P.G., Hutchinson, R. and Paul, D.K., 1972. Plutonic xenoliths and their relation to the upper mantle. Phil. Trans. R. Soc. Lond., 271: 313-323.

Ito, K. and Kennedy, G.C., 1968. Melting and phase relations in the plane tholeiite-lherzolite-nepheline basanite to 40 kilobars with geological implications. Contrib. Miner. Petrol., 19: 177-211.

Jamieson, B.G. and Clarke, D.B., 1970. Potassium and related elements in tholeiitic basalts. J. Petrol., 11: 183-204.

Krishnamurthy, P., 1974. Deccan trap lavas from western India. Unpubl. Ph.D. Thesis, Univ. Edinburgh, 129 pp.

Krishnamurthy, P. and Cox, K.G., 1977. Picrite basalts and related lavas from the Deccan traps of western India. Contrib. Miner. Petrol., 62: 53-75.

Kuno, H. and Aoki, K., 1970. Chemistry of ultramafic nodules and their bearing on the genesis of basaltic magmas. Earth Planet. Interiors, 3: 273-301.

Maaloe, S. and Aoki, K.I., 1977. The major element composition of the upper mantle estimated from the composition of lherzolites. Contrib. Miner. Petrol., 63: 161-173.

Manson, V., 1967. Geochemistry of basaltic rocks: major elements. In: H.H. Hess and A. Poldervaart (Editors), Basalts. Interscience, New York, pp. 215-269.

McIver, J.R., 1975. Aspects of some high magnesia eruptives. Contrib. Miner. Petrol., 51: 99-118.

Mysen, B.O. and Holloway, J.R., 1976. Rare earth fractionation with controlled partial melting of peridotite. Carnegie Inst. Washington Yearbk., 75: 675-677.

Mysen, B.O. and Kushiro, I., 1976. Compositional variation of coexisting phases with degree of melting of peridotite under upper mantle conditions. Carnegie Inst. Washington Yearbk., 75: 546-555.

Nesbitt, R.W. and Sun, S.S., 1976. Geochemistry of Archean spinifex-textured peridotites and magnesian and low magnesian tholeiites. Earth Planet. Sci. Lett., 31: 433-453.

Nicholls, G.D., 1967. Geochemical studies in the ocean as evidence for the composition of the mantle. In: S.K. Runcorn (Editor), Mantles of the Earth and Terrestrial Planets. Interscience, New York, pp. 285-292.

Nockolds, S.R., 1954. Average chemical compositions of some igneous rocks. Bull. geol. Soc. Amer., 65: 1007-1032.

O'Hara, M.J., 1968. The bearing of phase equilibria studies in synthetic and natural systems on the origin and evolution of basic and ultrabasic rocks. Earth Sci. Rev., 4: 69-133.

O'Hara, M.J., 1970. Upper mantle composition inferred from laboratory experiments and observation of volcanic products. Phys. Earth Planet. Interiors, 3: 236-245.

Ringwood, A.E., 1966. The chemical composition and origin of the earth. In: P.M. Hurley (Editor), Advances in Earth Sciences. M.I.T. Press, Cambridge, Mass.

Stanton, R.L. and Bell, J.D., 1969. Volcanic and associated rocks of the New Georgia Group, British Solomon Island Protectorate. Overseas Geol. Mineral Resources, 10: 113-145.

Sun, S.S. and Nesbitt, R.W., 1977. Chemical heterogeneity of the Archean mantle, composition of the earth and mantle evolution. Earth Planet. Sci. Lett., 35: 429-448.

Viljoen, M.J. and Viljoen, R.P., 1969. A reappraisal of the granite-greenstone terrains of shield areas based on the Barberton model. Sp. Publ. geol. Soc. S.Afr., 2: 245-274.

Viswanathan, S., 1974. Contemporary trends in geochemical studies of early Precambrian greenstone-granite complexes. J. geol. Soc. India, 15: 347-379.

White, I.G., 1967. Ultramafic rocks and the composition of the upper mantle. Earth Planet. Sci. Lett., 3: 11-15.

Williams, D.A.C., 1972. Archean ultramafic, mafic and associated rocks, Mt. Monger, Western Australia. Geol. Soc. Austr., 19: 163-188.

GEOCHEMISTRY OF METAVOLCANICS FROM THE BABABUDAN SCHIST BELT; A LATE ARCHAEAN/EARLY PROTEROZOIC VOLCANO-SEDIMENTARY PILE FROM INDIA

Y.J. Bhaskar Rao and S.M. Naqvi
National Geophysical Research Institute, Hyderabad, India

Abstract

The lithological sequence in the Bababudan schist belt, Karnataka, represents a late Archaean/early Proterozoic volcano-sedimentary association. The metavolcanic suite, the dominant unit of the sequence, is mainly a bimodal (basalt/ basaltic andesite-rhyodacite) suite with low to moderate potash of tholeiitic association. It displays a mineralogy characteristic of the greenschist facies of regional metamorphism. The mafic metavolcanics in some places show nepheline normative character as a result of secondary enrichment of soda. These rocks show (significantly) low Cr, Co and Ni concentrations unlike most Archaean/Lower Proterozoic metatholeiites and Phanerozoic oceanic tholeiites. This anomaly may be attributed to the degree and nature of differentiation of the Bababudan melt involving an early segregation of the Mg-rich silicate phases.

The differences in the chemical characters of the Bababudan and the younger Chitradurga metavolcanic suite can be attributed to the differences in their environmental conditions of extrusion, implying that the respective parental melts had broadly comparable compositions. The chemical data point to the possibility of a rather coeval but spatially different evolution of the Bababudan and Chitradurga sequences in the geosynclinal piles of the Dharwar craton.

INTRODUCTION

Geochemical studies on mafic and ultramafic rock associations have assumed great significance as they impart information on the physico-chemical conditions under which they were emplaced and help in our understanding of the processes of crustal evolution. The greenstone belts of the Dharwar craton preserve mafic and ultramafic rocks pertaining to a wide span of time from the early Archaean to the late Proterozoic, and a systematic study of these will assist in the elucidation of the early history of the earth.

The Bababudan schist belt, in the west central part of southern Karnataka, preserves an interesting late Archaean/early Proterozoic volcano-sedimentary pile. The metavolcanics of this sequence are the largest single and (probably) the thickest of the metavolcanic suites in the craton. Previous work on the geology of the region (summarised in Pichamuthu, 1935, 1974) concentrated on aspects of

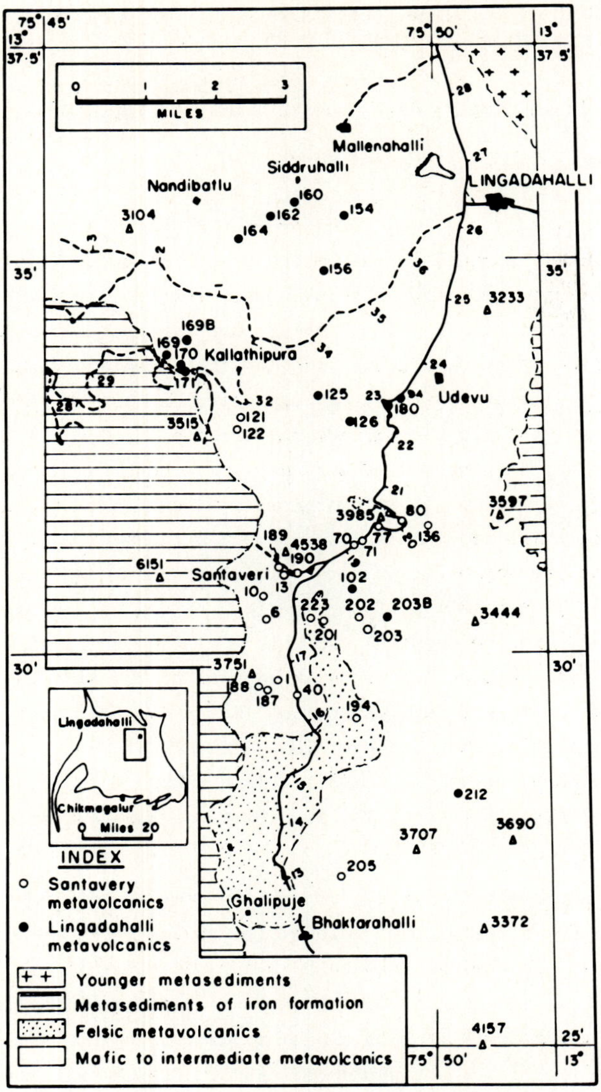

Fig. 1. A geological outcrop map of the Lingadahalli and Santavery areas in the eastern part of Bababudan schist belt (outline of the belt shown in the inset), Karnataka, India, with sample locations.

the iron formations in the belt. The geochemistry of associated metavolcanics, generally referred to as the 'Lingadahalli (LH) and Santavery (SV) traps', has not been studied in detail. In an effort to bridge this gap in our knowledge, samples were collected from the region around Lingadahalli and Santavery and studied for their petrography and geochemistry (Fig. 1).

We present our analyses here and compare them with those of metavolcanics from the other Archaean and early Proterozoic schist belts of the Dharwar craton and with their analogues from other shield areas. With the help of data on basaltic rocks from different geological settings and ages, an attempt is made to understand the late Archaean/early Proterozoic crustal development in the Indian Peninsula.

GENERAL GEOLOGY

The stratigraphic position of the Bababudan sequence in the Dharwar stratigraphy was discussed by several workers, but some of the recent work, applying modern concepts to the problem of greenstone belts (Radhakrishna, 1967; Srinivasan and Sreenivasa, 1972, 1975; Swami Nath et al., 1976; Radhakrishna and Vasudev, 1977) are pertinent to our objective. There is general agreement with regard to a late Archaean/early Proterozoic position of the Bababudans in the Dharwar craton. Swami Nath et al. (1976) consider the sequence represents a 2600-2700 my old shallow water platformal association constituting the lower part of their "Dharwar supergroup". A generalised lithological succession in the Bababudan belt after Swami Nath et al. (1976) is:

		Approx. thickness m
Chemical precipitates and euxetic shales	Chloritic argillites Iron formations Chloritic argillites	250
Mainly volcanic sequence	Mafic to felsic metavolcanics	600
Mafic platformal sequence	Amygdaloidal metabasites interbedded with current bedded and ripple marked quartzites, chloritic schist	350
	Basal oligomictic conglomerate (pyritiferous and uraniferous)	

- - - - - - - - - - - U N C O N F O R M I T Y - - - - - - - - - - - - -

Granites and migmatites

The succession commences with a quartz-pebble oligomictic conglomerate horizon exposed extensively along the southeastern margin of the Bababudan schist belt. This important basal conglomerate horizon in the Dharwar craton contains detrital pyrite and uraninite suggestive of an anoxygenic atmospheric condition during its deposition. It is closely associated with the other rock types of the platformal sequence that include several cyclically repeated volcanic and sedimentary horizons grading upwards in the succession to an essentially volcanic sequence comprising mafic-felsic metavolcanics (LH and SV traps). The metavolcanics are overlain by a metasedimentary assemblage of mainly iron formations (banded ferruginous quartzites and quartzites) that seem to occupy certain isolated basins on the 'metavolcanic platform'. Carbonaceous metasediments are conspicuous by their absence.

The sequence suffered at least two phases of deformation. Several phases of igneous intrusion are represented by dykes and differentiated sills.

Srinivasan and Sreenivas (1972, 1975), on the other hand, consider the sequence represents the pregeosynclinal phase of the 'Dharwars', and suggest that the LH and SV traps constitute the pregeosynclinal mafic platforms on which the metasediments were laid in a subaerial to shallow marine environment.

GENERAL CHARACTERISTICS OF THE METAVOLCANICS

The LH and SV metavolcanics of the main volcanic sequence (Fig. 1) vary from fine-grained massive to medium-grained granular and are for the most part amygdaloidal. Few instances of columnar jointing (predominantly six-sided) are encountered between Lingadahalli and Kallathipura. Pillow and variolitic structures were not observed in the area investigated but one instance of pillow lava was reported c. 12 miles SW of Santavery (Pichamuthu, 1957). Several tuffaceous and a few phychloritic outcrops were encountered and an effort is underway to decipher the various flows. The areal extent and the physical characters mentioned above are suggestive of subaerial to shallow marine conditions of extrusion.

The rocks vary in composition from Mg-rich basalts to rhyodacites. Our data indicate that the suite is essentially a bimodal mafic (basalt/basaltic andesite)-felsic (rhyodacite) association. The dominant rock type is a basaltic andesite closely associated with the basaltic and metapyroxenite (picritic?) variants. The felsic component overlies the mafic flows at several places around Santavery and is generally micro-porphyritic with acidic plagioclase, quartz, chlorite and carbonate.

Mineralogically the whole suite is an assemblage characteristic of the greenschist facies of regional metamorphism, dominated by chlorite, quartz, magnetite (or ilmenite, hematite and pyrite), actinolite and epidotes (mainly epidote, clinozoisite). In places the metavolcanics suffered intense carbonitisation and saussuritisation. The mineralogy of mafic and intermediate rocks from a region around Santavery is similar to that of spilites: chlorite, plagioclase, quartz, carbonates, hematite, pyrite, etc. Porphyritic and relict ophitic textures are common and microvariolitic and crude fluidal textures are seen in a few sections.

PETROCHEMISTRY OF BABABUDAN METAVOLCANICS

We consider here major and trace element analyses of thirty-six mafic and intermediate rocks. Major element oxides were estimated by wet rapid methods (after Shapiro and Branock, 1962) and trace elements by emission arc spectrography. Details of the methods are summarised in Naqvi and Hussain (1972; Geochemistry Group 1977).

The variation in the concentration of some constituents is shown by histograms depicting percentage frequency distributions (Fig. 2). Samples from a region around Santavery (SV) and near Lingadahalli (LH) are represented separately following a tentative grouping, mainly to project the existing chemical dissimilarities among them. The SV samples show enriched Na_2O (up to 8 wt %) and depleted CaO contents relative to the LH samples. This character and the mineralogy of

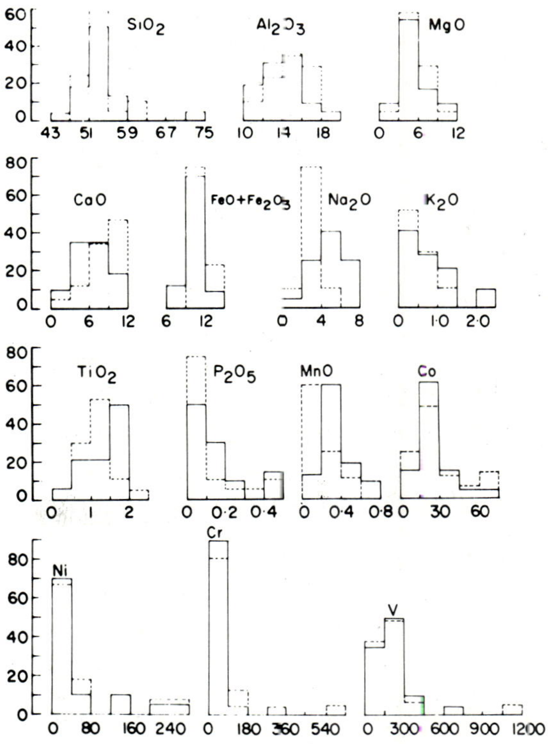

Fig. 2. Histograms for the Bababudan metavolcanics depicting percentage frequency distribution. The soda-enriched SV (continuous lines) and the sub-alkalic LH samples (broken lines) are represented separately.

the SV rocks conforms to those of spilites as suggested by Fiala (1974). Average analyses presented in Table 1 (along with the range) incorporate a consideration of SiO_2 wt % (CIPW normative composition, see Fig. 2), and the spilitic character (Na_2O wt % more than 4.70, and the typical mineralogy described in the previous section) of the metavolcanics. The analyses are included in the special publication, Geochemistry Group 1977, table 3a, pp. 8-11.

A majority (c. 80%) of the sub-alkalic basalts and basaltic andesites (Table 1, Fig. 2) have a moderate to high Al_2O_3 wt % (c. 13-17), moderate TiO_2 wt % (c. 0.60-1.50) and a low to moderate K_2O wt % (0.10-0.65). Total iron is less than 15 wt % but higher than MgO by a factor of c. 1.5-2.0 in most cases.

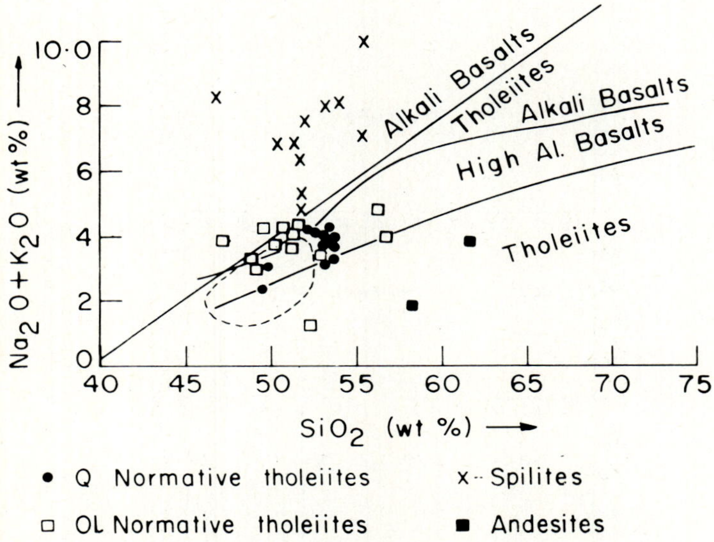

Fig. 3. Alkali-silica binary variation plot with fields of alkali and tholeiitic basalts after MacDonald and Katsura (1964) and of alkali, high-Al and tholeiitic basalts after Kuno (1967). The circular field includes metatholeiites from several Archaean/early Proterozoic greenstone belts of S. India, W. Australia, Canada and S. Africa.

In the alkali-silica binary variation plot (Fig. 3) all the subalkalic samples are tholeiites by the definition of MacDonald and Katsura (1964), and vary from tholeiites to high-Al basalts in composition by the definition of Kuno (1967). The samples show a distinct iron enrichment trend in the A-F-M diagram (Fig. 4) and are thus clearly non-calc-alkalic by the definition of both Kuno (1967) and Irvine and Baragar (1972).

TABLE 1. Average composition (and range) of Bababudan (Lingadahalli and Santavery) metavolcanics

	Rocks with < 54% SiO_2						54 - 57% SiO_2		57-63% SiO_2
	Ol-normative rocks	Range	Q-normative rocks	Range	Spilites	Range	Spilites	Range	Andesites
SiO_2	50.62	47.18-53.98	51.78	49.21-54.23	51.42	46.44-53.80	55.61	55.36-56.21	59.63
TiO_2	1.16	0.75-1.62	1.40	0.66-1.87	1.37	0.69-1.87	1.72	1.44-1.89	1.14
Al_2O_3	15.16	13.58-17.26	14.17	11.23-17.27	14.52	13.24-18.53	13.08	11.94-14.31	14.57
Fe_2O_3	2.45	1.25-3.31	2.05	0.56-3.75	1.93	0.93-3.37	1.27	1.09-1.38	1.86
FeO	8.45	7.80-12.18	8.80	7.41-9.88	8.77	7.14-10.17	9.56	8.63-10.64	10.21
MgO	6.90	4.62-8.87	5.76	3.65-9.91	5.42	2.72-9.38	4.45	4.19-4.86	3.01
CaO	2.48	3.18-10.65	8.64	6.50-10.61	6.00	3.09-9.98	4.20	1.85-5.94	2.52
Na_2O	3.60	2.20-4.32	2.86	1.85-3.92	6.03	4.70-7.85	6.26	4.85-7.98	1.81
K_2O	0.31	0.10-1.20	0.62	0.20-1.30	0.68	0.10-2.35	1.20	0.92-1.43	1.29
P_2O_5	0.17	* Tr-0.64	0.24	* Tr-0.64	0.14	0.01-0.43	0.03	* Tr-0.08	0.04
MnO	0.37	0.17-0.49	0.38	0.23-0.58	0.33	0.23-0.53	0.25	0.17-0.33	0.12
Co	37	10-84	32	18-72	27	8-70	21	18-25	12
Cr	84	10-580	28	10-70	48	10-310	16	10-24	10
Cu	93	10-350	116	10-1000	189	71-800	46	10-130	10
Ni	63	10-270	40	10-240	56	10-240	50	10-145	161
V	200	10-600	164	47-300	253	10-1000	100	11-180	175

* Tr denotes concentration in traces
Only samples having trace element abundances much above the lower detection limit (10) are included in the average of trace elements
Major oxides in weight % and trace elements in ppm

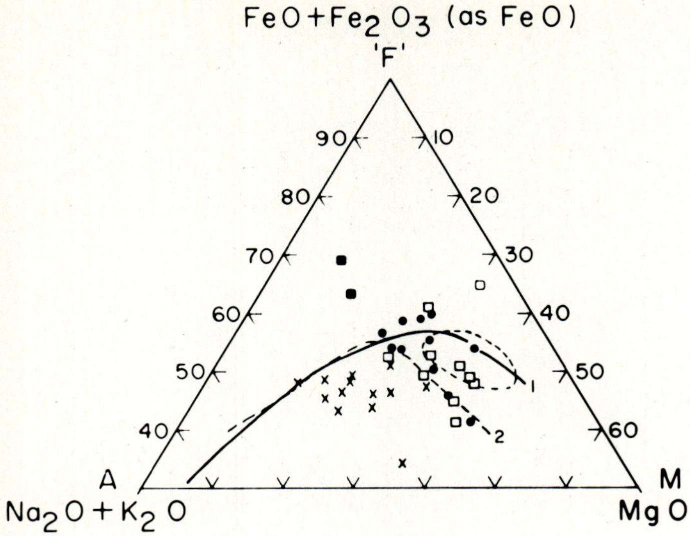

Fig. 4. A-F-M ternary variation diagram showing the non-calc-alkaline iron enrichment trend of the sub-alkalic Bababudan metavolcanics. Boundaries of calc-alkaline and non-calc-alkaline volcanic suites after Kuno (1967) (curve 1) and Irvine and Baragar (1972) (curve 2). Legend of sample symbols and description of the circular field as in Fig. 3.

The spilitic samples plot in the field characteristic of alkali basalts in the alkali-silica plot. They show alkali olivine basalt normative compositions (nepheline and olivine normative, Fig. 5). In some samples their Na_2O wt % is much higher (by up to c. 3 wt %) than those of average spilites and their Na_2O/K_2O values show a wide range of variation. Moreover, since these samples come from a specific region (around Santavery) that belongs to a stratigraphically higher position in the metavolcanics, one might be tempted to invoke from their chemistry the existence of flow(s) of alkali olivine basalt type in the Santavery region. Alkali basalt suites (mafic trachytes, trachyandesites, nepheline- and leucite-bearing lavas, etc.) have been described from near Kirkland Lake, Ontario, Canada, and in the Fig Tree Series, Barberton Mountainland, S. Africa (cited in Anhaeusser, 1973). Alkali basalts like these are rare in Archaean and lower Proterozoic greenstone belts and confirmation of their existence elsewhere would have intriguing tectonic implications.

K_2O in the metavolcanics is independent of enrichment in Na_2O (Fig. 6). A statistical test for the LH and SV samples was attempted following the approach of Narebski (1974) to the spilite problem. Our results are presented in Table 2

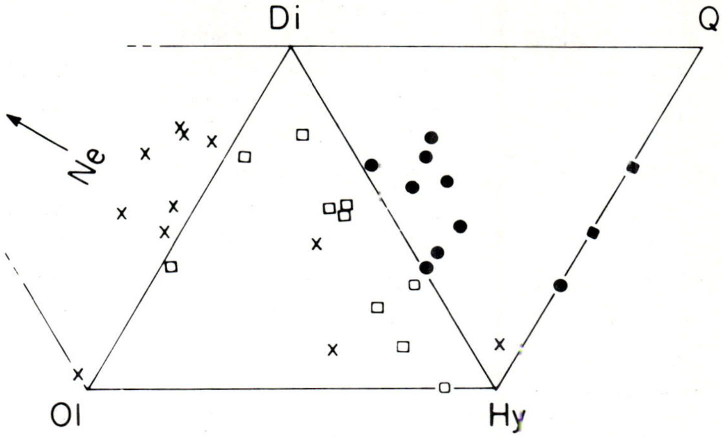

Fig. 5. The CIPW normative variation in the Bababudan rocks (from alkali olivine to saturated compositions). Legend of sample symbols as in Fig. 3.

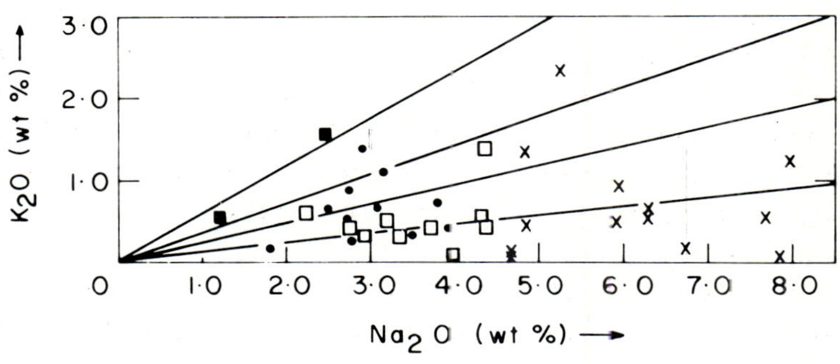

Fig. 6. Na_2O-K_2O binary variation in the Bababudan metavolcanics showing an enrichment in Na_2O independent of K_2O. Legend for samples as in Fig. 3.

which shows that most major element oxides have a normal distribution (except for K_2O, MnO and P_2O_5) and the traces approach a log-normal distribution pattern. CaO shows some bimodality.

TABLE 2. Coefficients of correlation between some elements in the Lingadahalli-Santavery metatholeiites (basalts and basaltic andesites N = 34)

Na_2O-SiO_2	-0.117	Na_2O-FeO/MgO	-0.359
Na_2O-K_2O	0.115	Na_2O-FeO/MnO	-0.579
Na_2O-Al_2O_3	-0.145	Na_2O-Fe_2O_3/FeO	-0.260
**Na_2O-CaO	-0.465	SiO_2-FeO /MgO	-0.026
Na_2O-MgO	0.063	SiO_2-MgO	-0.579
MgO-Cr	0.315	SiO_2-Fe_2O_3/FeO	-0.113
**CaO-FeO/MnO	-0.579	SiO_2-Al_2O_3	-0.231
SiO_2-K_2O	0.487	**CaO-Al_2O_3	-0.111
K_2O-MgO	-0.310		

The underlined coefficients are significant at 95% level
** Are not useful as CaO shows some bimodality

The correlation coefficients (Table 2) for the pairs SiO_2-MgO, MgO-Cr and SiO_2-K_2O, etc. (significant but low to moderate correlations) are as expected for a differentiated volcanic suite. A significantly good correlation between the pairs Na_2O-FeO/MnO and Na_2O-FeO/MgO, together with a low but significant negative Na_2O-MgO correlation, a negative Na_2O-SiO_2 correlation and a low Na_2O-K_2O correlation, suggests a secondary enrichment of Na_2O independent of K_2O. Due to the bimodality of CaO, the behaviour of Ca and Al in relation to Na in the process is difficult to interpret but a decrease in these elements, with the increase in Na, is apparent from the low and insignificant negative CaO-Al_2O_3 and Na_2O-CaO coefficients. Thin sections of the Na-rich rocks sometimes show two generations of plagioclase which are highly carbonatised in some cases. The plagioclases generally constitute a felt-like network. Microlite and fluidal textures are very rare.

Furthermore, there are very insignificant differences in the abundance levels and pattern of distribution of the ferromagnesian traces (Co, Cr, Ni, V, Cu) among the undersaturated (Ne normative, Na_2O enriched) and saturated rocks (Fig. 2). These features suggest that the SV and LH suites of metavolcanics suffered secondary soda enrichment and that the observed alkali olivine basalt character is a consequence of this process.

LATE ARCHAEAN/EARLY PROTEROZOIC VOLCANISM OF THE DHARWAR CRATON

A comparison was made of the composition of the Bababudan (BBN) metavolcanics (LH and SV rocks) with that of similar rocks from other Archaean and Lower Proterozoic greenstone belts within the Dharwar craton and from other shields (viz. W. Australia, Canada, S. Africa, amphibolites from Greenland) and with basaltic tholeiites from various Phanerozoic tectonic environments (viz. continental, oceanic-abyssal/ridge/island, island arc, etc.) (Ramachandra Rao, 1937; Engel et al., 1965; Wilson et al., 1965; Manson, 1967; Baragar and Goodwin, 1969; Viljoen and Viljoen, 1969; Jakes and Gill, 1970; Glikson, 1971, 1972; Jakes and White, 1971; Hallberg and Williams, 1972; Naqvi and Hussain, 1973; Naqvi et al., 1974; Arth and Hanson, 1975; Middlemost, 1975; Rivalenti, 1976). The BBN suite differs from most of the Archaean/early Proterozoic greenstone suites when the overall chemical variation is considered. However, the sub-alkalic BBN rocks have similar major element abundances, but the concentration of their ferromagnesian traces (especially Cr, Ni and Co) are significantly much lower than those in the average Archaean metatholeiites. The concentrations of these elements in the BBN rocks are lower than those in the Phanerozoic oceanic tholeiites and lie within the range of island arc and some continental tholeiite suites. The physical features and the chemistry of the BBN volcanic suite are thus similar to Phanerozoic continental plateau or flood basalts.

The Chitradurga (CHD) volcanic suites (Jogimardi, Mardihalli, etc.) are generally thought to represent the next major volcanic episode after that of the Bababudan in the Dharwar craton (Radhakrishna, 1967; Swami Nath et al., 1976; Radhakrishna and Vasudev, 1977; Naqvi et al., 1978); a comparative study of the compositional features of the two suites will help to evaluate the secular changes in chemistry of volcanic rocks from the late Archaean through the middle Proterozoic.

The (CHD) metavolcanics from the Jogimardi and Mardihalli suites vary from mafic to intermediate in composition. They are associated with metasediments of deep marine (eugeosynclinal) facies - argillites, greywackes, cherts, etc. They show calc-alkalic chemical affinities but are characterised by higher abundance of ferromagnesian traces (Co, Cr, Ni, V, Cu, etc.). The level of abundance of these elements in the CHD rocks is similar to that in Phanerozoic oceanic tholeiites (Naqvi and Hussain, 1973). The textural features of the rocks (Pichamuthu, 1930, 1957) resemble those of "quench tholeiites" (Pearce, 1974).

The sub-alkalic BBN metabasaltic rocks have a major element chemistry broadly comparable with that of the CHD metabasalts but differ on account of their low concentration of Co, Cr and Ni (Table 1). This feature is evident from Fig. 7 in which trace elements (Cr, Ni and V) are plotted agains total iron as FeO/MgO. In these, and in a Cr vs. Ni plot, the CHD metavolcanics plot in or close to the fields characteristic of abyssal tholeiites, while the BBN samples occupy fields of volcanic rocks from stable and active continental regimes (fields after Miyashiro and Shido, 1975).

Apart from differences in the abundance of ferromagnesian trace elements (mainly Cr, Co and Ni) among the BBN and CHD mafic and intermediate rocks, differences in the degree and nature of the crystallisation and differentiation undergone by their respective melts are obvious in Figs. 8 and 9. The depleted Co

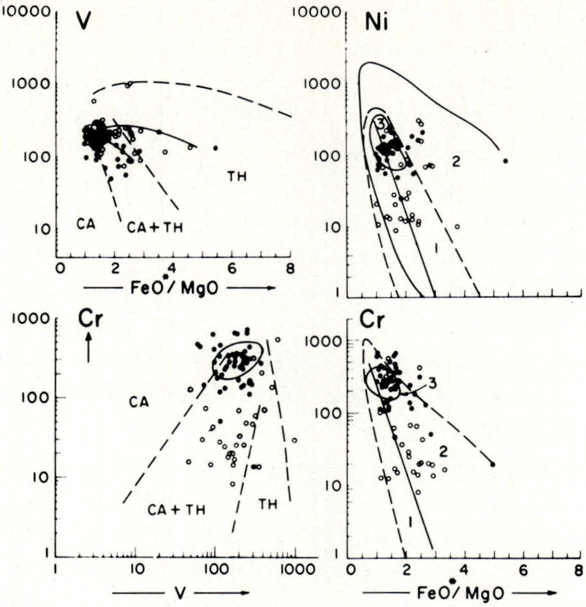

Fig. 7. Binary variation of Cr vs. V, and V, Ni and Cr vs. total iron as FeO/MgO, showing the depleted trace element concentrations of the Bababudan metavolcanics (unfilled circles), compared with younger Chitradurga metavolcanics (filled circles). The Chitradurga metavolcanics have trace element abundances similar to abyssal tholeiites, while the Bababudan rocks show abundances similar to tholeiitic basalts from continental areas. 1 = field of volcanic rocks from island arcs and active continental margins; 2 = field of volcanic rocks from stable continental and oceanic regions; 3 = field of abyssal tholeiites. Fields after Miyashiro and Shido (1975).

and Ni levels in the BBN rocks compared to the CHD ones in the Co-V-Ni ternary diagram (Fig. 8) suggests that the BBN melt probably underwent a relatively higher degree of differentiation prior to consolidation. The behaviour of SiO_2, FeO and TiO_2 of the CHD metavolcanics in Fig. 9 is suggestive of their mild calc-alkaline affinity relative to the BBN rocks (a steeper trend in SiO_2, and a gentle monotonic decrease in TiO_2 and FeO with increasing FeO/MgO; Miyashiro and Shido, 1975). The behaviour of these elements suggests that the melts of the two suites had significant differences in the volatile partial pressures and fugacity of oxygen etc. which contribute to diversity in trends of crystal differentiation, particularly the stages of separation of Fe- and Ti-rich oxide phases. The behaviour of MgO, CaO, Al_2O_3 and SiO_2 against fractionation indices F = FeO/(FeO+MgO) (Fig. 9) suggests that differentiation of both BBN and CHD melts was dominated by

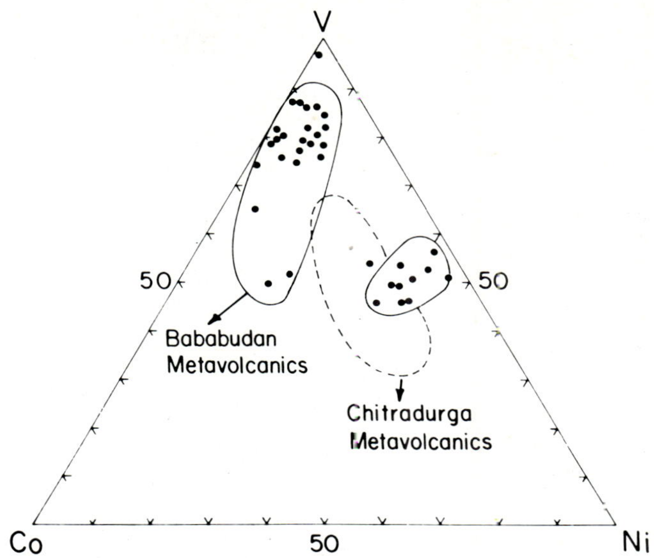

Fig. 8. V-Co-Ni ternary variation diagram showing the Co- and Ni-depleted character of the Bababudan metavolcanics relative to the Chitradurga rocks.

olivine separation which causes a depletion of MgO in the melt. As silica remains almost constant, the possibility of an orthopyroxene fractionation is not suggested. Al_2O_3 in the BBN case remains practically constant with increasing F, while it shows a slight increase in the CHD samples paralleled by a progressive increase in CaO.

A slight decrease of CaO in the sub-alkalic BBN tholeiites and the constancy of Al_2O_3 might suggest some clinopyroxene, plagioclase and olivine fractionation. The iron enrichment trend in the A-F-M plot and the constancy in silica might suggest differentiation under decreasing or low oxygen fugacity (Osborn, 1959). The fractionation trend of the BBN tholeiites can be regarded as a low pressure fractionation (Green and Ringwood, 1967).

DISCUSSION AND CONCLUSIONS

Volcanic rocks dominate the succession in the Bababudan sequence. Similar platformal-volcanic-iron formation sequences with a basal conglomerate are identified from several widely separated areas in the Dharwar craton and constitute the "Bababudan Supergroup" of Swami Nath et al. (1976). Metavolcanics in these sequences are closely associated with several types of metamorphosed

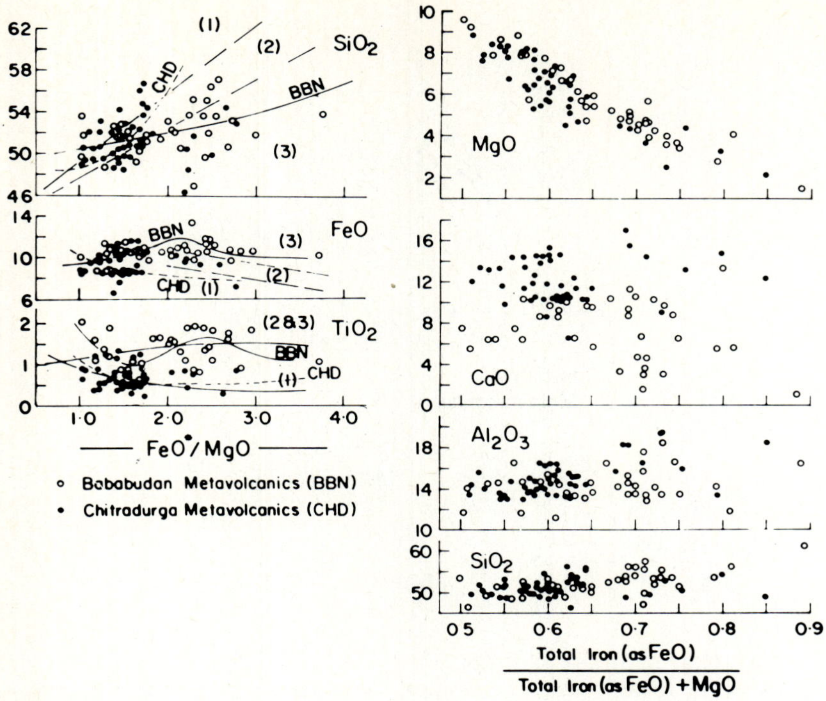

Fig. 9. Variation of SiO_2, FeO and TiO_2 vs. total iron as FeO/MgO and MgO, CaO Al_2O_3 and SiO_2 vs. the fractionation index (F). The Bababudan (BBN) metavolcanics (unfilled circles) have tholeiitic, and the Chitradurga (CHD) metavolcanics (filled circles) mild calc-alkaline, variation trends. The differentiation of both the melts is dominated by separation of olivine but in the BBN it is probably also accompanied by clinopyroxene. Fields in the left-hand side figure after Miyashiro and Shido (1975) (see Fig. 7).

ultrabasic rocks, i.e. meta-picrites underlying metavolcanic flows, in differentiated sills, and as dykes. The authors feel that some of these might represent the subvolcanic phases of the Bababudan volcanism. Such a meta-ultrabasic-volcanic association is rarely observed in, and is not a characteristic feature of, the younger Chitradurga group (ibid.) in which the metavolcanics are generally saturated and show calc-alkaline affinities (Naqvi and Hussain, 1973; Naqvi, 1976).

A high ferromagnesian trace element abundance in metavolcanics from Archaean greenstone belts in general is attributed to the high level of abundance of these elements in the Archaean mantle (Glikson, 1971, 1976; Naqvi and Hussain, 1973). The significantly low Cr, Co and Ni composition of the BBN metavolcanics relative to the younger CHD rocks is difficult to conceive on account of their older age, volume and the observed ultramafic association.

Trace element abundances in volcanic rock suites are a reflection of the magma type, and depend to a large extent on controlling factors such as composition and degree of partial melting of the parental source, depth of magma generation and crustal thickness, degree and nature of crystallisation differentiation of the melt, etc. Because most of these factors are related to the tectonic environment of lava extrusion, a consideration of this aspect of the BBN and CHD sequences is relevant.

The volcano-sedimentary sequences in the BBN and CHD belts have dissimilarities in the lithofacies of their volcanic and sedimentary rocks. The lithologies (discussed in the previous sections) are suggestive of a shallow marine, stable environmental condition for the BBN and a deep marine (eugeosynclinal) environment for the CHD sequence. This feature is also obvious from the mineral facies of the associated iron formation (oxide-type facies in the BBN belt and sulphide type in the CHD belt). The existence of a persistent basal conglomerate and thick beds of orthoquartzites underlying the metavolcanic flows in the BBN sequence suggest that the metavolcanics were extruded over a fairly developed crustal column. The CHD metavolcanics were shown to be emplaced over a thin oceanic type of crust (Naqvi and Hussain, 1973) and are believed to be younger than the metavolcanics in the BBN belt (Swami Nath et al., 1976). It is also possible that both the suites were coeval: the Bababudan suite represents a marginal, platformal association and the Chitradurga a deep oceanic type. This possibility appears to be supported by the available chemical and structural data (Naqvi et al., 1978; see Geochemistry Group, 1977). The relatively thicker crustal condition for the BBN melt results in a relatively higher degree of differentiation. The depleted Cr, Co and Ni abundances in the BBN rocks are probably due to separation of the early crystallising Mg-rich silicate phases (olivines and clinopyroxenes) from the tholeiitic melt. However, in the absence of reliable REE and isotopic data, the other possibility of the Bababudan sequence being older than the Chitradurga one cannot be ruled out.

ACKNOWLEDGEMENTS

The authors acknowledge the help rendered by Mr B.K. Verma and Mr T. Gnaneshwar Rao in processing the data on a PDP 11-40 computer. Thanks are due to Mr S.M. Ahmad and Mr B. Govinda Rajulu for help.

The authors thank the Director, National Geophysical Research Institute, Hyderabad, for permission to publish this paper. The work of the first author is supported by a C.S.I.R. Fellowship.

REFERENCES

Anhaeusser, C.R., 1975. Precambrian tectonic environments. In: F.A. Donath (Editor), Annual Review of Earth and Planetary Sciences, U.S.A., 5: 31-55.

Arth, J.G. and Hanson, G.N., 1975. Geochemistry and origin of the early Precambrian crust of northeastern Minnesota. Geochem. Cosmoch. Acta, 39: 325-362.

Baragar, W.R.A. and Goodwin, A.M., 1969. Andesites and Archaean volcanism in the Canadian Shield. Oregon Dept. Geol. Min. Ind., pp. 121-142.

Engel, A.E.J., Engel, C.G. and Havers, R.G., 1965. Chemical characteristics of oceanic basalts and upper mantle. Bull. geol. Soc. Am., 76: 719-734.

Fiala, F., 1974. Some notes on the problem of spilites. In: G.C. Amstutz (Editor), Spilites and spilitic Rocks. Springer Verlag, 9 pp.

Glikson, A.Y., 1971. Primitive Archaean element distribution patterns; chemical evidence and geotectonic significance. Earth Planet. Sci. Lett., 12: 309-320.

Glikson, A.Y., 1972. Early Precambrian evidence of a primitive ocean crust and island nuclei of sodic granites. Bull. geol. Soc. Am., 83: 3323-3344.

Glikson, A.Y., 1976. Stratigraphy and evolution of primary and secondary greenstones; significance of data from shields of the southern hemisphere. In: B.F. Windley (Editor), The Early History of the Earth. Wiley, London, pp. 257-279.

Goodwin, A.M., 1968. Archaean protocontinental growth and early history of the Canadian Sbield. Proc. 23rd int. geol. Congr., 1: 69-89.

Green, D.H. and Ringwood, A.E., 1967. The genesis of basaltic magmas. Contr. Mineral. Petrol., 15: 103-190.

Hallberg, J.A. and Williams, D.A.C., 1972. Archaean mafic and ultramafic rock associations in the eastern gold fields of Western Australia. Earth Planet. Sci. Lett., 15: 191-200.

Irvine, T.N. and Baragar, W.R.A., 1972. A guide to chemical classification of the common volcanic rocks. Can. J. Earth Sci., 8: 523-548.

Jakes, P. and Gill, J.B., 1970. Rare earth elements and the island arc tholeiitic series. Earth Planet. Sci. Lett., 9: 17-28.

Jakes, P. and White, A.J.R., 1971. Composition of island areas and continental growth. Earth Planet. Sci. Lett., 12: 274-280.

Kuno, H., 1967. Differentiation of basaltic magmas. In: H.H. Hess and A. Poldervaart (Editors), Basalts - a Poldervaart Treatise on Rocks of Basaltic Composition. Wiley, New York, Vol. 2: 623-688.

MacDonald, G.A. and Katsura, T., 1964. Chemical composition of Hawaiian lavas. J. Petrol., 5: 82-133.

Manson, V., 1967. Geochemistry of basaltic rocks - major elements. In: H.H. Hess and A. Poldervaart (Editors), Basalts - a Poldervaart Treatise on Rocks of Basaltic Composition. Wiley, New York, Vol. 1: 215-269.

Middlemost, E.A.K., 1975. The basalt clan. Earth Sci. Rev., 11: 337-364.

Miyashiro, A. and Shido, F., 1975. Tholeiitic and calc-alkaline rocks in relation to the behaviours of titanium, vanadium, chromium and nickel. Am. J. Sci., 275: 265-278.

Naqvi, S.M., 1976. Physico-chemical conditions during the Archaean as indicated by Dharwar geochemistry. In: B.F. Windley (Editor), The Early History of the Earth. Wiley, Lond., pp. 289-298.

Naqvi, S.M. and Hussain, S.M., 1972. Petrochemistry of early Precambrian metasediments from the central part of Chitaldrug schist belt, Mysore, India. Chem. Geol., 10: 109-135.

Naqvi, S.M. and Hussain, S.M., 1973. Relation between trace and major element composition of Chitaldrug metabasalts, Mysore, India, and the Archaean mantle. Chem. Geol., 11: 17-30.

Naqvi, S.M., Divakara Rao, V. and Hari Narain, 1974. Protocontinental growth of the Indian shield and the antiquity of its rift valleys. Precamb. Res., 1: 345-398.

Naqvi, S.M., Divakara Rao, V. and Hari Narain, 1978. The primitive crust; evidence from the Indian shield. Precamb. Res. (in press).

Narebski, W., 1974. A statistical study of specific petrochemical features of some spilitic rock series. In: C.G. Amstutz (Editor), Spilites and Spilitic Rocks. Springer Verlag, pp. 127-161.

Osborn, E.F., 1959. Role of oxygen pressure in the crystallisation and differentiation of basaltic magmas. Am. J. Sci., 257: 609-647.

Pearce, T.H., 1974. Quench plagioclase from some Archaean basalts. Can. J. Earth Sci., 11: 715-719.

Pichamuthu, C.S., 1930. On the trap rocks of Chitaldrug schist belt. J. Mysore Univ., 14(2): 1-13.

Pichamuthu, C.S., 1935. The iron formations and the associated rocks of the eastern Bababudans, Kadur district, Mysore. J. Mysore Univ., 8.

Pichamuthu, C.S., 1957. Pillow lavas from Mysore state, India. Bull. Mysore Geol. Assoc., 15.

Pichamuthu, C.S., 1974. On the banded iron formations of Precambrian age in India. J. geol. Soc. India, 15: 1-31.

Radhakrishna, B.P. and Vasudev, V.N., 1977. The early Precambrian of the South Indian Shield. J. geol. Soc. India, 18: 525-541.

Ramachandra, Rao, M.B., 1937. The petrology of hornblendic rocks from the Kolar schist belt. Mysore Geol. Dept., 16: p.103.

Rivalenti, G., 1976. Geochemistry of metavolcanic amphibolites from South-West Greenland. In: B.F. Windley (Editor), The Early History of the Earth. Wiley, Lond., pp. 213-223.

Shapiro, L. and Brannock, W.W., 1962. Rapid analysis of silicate, carbonate and phosphate rocks. Bull. U.S. geol. Surv., 1144-A, 56 pp.

Srinivasan, R. and Sreenivas, B.L., 1972. Dharwar stratigraphy. J. geol. Soc. India, 13: p. 75.

Srinivasan, R. and Sreenivas, B.L., 1975. A reappraisal of some aspects of the geology of the Dharwars of Mysore State, India. Misc. Publ. geol. Surv. India, 23: 49-69.

Swaminath, J., Ramakrishnan, M. and Viswanatha, M.N., 1976. Dharwar stratigraphic model and Karnataka craton evolution. Rec. geol. Surv. India, 107(2): 149-175.

Viljoen, M.J. and Viljoen, R.P., 1969. The geology and geochemistry of the lower Onwerwacht group and a proposed new class of igneous rocks. In: The Upper Mantle Project, Sp. Publ. geol. Soc. S.Afr., 2: 55-86.

Wilson, H.D.B., Andrews, P., Moxham, R.L. and Randall, K., 1965. Archaean volcanism in the Canadian Shield. Can. J. Earth Sci., 2: 161-175.

Geochemistry Group, 1977. Archaean geochemistry. Contrib. Prog. N.G.R.I. geophys. Res. Bull. Sp. Iss., Nov. 1977, Vol. 15.

GEOCHEMISTRY OF ARCHAEAN METASEDIMENTS; EVIDENCE FOR PROMINENT ANORTHOSITE-NORITE-TROCTOLITE (ANT) IN THE ARCHAEAN BASALTIC PRIMORDIAL CRUST

S.M. Naqvi
National Geophysical Research Institute, Hyderabad, India

Abstract

Siliceous-aluminous schists are common in the lower succession of the greenstone belts of India, Africa and other shield areas; they overlie basic-ultrabasic flows. The presence of associated bedded barites and secondary fuchsite quartzite convincingly corroborates their sedimentary origin. In the Dharwar craton they are found in the Sargur group as sillimanite-kyanite-staurolite-micaceous-corundum-quartz schists and in the Bababudan-Chitradurga groups as sericitic phyllites. The Sargur group metasediments vary in composition from highly Al (K + Ti) to Al (Mg + Fe + Ti + Cr + Ni)-rich varieties. They are depleted in Si and K-group elements like Rb and Sr, and enriched in Al, Ti, Mg, Fe, Cr and Ni compared with Bababudan-Chitradurga group metasediments and surrounding tonalitic gneisses. The evidence of a mixed source provenance (tonalites, trondhjemites, basalts and sediments) is preserved in the composition of the sediments of the Bababudan and Chitrudurga groups. A similar source is not expected to provide the material for the Sargur group sediments which require a source enriched in Al, Mg, Fe, Ca, Co, Cr and Ti, and depleted in Si and K-group elements. Archaean crust, made up of 60% low-K tholeiites, 30% high-Ti and Al anorthosites and 10% ultramafics like norite, troctolite and peridotite, may give rise to such sediments. Data from the silicate planets indicate their identical initial evolutionary history. The lunar crust, whose evolution was arrested c. 4-3 by ago, is mainly made up of basalts and anorthosites. Therefore, it is possible that the results of the early exogenic process at the lunar type of earth's crust are preserved as meta-sediments in the lower parts of the supracrustals of great antiquity. Absence of primary quartzites in the early-middle Archaean, their relatively low to moderate abundance in the early Proterozoic and their prominence since the middle Proterozoic probably suggests that the change from an ANT-rich basaltic primordial crust to a quartz-rich granitic crust was between 4.0 and 2.0 by ago.

INTRODUCTION

During the last few years geochemical research on Precambrian rocks has been confined, with few exceptions, to basic, ultrabasic, gneissic and granulitic rocks;

metasediments have received little attention. In the opinion of many geologists working in Precambrian terrains (e.g. Glikson, 1971; Pettijohn et al., 1972; Ronov, 1972; Turner and Walker, 1973; Veizer and Hoefs, 1976) sediments are equally useful in understanding the early history of crustal evolution. However, the scales seem to be heavily weighted against sedimentologists, at least in the early Precambrian, because of severe deformation and protracted erosion. In Precambrian rocks selective loss of certain rock types has had a most profound effect: if estimates of rates of deposition are roughly correct, c. 90% of Precambrian sediments have been removed and reworked since their deposition. The residual sedimentary record must therefore be interpreted with great care to avoid mistaking selective loss for a primary record (Garrels et al., 1971, pp. 119-120). Nevertheless, this important source of information cannot be ignored and during the past few years geochemists of the NGRI have approached problems of crustal evolution from a sedimentological standpoint (Naqvi and Hussain, 1972; Satyanarayana et al., 1973; Naqvi, 1976a,b, 1977; Naqvi et al., 1978a).

Siliceous and aluminous schists are commonly found in greenstone belts, generally associated with mafic and ultramafic units. All the Archaean sedimentary series of the Aldan Shield overlie basic schists and their lower parts are represented by quartzites and high-alumina gneisses and locally by corundum rocks (Glukhovskiy and Pavlovskiy, 1973). In the Barberton Mountain Land siliceous aluminous schistose rocks are found stratigraphically above basic-ultrabasic volcanic units and are made up of quartz, sillimanite, andalusite, staurolite, chloritoid, sericite, ilmenite and various accessories (Anhaeusser, 1972). In the Archaean of Karnataka, India, siliceous aluminous schists occur in amphibolite facies supracrustal sequences designated as the Sargur group by Viswanatha and Ramakrishnan (1975), and in low-grade schistose groups named the Bababudan and Chitradurga greenstones of Karnataka they were originally termed G.R. clays by Sampat Iyengar (1905).

These siliceous and aluminous schists are thought to be siliceous tuffs and agglomerates by Anhaeusser (1972), Archaean metabauxite by Martin (1968) and Salop (1968), and metasediments by Fyfe (1973) and Ramakrishnan et al. (1976). The followers of the school of chemical uniformitarianism regard them as shallow water detrital sediments and cite them as evidence for a widespread granitic crust during the early Archaean (Swaminath et al., 1976). The thin layers of impure quartzite, bedded barite, fuchsite quartzite and limestone (marble) in them are sufficiently convincing to concede their sedimentary character. However, this does not necessarily imply that they provide strong evidence for the existence of a predominantly granitic crust during the early Archaean. In view of the global importance (Windley, 1977, p. 26) of these rocks, this paper describes the geochemistry of their Indian representatives in terms of the concept of material

TABLE 1. Geological setting of the siliceous aluminous schists of the Dharwar craton following G.S.I. classification of greenstone belts

	Schist belt	Main rock type	Underlain by	Overlain by	% of quartz and sedimentary textures and structures	Associated rocks
Chitradurga Group 2.8-2.6 by	Chitradurga	Sericitic phyllites	Basic flows of quartz and olivine normative tholeiites	Pyritiferous chert	Greywackes form major portion of schist belt	Lutites (chlorite schists) and greywacke conglomerates
Bababudan Group 2.8-2.6 by	Chitradurga	Sericitic phyllites	"	Magnetite quartzites (B.M.Q.)	10-15%, current bedding, ripple marks and other features	Chloritic schist, current bedded quartzites, limestones and basal conglomerates
	- - - VAST TONALITIC-TRONDHJEMITIC ACTIVITY - - -					
Sargur Group 3.2-3.5 by?	Holenarasipur	Kyanite-staurolite garnet micaceous, schists, amphibolites, chlorite and chloritoid magnetite schists	Tonalite*, ultramafic and mafic komatiitic flows**	Magnetite-grunerite garnet-quartzites	3-5%, no detrital sedimentary textures or structures yet found	Fuchsite quartzites, sericitic quartzites
	Sargur	Sillimanite garnet corundum quartz schist	"	"	"	Limestones, fuchsite quartzite, barites interbedded with fuchsite quartzite

* Ramakrishnan et al., 1971

** Naqvi et al., 1978b

balance. This method, first introduced by Mead (1907), has proved useful in evaluating the provenance of various sedimentary sequences (Brotzen, 1966; Horn and Adams, 1966; Garrels et al., 1971; Sibley and Wilband, 1977). Here it is used to propose a tentative model for secular change in the predominant composition of the Archaean crust based on analyses of sediments of various relative ages from Peninsular India.

GEOLOGY

At the outset it should be stated that I disagree with many local details of the stratigraphic sequence proposed for the Archaean of Karnataka by Ramakrishnan et al. (1976). However, because their nomenclature and sequence has no great effect on the general observations made here, their terminology is used to avoid at least semantic confusion amongst readers. The age of the oldest high-grade supracrustal sequences (Sargur group) in Karnataka, relative to the enveloping gneiss-tonalite complex, has not been resolved by stratigraphic or radiometric means. They are older than the major deformation and plutonic activity in the gneisses, which definitely predate the deposition of the lower grade supracrustals (Bababudan and Chitradurga groups) (Radhakrishnan, 1976; Naqvi et al., 1978b). The exact stratigraphic position of the siliceous aluminous metasediments in the Sargur, particularly their relation to the dominant basic-ultrabasic unit, is also controversial (Naqvi et al., 1978b). The local stratigraphic position of these rocks will not alter their provenance, and therefore the succession proposed by Viswanatha and Ramakrishnan (1975) is followed here for the sake of brevity. The relevant geological information about these metasediments is given in Table 1. The metasediments are intruded by a large number of pegmatites and quartz veins which have increased the potash content of their country rocks. In places kyanite concentrated in crystalline aggregates around these pegmatites and veins is economic.

In the Bababudan group they (G.R. clays) occur above tholeiites and below an oxide facies iron horizon, and in the Chitradurga group above tholeiites and below a sulphide facies iron horizon (see Naqvi and Hussain, 1972). In the Chitradurga group they are interbedded with chert. The layers of chert and phyllite vary from a few centimetres to 1 m. In between the Sargur and Bababudan groups vast tonalitic activity has taken place (Radhakrishna, 1976). Thus, the source region of the Bababudan group was definitely more sialic than that of the older Sargur group.

The Sargur group siliceous-aluminous schists contain layers made up of different proportions and combinations of sillimanite kyanite, staurolite, corundum, garnet quartz, muscovite, biotite, chloritoid, amphiboles, sericite, chlorite, fuchsite, magnetite, ilmenite and graphite. The entire mineral paragenesis of the different layers is derived from pelites, semipelites, psammites and quartzites

TABLE 2. Average chemical composition of the Archaean siliceous-aluminous sediments from the Sargur Group

	METAPELITES				CHLORITE SCHISTS		QUARTZITES		IRONSTONES		GNEISSES	PELITE	SHEBA SHALES
	1	2	3	4	5	6	7	8	9	10	11	12	13
SiO_2	52.31	54.21	70.85	67.32	44.70	56.04	88.00	70.85	47.87	48.10	68.28	61.54	54.11
TiO_2	1.42	1.62	0.59	0.48	0.46	0.54	Tr	0.12	1.57	0.69	0.34	0.83	1.00
Al_2O_3	23.89	24.16	12.71	14.58	18.10	14.51	5.58	18.81	16.02	6.69	12.52	16.95	17.54
Pe_2O_3	1.38	1.57	6.66	6.38	1.79	7.15	0.38	0.53	1.16	27.52	0.98	2.36	8.88
FeO	5.41	3.32			10.72		0.98	2.40	12.97	11.52	3.26	3.90	
MgO	4.97	5.08	1.80	1.83	12.45	5.11	0.72	1.80	5.85	2.38	3.10	2.52	6.97
CaO	4.41	2.43	1.74	2.04	5.16	7.29	1.95	3.02	10.78	2.33	5.18	1.76	0.22
Na_2O	2.55	2.10	0.84	0.91	1.37	3.94	0.30	3.20	2.72	1.00	4.85	1.84	2.14
K_2O	1.58	1.86	2.08	2.13	0.08	1.29	0.30	0.40	1.35	0.20	1.40	3.45	2.72
MnO	0.16	0.07	0.05	0.03	0.12	0.39	-	-	0.31	0.05	0.16	-	0.05
P_2O_5	0.04	0.09	0.09	0.07	0.07	0.06	-	-	0.17	0.21	0.04	-	
Co	26.40	30.87	85	62	88	69	BDL	12	BDL	BDL	4.17	-	-
Cr	1000	1000	75	42	600	120	270	380	24	353	48.25	-	-
Ni	256.60	203	142	95	205	76	140	240	26	BDL	24.00	-	-
V	143	165	122	30	205	142	62	68	77	BDL	78.26	-	-

1 = Average of 10 metapelites from Holenarasipur
2 = Average of 12 metapelites from Sargur
3 = Average of 4 metapelites from Chitradurga (sulphide facies)
4 = Average of 2 metapelites from Chitradurga (oxide facies)
5 = Average of 5 chlorite schists from Holenarasipur
6 = Average of 5 chlorite schists from Chitradurga
7 = Average of 2 quartzites from Holenarasipur
8 = Impure quartzite from Sargur
9 = Garnet-grunerite-quartz rock from Holenarasipur
10 = Quartz magnetite rock from Sargur
11 = Average of 12 gneisses from Holenarasipur
12 = Average pelite (Shaw, 1956)
13 = Average Sheba Shales (Condie et al., 1970)

Viswanatha and Ramakrishnan, 1975; Ramakrishnan et al., 1976; Swaminath et al., 1976). Thin quartzite (?) layers found between the pelites and the ultramafics vary from sericitic quartzite to fuchsitic quartzite. Nowhere do these so-called quartzites in the Sargur group exhibit any sedimentary structures such as current bedding and ripple marks, which are quite common in the quartzites of the younger Bababudan group (see Janardhan and Srikantappa, 1975; Nair et al., 1975; Vasudevan and Rao, 1975). In places boudins, rods and mullions in the quartzites between more fuchsite layers result in a pseudo-conglomeratic appearance (Rama Rao, 1962; Srinivasan and Sreenivas, 1968; Naqvi et al., 1978a). The quartz grains of the quartzites vary from very fine to coarse and do not show that intricate boundary texture which is characteristic of metamorphosed mature sandstones. These grains show inclusions of fuchsite and apatite. Zircon and other detrital heavy minerals are absent, whereas zircons are commonly found in the mature detrital sedimentary quartzites of the Bababudan group. The percentage of these quartzites (?) in the Sargur group varies between 3 and 5% of the total sedimentary sequence (Naqvi et al., 1978b); in fact their percentage is so low that in the earlier works of Sampat Iyengar (1905), Rama Rao (1962) and Iyengar (1971), and the recent maps by Ramakrishnan and Viswanatha (unpublished maps of GSI) and Hussain and Naqvi (see Geochemistry Group NGRI Sp. Publ., 1977), they were difficult to show on a 1"- or 2"-to-a-mile map. In the Bababudan group the quartzites (mature sandstones) increase to 10-15%. In the Chitradurga group orthoquartzites are replaced by greywackes which exhibit graded bedding and form the predominant part of the schist belt (Naqvi, 1977).

GEOCHEMISTRY

The averages of the siliceous-aluminous rocks of Karnataka from the Sargur, Bababudan and Chitradurga groups are given in Table 2 and graphically represented in Figs. 1-5. It can be seen that the metasediments of the Sargur group range in composition from highly Al (K+Ti)-rich to Al (Mg+Fe+Ti, Ca+Ni and Cr)-rich varieties (Fig. 1). The abundance level of TiO_2 is very high (2.75%). The SiO_2 of pelites and semipelites reaches 60% and the quartzites are generally impure, having SiO_2 up to 90%. These rock types show a very wide range of abundance of CaO, Na_2O and K_2O (Fig. 1) and a very high abundance level of ferromagnesian traces (Fig. 2). The Co content varies from less than 10 to 400 ppm, Ni is between 180 and 500 ppm, Cr ranges from 200 to more than 1000 ppm, and V varies from 50 to 400 ppm (Fig. 2). The Rb content of the majority of the samples is less than 10 ppm, whereas Sr varies from less than 10 to 250 ppm (Fig. 3). In the case of SiO_2, Al_2O_3, TiO_2, MgO, Ni, Cr, Co and a few other constituents, the composition of the Sargur group siliceous-aluminous sediments differs remarkably from the composition of the siliceous-aluminous sediments of the Bababudan and Chitradurga groups (Figs. 1-3). The sericitic phyllites, known as GR clays in the classical literature of the Dharwar

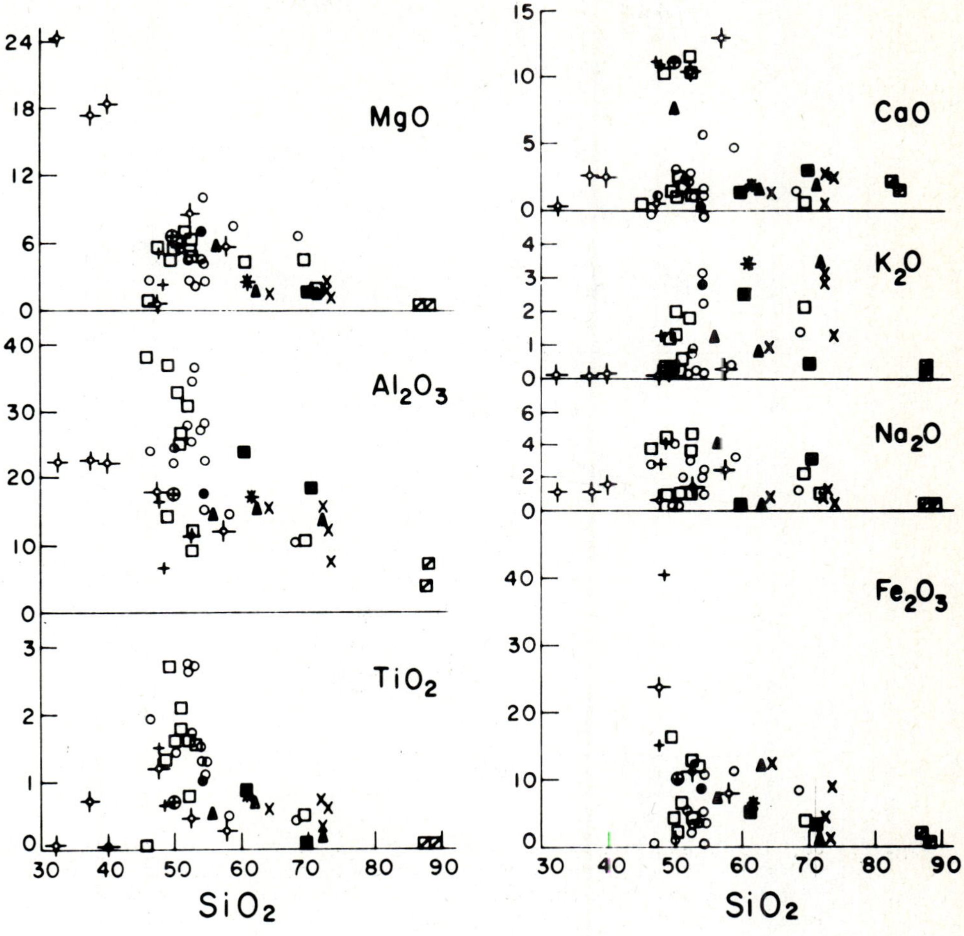

□ Holenarsipur Metapelites; + Holenarsipur Ironstone; ✧ Holenarsipur Chlorite Schist; ⧄ Holenarsipur Quartzites; ○ Sargur Metapelites; ■ Sargur Quartzites; ▲ Chitradurga Pelite (Lower Horizon with Oxide Facies); X Chitra durga Pelite (Upper Horizon-Sulfide Facies); ▲ Chitradurga Chlorite Schist; ∗ Average Pelite (Shaw, 1956); ● Sheba Shales (Condie, 1970); ⊕ Average Archaen Crust; △ Gneisses.

Fig. 1. Different wt % components of the siliceous-aluminous sediments of the Sargur, Bababudan and Chitradurga groups plotted against their SiO_2 wt % contents. Note the differences in composition of the metasediments of the Sargur and younger groups. Depletion in SiO_2 and K_2O and enrichment in Al_2O_3, TiO_2, MgO, Fe_2O_3, CaO and Na_2O are noticeable.

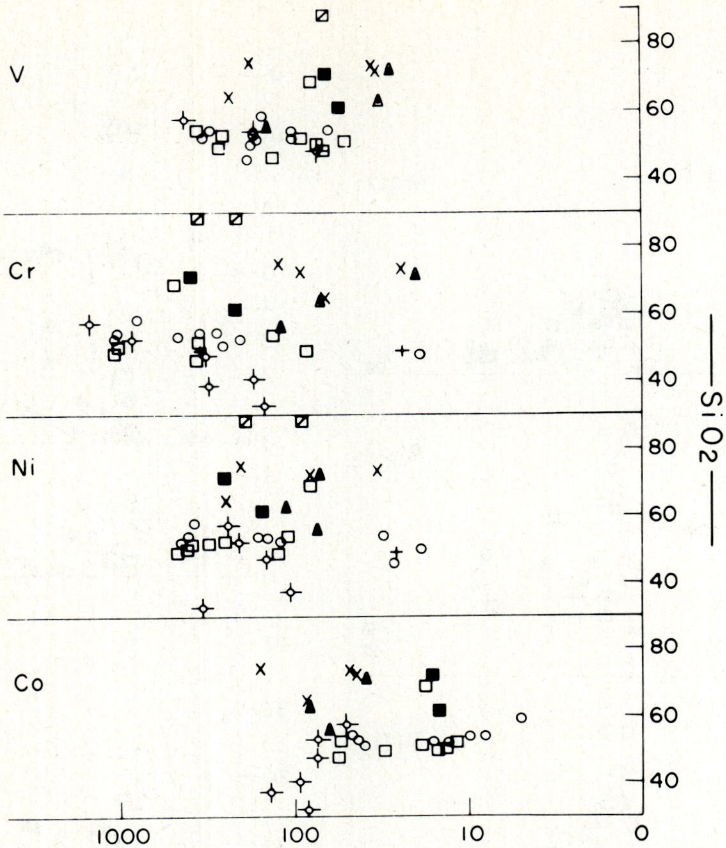

Fig. 2. Co, Ni, Cr and V contents (ppm) plotted against SiO_2 wt % in the meta-sediments of the Sargur and younger groups of the Dharwar greenschists (Bababudan and Chitradurga). Note the enrichment of the Sargur group sediments in Ni and Cr (the symbols are the same as in Fig. 1).

craton (Sampat Iyengar, 1905), of the Bababudan and Chitradurga groups have relatively higher SiO_2, K_2O and lower Al_2O_3, TiO_2, MgO, CaO, Ni and Cr than those of the Sargur group (Figs. 1-3). Similarly, the composition of the Sargur group pelites does not resemble the composition of the average pelite of Shaw (1956). These differences are more pronounced in the MgO/Al_2O_3, TiO_2/Al_2O_3 plots and ratios (Fig. 4). The K/Rb and Sr/Ca plots (Fig. 5) bring out their differences with the surrounding gneisses and granites.

The major and trace element data are probably sufficient to demonstrate the compositional differences between the siliceous-aluminous schists of the Sargur group on the one hand, and the Bababudan and Chitradurga groups on the other.

The Sargur group sediments are also remarkably different in their overall compositions from the tonalites surrounding them (for the composition of tonalite, see Naqvi et al., this volume). The most important inference from these data appears to be that both the composition and abundance of the Sargur-type metasediments are unique.

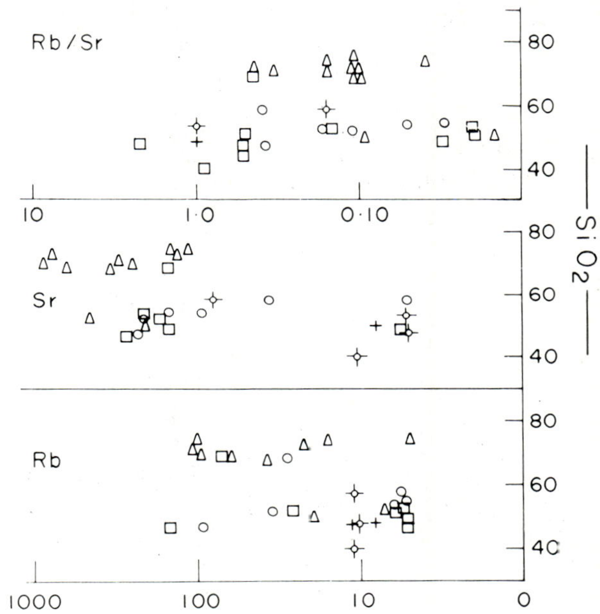

Fig. 3. Distribution of Rb and Sr and Rb/Sr in the metapelites of the Sargur group and the surrounding tonalitic gneisses. Note the differences between the Sargur group metasediments and the tonalitic gneisses. The data show that both rock types are depleted in Rb and Sr (the symbols are the same as in Fig. 1).

SOURCE

The compositional differences between metasediments of different ages may be interpreted as a reflection of the differences in their respective source areas. The presence and probably the abundance of tonalites and trondhjemites in the source area of the Bababudan and Chitradurga group metasediments are fairly established by the abundance of tonalitic and trondhjemitic pebbles in the conglomerate and of detrital quartz and plagioclase in the greywackes (Naqvi and

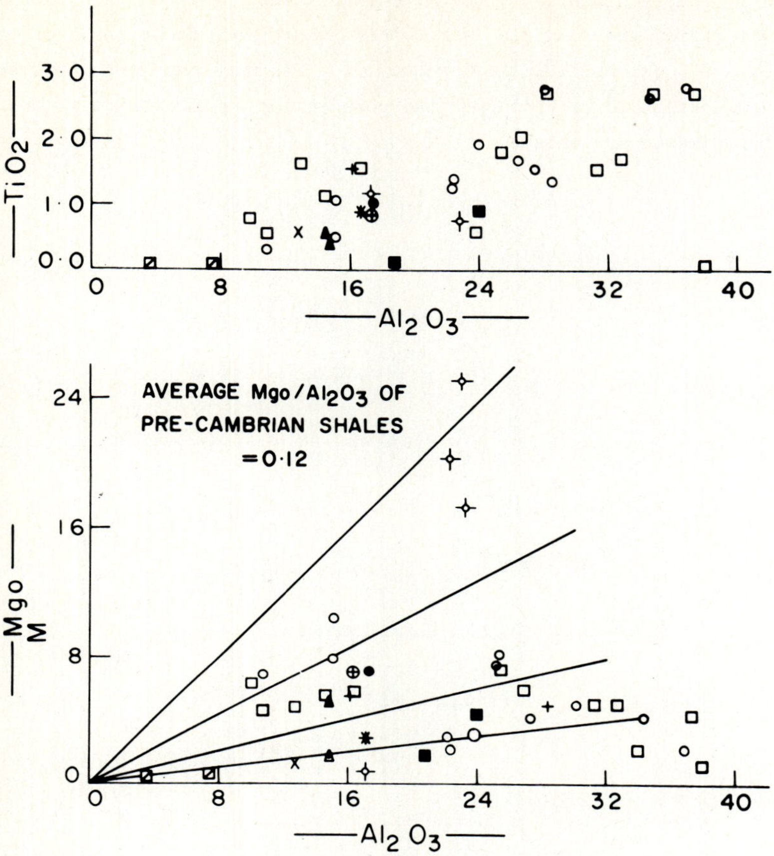

Fig. 4. Al_2O_3-TiO_2 and Al_2O_3-MgO plots. Note the difference in the ratios and composition of the Sargur group metasediments from the Holenarasipur and Sargur schist belt and the Bababudan-Chitradurga group. Note the sympathetic positive relationship between TiO_2 and Al_2O_3 and the higher MgO/Al_2O_3 ratios of most of the Sargur group metasediments. (The symbols are the same as in Fig. 1.)

Hussain, 1972; Naqvi et al., 1978a). Therefore, a mixed source area made up of. tonalites, trondhjemites, basalts, shales and quartzite has been proposed for the Bababudan-Chitradurga group (Naqvi et al., 1978a). A similar source area could not provide the material for the Sargur metasediments for which a source enriched in Al, Ti, Mg, Fe, Cu, Ni, Cr and Co and depleted in Si, K, Rb and Sr is necessary. If the Sargur group was laid down on, and derived from, a well-granitised and stabilised crust, the result should have been similar to the Bababudan and Chitradurga groups.

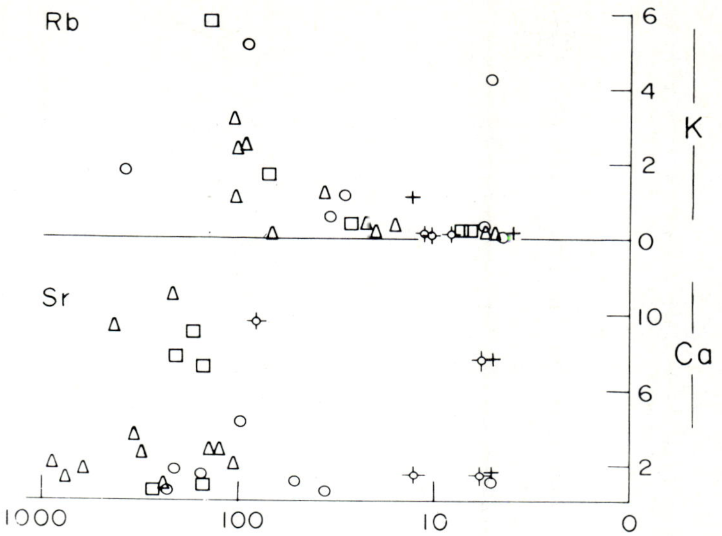

Fig. 5. Ca/Sr and K/Rb plot of the Sargur group metasediments and surrounding gneisses. Note the low level of the K and Rb contents of the metasediments relative to the gneisses. A few metapelite samples collected from near the pegmatites show higher abundances of K and Rb. The Sr content of the gneisses is generally higher relative to the Sargur group metasediments. (The symbols are the same as in Fig. 1.)

A geochemical balance between sediments and igneous activity has been established by many workers from Mead's (1977) time onward. Recently Sibley and Wilband (1977) have shown that since 1.5 by such a balance between sediments and preceding igneous activity has been maintained. They also found that an average igneous rock (2/3 granodiorite, 1/3 tholeiite) will give rise to 24-26% sandstones in geosynclines and platforms. A similar qualitative (with a low sandstone and high greywacke abundance) balance can be envisaged between the sediments of the Bababudan and Chitradurga groups (Naqvi et al., 1978a). The data on the greywackes and associated lutites from different shield areas (Nanz, 1953; Macpherson, 1958; Donaldson and Jackson, 1965; Condie et al., 1970; McGlynon and Henderson, 1970; Glikson, 1971; Naqvi and Hussain, 1972) strongly support the view that a qualitative geochemical balance was maintained between Archaean/early Proterozoic sediments and tonalitic-basaltic igneous activity. If this concept of material balance (Horn and Adams, 1966; Garrels et al., 1971) is correct, the absence of 24-26% of sandstones from the Archaean in general and the Sargur group of Karnataka in particular should be explained. The low abundance of limestones in the Sargur, Bababudan and Chitradurga groups may be explained by selective loss of limestone and its recycling (Garrels et al., 1971). The same selective loss of limestone will increase the relative abundance and proportion of sandstones (being the most resistant rocks) in the residual record of the Archaean. Ronov and Migdisov (1971) clearly demonstrated that the relative proportion of abundances between different sedimentary rock types are the function of age, and that mature sandstones (orthoquartzites) were present in minor quantities during the Archaean whilst the so-called Archaean quartzites are often secondary. The quartzites of the Sargur group contain fuchsite, layers of barite, and their constituent quartz grains contain minute fuchsite inclusions in some cases giving rise to the semi-precious gemstone aventurine. Interbedded barite also contains fuchsite and quartz. Fuchsite quartzite, in view of the work of Whitemore et al. (1964), cannot be regarded as detrital. Fuchsites (chrome mica) are invariably the product of solutions of magmatic derivations: the chrome was introduced into the solutions either as a result of the original fractionation of volatile magma constituents or through leaching of mafic-ultramafic rocks. Whitemore et al. (1964) and Gajer (1963), as quoted by Anhaeusser (1972), have also given several examples where it appears that chromium-bearing solutions generally emanate from a mafic-ultramafic magma. Intimately associated chemical precipitation is evident from barite-rich layers interbedded with the quartzites. Chemical precipitation to produce sedimentary barite and high energy winnowing to produce quartz sandstones can hardly be expected in the same sedimentary environment; the quartzites may therefore be recrystallised chemical precipitates, i.e. originally cherts. This is supported by the absence of zircons and other heavy mineral grains and by differences in the textures from those expected in primary quartzites like those of the Bababudan group. It can therefore be argued that the metasediments of the Sargur group are pelites, cherts, limestones and iron-stones and the source area for these sediments should be depleted in SiO_2, K_2O and K-group elements and enriched in TiO_2, Al_2O_3, MgO, FeO, Ni, Cr and Co. The titanium enrichment of the source area is further demonstrated by the titaniferous magnetite deposits of the Nuggihalli schist belt, another member of the Sargur group, where pillows are exhibited by underlying ultramafic units (Varadarajan, 1970; Srinivasan and Sreenivas, 1972; Vishwanathan, 1974).

MODEL

If a source area consisting of 60% low-K tholeiites, 30% high Ca, Ti anorthosites and 10% ultramafic material like norite, troctolite and peridotite is assumed, and if the concept of material balance (see Garrels et al., 1971) is accepted, pelites, limestones, cherts and ironstones chemically resembling those of the Sargur group can be derived. Table 3 shows that an average Archaean crust

TABLE 3. Composition of Archaean crust and sediments

	ARCHAEAN	CRUST	LIMESTONES DOLOMITES		SHALES	PELITES	IRONSTONES		CHERTS		OCEAN	AV. SARGUR GROUP PELITE
SiO_2	50.13	501	--	-	368.00	54.98	65.00	48.63	68	94.44	-	53.26
TiO_2	0.76	7.6	-	-	7.60	1.36	-	-	-	-	-	1.52
Al_2O_3	17.45	175	-	-	170.00	25.40	3.00	2.24	2	2.78	-	24.02
MgO	7.31	73	28	12.90	35.00	5.23	4.00	2.29	-	-	6.00	5.02
CaO	11.35	114	79	36.40	29.00	4.33	4.00	2.29	1	1.39	-	3.42
Fe_2O_3 + FeO	9.48	95	2	0.92	34.00	5.08	55.00	41.15	1	1.39	3.00	5.83
Na_2O	2.99	30	1	0.46	22.50	3.36	1.50	1.12	-	-	5.00	2.32
K_2O	0.19	2	-	-	2.00	0.30	-	-	-	-	-	1.72
MnO	0.22	2	-	-	0.84	0.13	1.00	0.75	-	-	0.14	0.11
P_2O_5	0.14	1	-	-	0.35	0.05	0.15	0.11	-	-	0.50	0.06
CO_2	-	-	107	49.31	-	-	-	-	-	-	-	-
TOTAL	100.00	1000	217	99.99	669.29	100.00	133.65	99.98	72	100.00	14.64	97.28
%	-	-	11	-	67	-	13	-	7	-	1.50	-

of 30% anorthosite, 60% low-K tholeiite, and 10% ultramafics can produce on weathering 67% shales-pelites, 11% limestone, 13% ironstones and 7% cherts. If the excess K_2O of the Sargur pelites is attributed to later pegmatitic activity and K reactivation, as can be demonstrated in the field, the composition of these pelites is quite similar to the pelites from the average Archaean crust.

Lowman (1976) suggested that the initial stages in the crustal evolution of all silicate planets were almost identical. The evolution of the lunar crust appears to have been arrested c. 4.0-3.0 by ago whereas the earth's crust has witnessed rapid changes in the hydrosphere, atmosphere and related exogenic processes. The lunar crust consists of basalts and ANT suites. Therefore it may be possible that the lunar type of the earth's primordial basic crust provided material and solutions for the Sargur group depositories.

The absence of quartzites and conglomerates (both basalt and greywacke type with tonalitic pebbles) from the Sargur group, and the relatively lower abundance of quartzites and basal conglomerates from the Bababudan and Chitradurga groups, indicate that free plutonic quartz was probably not available for weathering during pre-Sargur (3.2-3.5) times in the Indian Peninsula. After the post-Sargur tonalite-trondhjemite activity free plutonic quartz became available. However, tonalites and trondhjemites (with 30% quartz) are incapable of yielding vast amounts of mature quartzites and basal conglomerates. Thus in the Bababudan and Chitradurga groups arkoses, shales, greywackes and lutites are more prominent than sandstones (Naqvi, 1977; Naqvi et al., 1978a).

These secular changes in the composition of the sedimentary component of Archaean to Proterozoic rocks are illustrated when elemental and oxide abundances are plotted against age (Engel et al., 1974; Naqvi, 1976b; Veizer and Hoefs, 1976; Radhakrishna and Vasudev, 1977). The ferromagnesian elements show a decrease with age (Naqvi, 1976b). The study of the composition of conglomerate pebbles suggests that during the Bababudan-Chitradurga geosynclinal activity the emplacement of K-rich granites took place (Naqvi et al., 1978a). Late emplacement of pegmatites and quartz veins between 2.5-2.1 by made the crust rich in free plutonic quartz; thus the debris of middle Proterozoic depositories like the Cuddapah and Vindhyans became quartz dominated, and mature sandstones, shales and limestones became the prominent sediments in all environments. Therefore, on the basis of the data presented here and published earlier by the author and his coworkers, a qualitative balance between the sediments and igneous activity of the Precambrian can be envisaged (Table 4). It appears that the late Archaean/ early Proterozoic was a transitional era. In view of these changes from the Archaean to the middle Proterozoic in the sedimentary shell of the crust, a tentative model is proposed which suggests that ANT-rich primordial basaltic crust was progressively changed into sialic, quartz-rich crust during the period

TABLE 4. Qualitative geochemical balance during the Precambrian era

Time	Igneous activity	Sediments
Upper Proterozoic	Granodiorites Granites Basalts	Shales Sandstones Limestones
Middle Proterozoic	Tonalites Granites Basalts Pegmatites Quartz veins	Shales Sandstones Limestones
Early Proterozoic	Basalts Trondhjemites Tonalites	Greywackes Shales Sandstones Chert Ironstones Limestones
Archaean	Basalts ANT Peridotites	High Al, Mg, Ti pelites-shales Chert Ironstones Limestones

4.0-2.0 by. This tentative model needs to be more thoroughly substantiated by reliable radiometric ages and better geochemical characterisation of the differences between metamorphosed cherts and detrital quartzites. However, I maintain that it is a necessary development in beginning to explain in a better way the relative abundances and compositional trends of the sediments of the early history of the earth.

ACKNOWLEDGEMENTS

I am extremely grateful to Dr Hari Narain, Director National Geophysical Research Institute, Hyderabad, for permission to publish this paper, and to Drs S.A. Drury and A.K. Baksi for the review of the manuscript and suggestions for its improvement. The help of Shri K. Chandrashekar and S.H. Jafri during the preparation of this paper is acknowledged. The analytical work for this paper was done with the help of Shri S.M. Ahmad and B. Udayraj. The Rb and Sr data were borrowed from the Working Group Data.

REFERENCES

Anhaeusser, C.R., 1972. The geology of the Jamestown Hills area of the Barberton Mountainland, South Africa. Trans. geol. Soc. S.Afr., 75: 225-263.

Brotzen, O., 1966. The average igneous rock and geochemical balance. Geochim. Cosmoch. Acta, 30: 863-868.

Condie, K.C., Macke, E.J. and Reimer, O.T., 1970. Petrology and geochemistry of early Precambrian greywackes from the Fig Tree Group, South Africa. Bull. geol. Soc. Am., 81: 2759-2776.

Donaldson, J.A. and Jackson, G.D., 1965. Archaean sedimentary rocks of North Spit Lake area, northwestern Ontario. Can. J. Earth Sci., 2: 622-647.

Engel, A.E.G., Itson, S.P., Engel, C.G., Stickney, D.M. and Cray, E.J., 1974. Crustal evolution and global tectonics: a petrogenic view. Bull. geol. Soc. Am., 85: 843-858.

Fyfe, W.S., 1973. Granulite facies, partial melting and the Archaean crust. Phil. Trans. R. Soc. Lond., A273: 457-462.

Garrels, R.M., Mackenzie, F.T. and Siever, R., 1971. Sedimentary cycling in relation to the history of the continents and oceans. In: C.E. Robertson (Ed.), The Nature of Solid Earth. pp. 93-124.

Geochemistry Group, 1977. Archaean Geochemistry. Contributions and Programmes of N.G.R.I. Geophys. Res. Bull., 15, Sp. Iss., 54 pp.

Glikson, A.Y., 1971. Archaean geosynclinal sedimentation near Kalgoorlie, Western Australia. Sp. Publ. geol. Soc. Austr., 3: 343-360.

Glukhovskiy, M.Z. and Pavlovskiy, Ye.V., 1973. Early stages of the evolution of the earth. Geotectonics, 2: 61-63.

Horn, M.K. and Adams, J.A.S., 1966. Computer-derived geochemical balance and element abundances. Geochim. Cosmoch. Acta, 30: 279-297.

Iyengar, B.R.C., 1971. Precambrian rocks in the Holenarsipur, Hassan district, Mysore State (South India). J. Ind. Geosci. Ass., 13: 31-44.

Janardhan, A.S. and Srikantappa, C., 1975. Geology of the northern parts of the Sargur schist belt between Mavinhalli and Dodkanya, Mysore district, Karnataka. Ind. Mineral., 16: 66-75.

Lowman, Jr., P.D., 1976. Crustal evolution in the silicate planets: implications for the origin of continents. J. Geol., 84: 1-26.

Martin, H., 1968. Problems of age relations and structure in some metamorphic belts of southern Africa. Sp. Pap. Geol. Ass. Can., 5: 17-26.

Macpherson, H.G., 1958. A chemical and petrographic study of Precambrian sediments. Geochim. Cosmoch. Acta, 14: 73-92.

McGlynon, J.C. and Henderson, J.H., 1970. Archaean volcanism and sedimentation in the Slave structural province. Geol. Surv. Can. Pap., 70-40: 31-44.

Mead, W.J., 1907. Redistribution of elements in the formation of sedimentary rocks. J. Geol., 15: 238-256.

Nair, M.M., Vidyadharan, K.T., Power, S.D., Sukumaran, P.V. and Murthy, Y.G.K., 1975. The structural and stratigraphic relationship of the schistose rocks and associated igneous rocks of the Tellicherry-Manantoddy area, Cannanore district, Kerala. Ind. Mineral., 16: 89-100.

Nanz, R.H., 1953. Chemical composition of Precambrian slates with notes on the geochemical evolution of lutites. J. Geol., 61: 51-64.

Naqvi, S.M., 1976a. Physico-chemical conditions during the Archaean as indicated by Dharwar geochemistry. In: B.F. Windley (Ed.), The Early History of the Earth. Wiley, Lond., pp. 289-298.

Naqvi, S.M., 1976b. Distribution of elements in the crust and mantle during the Archaean. Abstr. 25th Int. geol. Congr., Sydney.

Naqvi, S.M., 1977. Archaean sedimentation of the Dharwars in the central part of the Chitradurga schist belt, Karnataka, India. Bull. Geophys. Res., 15: 17-30.

Naqvi, S.M. and Hussain, S.M., 1972. Petrochemistry of early Precambrian metasediments from the central part of the Chitaldrug schist belt, Mysore, India. Chem. Geol., 10: 109-135.

Naqvi, S.M., Divakara Rao, V., Rogers, J.J.W., Satyanarayana, K., Hussain, S.M. and Narayana, B.L., 1978a. The petrochemistry and geologic implications of conglomerates from the Archaean geosynclinal piles of southern India. Can. J. Earth Sci. (in press).

Naqvi, S.M., Divakara Rao, V. and Hari Narain, 1978b. The primitive crust; evidence from the Indian Shield. Precamb. Res. (in press).

Naqvi, S.M., Viswanathan, S. and Viswanatha, M.N. (this vol.) Geology and geochemistry of the Holenarasipur schist belt and its place in the evolutionary history of the Indian Peninsula.

Pettijohn, F.J., Potter, P.E. and Siever, R., 1972. Sand and Sandstone. Springer Verlag, Berlin. 618 pp.

Radhakrishna, B.P., 1976. Mineralisation episodes in the Dharwar craton of the Peninsular India. J. geol. Soc. India, 17: 79-88.

Radhakrishna, B.P. and Vasudev, V.N., 1977. The early Precambrian of the southern Indian Shield. J. geol. Soc. India, 18: 525-541.

Ramakrishnan, M., Viswanatha, M.N. and Swami Nath, J., 1976. Basement-cover relationships of Peninsular gneiss with high-grade schists and greenstone belts of southern Karnataka. J. geol. Soc. India, 17: 97-111.

Rama Rao, B., 1962. A handbook of the geology of Mysore State, southern India. Bangalore Printing and Publ. Co., 264 pp.

Ronov, A.B., 1972. Evolution of the rock composition and geochemical processes in the sedimentary shell of the earth. Sedimentology, 19: 157-172.

Ronov, A.B. and Migdisov, A.A., 1971. Geochemical history of the crystalline basement and the sedimentary cover of the Russian and North American platform. Sedimentology, 16: 137-185.

Salop, J.J., 1968. Precambrian of U.S.S.R. In: Geology of the Precambrian. Sect. 4, 23rd Int. geol. Congr.

Sampat Iyengar, P., 1905. Report on the Survey work in the Chitaldrug district. Rec. Mysore Geol. Dept., 6(2): 57-92.

Satyanarayana, K., Divakara Rao, V., Naqvi, S.M. and Hussain, S.M., 1973. On the provenance of Dharwar schists, Shimoga, Mysore. Bull. Nat. Geophys. Res. Inst., 11: 35-41.

Shaw, D.M., 1956. Geochemistry of pelitic rocks. Pt III: Major elements and general geochemistry. Bull. geol. Soc. Am., 67: 919-934.

Sibley, D.F. and Wilband, J.T., 1977. Chemical balance of the earth's crust. Geochim. Cosmoch. Acta, 41: 545-554.

Srinivasan, R. and Sreenivas, B.L., 1968. Sedimentation and tectonics in the Dharwars (Archaean), Mysore State, India. Ind. Mineral., 1-2: 47-58.

Swami Nath, J., Ramakrishnan, M. and Viswanatha, M.N., 1976. Dharwar stratigraphic model and Karnataka craton evolution. Rec. Geol. Surv. India, 107(2): 149-175.

Turner, C.C. and Walker, R.G., 1973. Sedimentology, stratigraphy and crustal evolution of Archaean greenstone belts near Sioux Lookout, Ontario. Can. J. Earth Sci., 10: 817-845.

Varadarajan, S., 1970. Emplacement of chromite-bearing ultramafic rocks of Mysore State, India. Proc. Symp. U.M.P. 2nd, Hyderabad, pp. 441-454.

Vasudevan, D. and Rao, T.M., 1975. The high-grade schistose rocks of the Nellore schist belt, Andhra Pradesh, and their geologic evolution. Ind. Mineral., 16: 43-47.

Veizer, J. and Hoefs, J., 1976. The nature of O^{18}/O^{16} and C^{13}/C^{12} secular trends in sedimentary carbonate rocks. Geochim. Cosmoch. Acta, 40: 1387-1395.

Viswanatha, M.N. and Ramakrishnan, M., 1975. The pre-Dharwar supracrustal rocks of the Sargur schist complex in southern Karnataka and their tectonometamorphic significance. Ind. Mineral., 16: 48-65.

Viswanathan, S., 1974. Basaltic komatiite occurrences in the early Precambrian of India. Proc. 61st Ind. Sci. Congr., Pt. 3, Abstr., p. 161.

Whitemore, D.R.E., Berry, L.G. and Hawley, J.E., 1964. Chrome micas. Amer. Miner., 31: 1-21.

Windley, B.F., 1977. The Evolving Continents. Wiley, Lond., 385 pp.

ANORTHOSITES AND ULTRAMAFICS

REE DISTRIBUTIONS IN BASALTIC ANORTHOSITES FROM THE HOLENARASIPUR GREENSTONE BELT, KARNATAKA, SOUTH INDIA

S.A. Drury
Open University, Milton Keynes, U.K.

S.M. Naqvi and S.M. Hussain
National Geophysical Research Institute, Hyderabad, India

Abstract

Extremely fine-grained meta-anorthosites to anorthositic meta-basalts are found as sheets and lenses amongst meta-peridotitic komatiites, serpentinites and amphibolites of the c. 3.5 by Holenarasipur greenstone belt in Karnataka State. Their REE patterns show slight enrichment in LREE relative to HREE, La varying between 12 and 7 times and HREE between 3 and 4 times chondritic abundances, and the average $(Ce/Yb)_N$ ratio is 2.1. Their Eu content is variable, the Eu/Eu* ratio ranging from 1.2-9.0 and showing a strong negative correlation with modal amphibole. In such respects these supracrustal anorthositic rocks differ greatly from high-grade rocks of similar major element chemistry from West Greenland and NW Norway, but show a remarkable resemblance to lunar highland basalts. This resemblance is confirmed by close textural and other geochemical similarities. The relatively unfractionated nature of the anorthositic basalts suggests that they may represent anorthositic basalt melts within which small amounts of plagioclase accumulation has occurred. The origin of such melts is restricted by the REE data to two possibilities: (a) by small degrees of partial melting of very hydrous mantle peridotite; (b) by large amounts of olivine and orthopyroxene fractionation from peridotitic komatiite magmas.

INTRODUCTION

The occurrence of supracrustal Archaean basaltic anorthosites in association with komatiites of a c. 3.5 by greenstone belt in South India (Radhakrishna, 1976; Ramakrishna et al., 1976) was as unexpected as the original discoveries of the komatiite suite of lavas (Viljoen and Viljoen, 1969) and the anorthositic component of the early lunar crust (Wood et al., 1970).

Many petrologists maintain that magmas of anorthositic composition cannot form by partial melting of mantle or crustal materials (e.g. Phinney, 1969; Ryder, 1974), the Proterozoic massif-type anorthosites (An_{35-60}) (Green, 1969) and the high-grade Archaean stratiform anorthosites (An_{90}) (Windley and Smith, 1976) being regarded as cumulates from basic magmas. However, studies of the system diopside-anorthite-silica indicate that anorthositic magmas may form by partial melting of wet basic materials at pressures greater than 20 kb at temperatures around 1350°C (Yoder, 1968). Experimental melting of amphibolite at low pressures (5 kb) yields anorthositic liquids (Yoder and Tilley, 1962). Rapid adiabatic rise of such magmas would form layered anorthositic intrusions and possibly anorthositic lavas. Murthy (1976) suggested that hydrous, high temperature conditions in a primitive mantle with a significant component of late-accreted refractory Ca- and Al-rich material would have promoted the formation of anorthositic magmas at an early stage in the earth's history.

In this paper the REE distributions in anorthositic basalts from South India are compared with those in a variety of terrestrial and lunar anorthositic rocks, and with those produced by petrogenetic models representing: (a) fractional crystallisation of basaltic and komatiitic magmas; (b) partial melting of basalt, amphibolite and mantle peridotite; and (c) partial melting of tonalite to yield adamellitic melts and anorthosite residues. Together with textural and other features, these data are used to suggest a framework for the origin of the supracrustal anorthositic basalts.

ANALYSED ROCKS

1. Geological Setting

The Holenarasipur greenstone belt, from which the anorthositic basalts were collected, has three arms representing tight synforms of supracrustals separated by trondjhemitic to tonalitic gneisses and intrusions (Rama Rao, 1940). The lower part of the belt comprises basal tremolite-actinolite-talc schists succeeded by meta-peridotitic komatiites, serpentinites, basaltic komatiites (high-MgO basalts) and garnet amphibolites, together with minor metasediments (kyanite-staurolite schists and iron formations). The whole sequence has been subject to almandine amphibolite facies metamorphism, the presence of kyanite suggesting an intermediate- to high-pressure facies series. The marginal zones of the belt have been invaded by several generations of pegmatites related to tonalitic plutons which, together with the absence of sedimentological evidence for an older eroded gneissic basement, suggests that the surrounding 'Peninsular Gneisses' are younger. However, without radiometric dates and a detailed tectonic chronology this is a matter of opinion only - the antecedence of the gneissic component of the 'Peninsular Gneisses' has been postulated by Ramakrishna et al. (1976).

Deformation in the belt is variable, ranging from intense in marginal areas of ultrabasic phyllonite to low when pillows may be observed in the meta-peridotitic komatiites. Spinifex textures are absent as a result of complete recrystallisation of the ultramafic rocks. However, the alternation of ultramafics with metacherts confirms the subaqueous nature of the igneous activity.

The anorthositic rocks are found in outcrops of weathering-prone ultramafics; hence exposure is poor and their relations with the enclosing ultramafics are largely obscured. The anorthositic rocks occur as up to 20 m wide pods, lenses

TABLE 1: Major and rare earth element analyses and modal analyses for anorthositic rocks from the Holenarasipur greenstone belt. Samples 5-8 are fine grained and show evidence of high temperature annealing; samples 1-4 are partly granulated, medium- to fine-grained rocks.

	1	2	3	4	5	6	7	8
NGRI No.	S299	S246	S302	S306	S309	S307	S244A	S243B
SiO_2	46.86	49.88	46.19	45.83	50.02	47.99	51.19	50.24
TiO_2	0.37	0.30	0.36	0.26	0.44	0.40	0.35	0.36
Al_2O_3	26.34	26.61	28.12	26.14	26.92	29.28	25.96	25.17
Fe_2O_3	0.62	0.31	0.21	1.73	0.11	0.56	0.37	0.07
FeO	3.36	2.74	1.60	2.88	1.88	0.94	2.62	2.68
CaO	17.39	15.02	16.03	15.25	14.50	14.69	14.19	13.69
MgO	1.59	2.66	4.22	5.04	4.03	2.78	2.44	6.08
Na_2O	2.20	2.20	2.00	2.35	2.20	2.15	2.57	1.50
K_2O	0.09	0.05	0.05	0.09	0.05	0.09	0.05	0.05
MnO	0.13	0.16	0.68	0.07	0.09	0.02	0.11	0.12
P_2O_5	0.33	0.31	0.31	0.35	0.25	0.43	0.61	0.66
H_2O	0.18	0.21	0.23	0.30	0.25	0.22	0.30	0.24
TOTAL	99.46	100.45	100.00	100.29	100.24	99.45	100.76	100.85
RARE EARTH ELEMENTS								
La	3.37	3.11	3.68	2.26	2.91	2.79	3.33	3.61
Ce	6.94	7.00	5.60	4.75	5.85	5.20	7.23	7.38
Nd	3.93	3.68	2.56	2.58	3.71	3.00	3.95	4.13
Sm	1.06	1.01	0.65	0.81	0.98	0.90	0.91	0.89
Eu	0.54	0.50	0.38	0.34	0.75	0.90	1.81	2.77
Gd	1.42	0.85	0.86	1.00	1.29	0.60	1.00	1.05
Tb	0.22	0.20	0.15	0.14	0.21	0.20	0.18	0.19
Tm	0.16	0.15	0.10	0.14	0.15	0.13	0.10	0.12
Yb	1.00	0.90	0.64	0.52	0.88	0.82	0.72	0.74
Lu	0.15	0.14	0.11	0.50	0.13	0.12	0.11	0.11
REE	18.80	17.54	14.70	12.60	16.90	14.60	19.40	21.00
$(Fe/Yb)_N$	1.80	1.90	2.10	2.30	1.70	1.60	2.50	2.50
Eu/Eu*	1.30	1.40	1.60	1.20	1.90	2.80	6.00	9.00
MODAL ANALYSES								
Qtz	-	0.80	-	-	1.20	0.02	2.90	2.35
Plag	74.60	66.40	81.75	71.96	76.78	89.71	75.50	80.41
Amph	24.00	31.40	15.90	27.27	18.90	7.90	12.40	8.29
Garnet	-	-	-	-	0.20	-	5.90	6.05
Mafics	1.17	0.70	0.50	0.26	1.70	0.90	1.90	1.61
Altn. & Acc.	0.14	0.60	1.80	0.50	1.20	1.40	1.25	1.29

and sheets parallel to the foliation in deformed ultramafics. The limited exposure shows no mineral layering within the anorthositic rocks, but there is irregular variation in both their grain size and mafic index.

2. Petrography

The analysed rocks range from anorthosites, through basaltic anorthosites to anorthositic basalts, and consist of anorthite, low-Mg tschermakitic hornblende, almandine-grossular garnet and minor apatite, sphene, magnetite, spinel, scapolite and quartz. The grain size varies from 0.02-1.0 mm. The medium- to fine-grained material exhibits irregular plagioclase grains surrounded by very fine-grained granular plagioclase in what appears to be a cataclastic texture; a gradation to entirely very fine-grained material can be demonstrated. In this the feldspar grains exhibit a polyhedral granular texture and only occasionally show twinning. Texturally there is a most striking similarity to anorthositic impactites from the lunar highlands (Taylor, 1975, Fig. 5.5), analogues of cataclastic anorthosites and granulitic anorthosites produced by prolonged high-temperature subsolidus annealing being present. It is worth noting that under almandine amphibolite conditions and in the presence of Mg, Fe and H_2O from the ultramafics, tectonic cataclasis of anorthosites is expected to produce zoisite-epidote-rich rocks, commonly found in deformed Archaean high-grade anorthosites (Watson, 1969). In fact the feldspars are not even clouded. The hornblendes in the fine-grained material exhibit textural equilibration with plagioclase and define a weak planar fabric. The garnets occur as euhedral porphyroblasts in the fine-grained variety and enclose minute grains of plagioclase and hornblende. They grew after annealing of the polyhedral plagioclase-rich matrix, probably under almandine amphibolite facies conditions.

3. Geochemistry

The compositional range of the samples is encompassed by the fields of anorthosite to anorthositic gabbro (Table 1). There are both olivine- and quartz-normative samples. The garnet-bearing samples are quartz-normative. The suite is broadly comparable with typical high-grade Archaean anorthosites from Fiskenaesset and Sittampundi, except for higher Na_2O, SiO_2 and TiO_2, and shows a closer resemblance to lunar anorthositic rocks, particularly 'highland basalts', except for higher SiO_2, MgO and H_2O contents. They plot together with the lunar rocks on CaO-Al_2O_3, Al_2O_3-2Ti-Cr and AFM diagrams. The plagioclase compositions fall within the range of lunar anorthositic plagioclases and are all in the anorthite range An_{92-96} (electron microprobe analyses, Open University). Details of the geochemistry will be published elsewhere.

RARE EARTH ELEMENT ANALYSES

1. Analytical Technique

Eight samples were analysed for La, Ce, Nd, Sm, Eu, Gd, Th, Tm, Yb and Lu by instrumental neutron activation using a method described by Paul et al. (1974). Irradiation was carried out at the University of London Reactor Centre, Ascot, and activities were detected by instrumentation at the Open University and converted to element abundances using the peak analysing programme SAMPO at the University of London Computer Centre.

Because of very low levels of interfering elements, such as Ba and Th, excellent resolution for the relatively low abundances of REE was achieved (detection limits, precision and comparability of REE analysis with BCR-1 are given in Table 2).

TABLE 2: Lower limits of detection (1), analytical error (2), and comparability between Open University values (3) and accepted values for BCR-1 for REE

	1 (ppm)	2 (%)	3 (ppm)	4 (ppm)
La	2.00	10	21.10	26.00
Ce	0.60	7	51.20	53.90
Nd	1.00	7	28.20	29.00
Sm	0.70	7	6.98	6.60
Eu	0.01	4	2.07	1.94
Gd	0.80	10	7.80	6.60
Tb	0.03	4	1.01	1.00
Tm	0.06	10	0.50	0.60
Yb	0.04	4	3.31	3.36
Lu	0.03	10	0.54	0.55

2. Results

Table 1 lists major element and REE analyses, and modal analyses for the 8 anorthosites to anorthositic basalts. Their REE abundances, normalised relative to the average of 10 ordinary chondrites (Nakamura, 1974), are plotted on Fig. 1 together with the fields of high-grade anorthositic rocks from West Greenland (O'Nions and Pankhurst, 1974; Henderson et al., 1976) and NW Norway (Green et al., 1972) and lunar anorthositic gabbros (highland basalts) (Hubbard et al., 1974).

The REE patterns in Holenarasipur anorthositic rocks show the following features:

(a) Slight enrichment in LREE relative to HREE, with an average $(Ce/Yb)_N$ ratio of 2.1 (Fig. 1a,b). In this respect they differ from the highly fractionated meta-anorthositic rocks from the Archaean of NW Norway (Green et al., 1972). Meta-anorthosites from West Greenland, however, also display low $(Ce/Yb)_N$ ratios ranging from 0.5-14.5 with an average of 3.6 (Fig. 1c).

(b) Bar Eu, only a small range in the abundances of each REE. The overall pattern indicates La at between 12 and 7 times and HREE between 3 and 4 times chondritic abundances. This contrasts with the very high variability of REE in West Greenland meta-anorthosites (Fig. 1c).

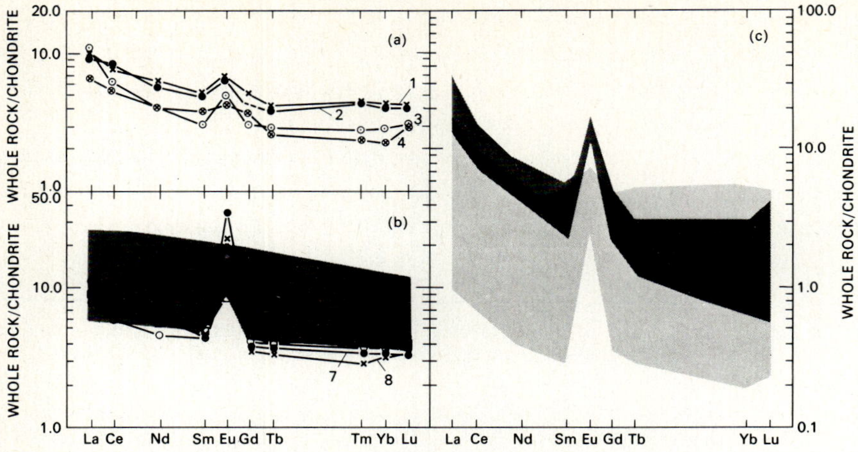

Fig. 1. Chondrite-normalised REE plots for (a) partly granulated medium- to fine-grained, Holenarasipur anorthositic rocks, (b) annealed fine-grained anorthositic rocks from Holenarasipur showing field of lunar anorthositic rocks (dark tone) (Hubbard et al., 1974), and (c) high-grade Archaean meta-anorthosites from NW Norway (dark tone) (Green et al., 1972) and from the Fiskenaesset complex, West Greenland (light tone) (O'Nions and Pankhurst, 1974; Henderson et al., 1976).

(c) Variable Eu abundances, Eu/Eu* ranging from 1.2-9.0. As shown by Fig. 1, the very fine-grained examples have stronger positive Eu anomalies than the medium- to fine-grained variety. The Eu/Eu* ratios show a strong negative correlation with modal amphibole content but no significant correlations with modal feldspar, Al_2O_3 or Fe_2O_3/FeO ratios. The samples show much lower Eu/Eu* ratios than the majority of Archaean meta-anorthositic rocks.

(d) A very close resemblance to lunar 'highland basalts' in terms of ΣREE, $(Ce/Yb)_N$ and Eu/Eu* ratios (Fig. 1b), the average abundances of REE in the Holenarasipur material being very close to those for lunar highland basalt quoted by Taylor (1975).

DISCUSSION

There are four plausible explanations for the occurrence of the supracrustal basaltic anorthosites:

(1) Feldspathic cumulates from basaltic or peridotitic komatiite flows or sills.

(2) Residual melts of basaltic or peridotitic magma from which olivine and orthopyroxene were fractionated.

(3) Anorthositic basalt magma formed by partial melting of a basaltic or

peridotitic source under high P_{H_2O}.

(4) Anorthositic residual solids from partial melting of tonalites to yield adamellitic melts.

(Throughout this discussion phenocryst-matrix partition coefficients for REE are as compiled by Arth, 1976.)

(1) The field relations of the basaltic anorthosites as small bodies in large volumes of ultramafic material suggest a possible origin by flotation of feldspars after emplacement of thick, fluid ultrabasic flows or sills. However, the anorthositic rocks have higher total REE contents than Archaean peridotitic komatiites which have 2-3 times chondritic REE abundances and show depletion in LREE relative to HREE (Hawkesworth and O'Nions, 1977). The low phenocryst-matrix partition coefficients for REE in plagioclase should lead to their depletion in anorthositic rocks formed by fractional crystallisation of feldspars. Arndt (1976) has shown that plagioclase is the last mineral to crystallise from peridotitic komatiite melts and appears only at low pressures.

Accumulation of plagioclase from a basaltic source liquid could feasibly produce the REE patterns seen in the basaltic anorthosites. However, even if REE-enriched Archaean tholeiites are used as a starting liquid, impossibly high proportions (70-90%) of plagioclase crystals must be removed to give appropriate REE contents in a feldspathic cumulate. The only theoretically possible plagioclase fractionation mechanism is one involving parental liquids with the high REE contents of Archaean andesitic lavas (Condie, 1976). Such materials have not been reported from the older greenstone belts of the Indian Archaean.

(2) The phenocryst-matrix partition coefficients for REE in olivine and orthopyroxene in basaltic systems are less than unity, and neither selectively concentrates or is depleted in Eu (see Arth, 1976). Therefore residual melts after fractionation of olivine and orthopyroxene from basaltic melts should show relative enrichment in total REE and no significant Eu anomalies. Even the most REE-depleted Archaean basalts have REE with flat patterns greater than 6 times chondritic abundances so that calcic residual liquids in the basalt system would show much more abundant REE than the basaltic anorthosites.

Fractional crystallisation of various olivine- and pyroxene-bearing assemblages from the average of Rhodesian ultrabasic komatiites (Hawkesworth and O'Nions, 1977) is capable of producing the relative REE enrichment observed in the basaltic anorthosites. Such a process is also compatible with experimental data on crystallisation of komatiites (Arndt, 1976). However, since the REE patterns in these komatiites show enrichment in HREE, to produce the required $(Ce/Yb)_N$) ratios from them, large amounts of augite and/or hornblende must be present in the extracted crystals. This is also a prerequisite for ultrabasic komatiite magmas with flat REE patterns. For the LREE-enriched ultrabasic komatiites reported by

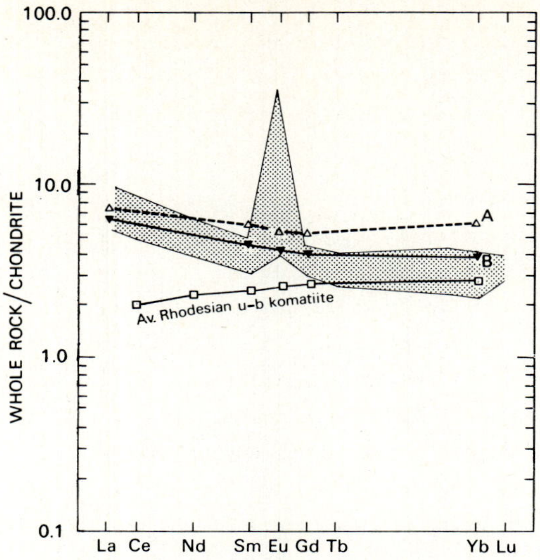

Fig. 2. Plots showing chondrite-normalised Ce, Sm, Eu, Gd and Yb for (A) residual liquid after 80% equilibrium fractional crystallisation of assemblage (ol_{40}, opx_{20}, cpx_{20}, hb_{20}) from average Rhodesian ultrabasic komatiite (Hawkesworth and O'Nions, 1977); and (B) 10% equilibrium modal melting of hornblende peridotite (ol_{50}, opx_{10}, cpx_{10}, hb_{20}, $oxide_{10}$) with chondritic REE. Light tone indicates field of Holenarasipur anorthositic rocks.

Nesbitt and Sun (1976), fractionation of olivine plus orthopyroxene alone can mimic the general REE pattern of the basaltic anorthosites. Fig. 2 shows the REE pattern derived by 80% fractional crystallisation of the assemblage (ol_{40}, opx_{20}, cpx_{20}, hb_{20}) from average Rhodesian ultrabasic komatiite.

(3) The REE patterns in calcic and aluminous melts formed by partial fusion in wet basaltic or peridotitic compositions are large determined by the mineralogy of the residual solids. In cases of small degrees of melting, which anorthositic materials must represent in such hypothetical models, this mineralogy would differ little from the modal composition of the source material. Because of the low $(Ce/Yb)_N$ ratios in the anorthositic rocks, very rigid limits must be placed on HREE-concentrating minerals such as garnet in the source materials, which can safely be assumed to have flat or only slightly fractionated chondrite-normalised REE patterns.

With source material of basaltic composition, assemblages with high proportions of olivine and pyroxene, which have partition coefficients well below 1.0 for basaltic matrices, ensure enrichment in REE, whereas the anorthositic basalts are depleted relative to basalts. Even basaltic source materials in the form of pure hornblendite are incapable of producing depleted REE patterns due to the low partition coefficients for REE between amphiboles and basaltic matrices. On these grounds, an origin by partial melting of wet basaltic rocks seems implausible, unless REE-depleted basalts prove to be uniquely present in the Indian Archaean.

Low degrees of melting of nearly anhydrous peridotitic source materials with near-chondritic REE patterns generate REE-enriched materials which fall in the envelopes of Archaean to Recent tholeiites, as a result of low bulk distribution coefficients for REE elements. Only by incorporating large amounts of amphibole into the parent mineralogy can bulk distribution coefficients be assembled which are capable of producing the 3-10 times enrichment observed in the basaltic anorthosites. Fig. 2 shows the REE pattern generated by 10% melting of a hornblende peridotite (hb_{20}, ol_{50}, opx_{10}, cpx_{10}) with chondritic REE abundances. Though not showing a positive Eu anomaly it has a $(Ce/Yb)_N$ ratio similar to that observed in the basaltic anorthosites. As suggested by the correlation between Eu/Eu* ratios and modal hornblende contents, the observed Eu anomalies could represent slight <u>in situ</u> fractionation of plagioclase.

(4) Green (1969) suggested that some anorthosites may be residues of partial melting of tonalites to yield adamellitic melts. All Archaean rocks of tonalitic composition that have been analysed for REE show very marked enrichment of LREE relative to HREE (e.g. Condie and Hunter, 1976). Such patterns would be preserved in feldspar-rich fractionates from tonalites due to the lack of significant differences in partition coefficients between LREE and HREE in plagioclase. This, together with a total lack of field evidence for relations between basaltic anorthosites, tonalites and derived melts, rules out such a petrogenetic model.

On the basis of REE patterns alone, the basaltic anorthosites could represent either residual melts after large-scale fractionation of hornblende- and/or augite-rich assemblages from peridotitic komatiite magma, or low degrees of partial melting of a hornblende peridotite source. Given the limited scope of this study and the poor exposure it is not possible to favour either mechanism.

Taking into account the close geochemical and textural similarities between the basaltic anorthosites and lunar anorthositic rocks, it is tempting to speculate that they share either a similar formative process or common precursors. It is widely held that the geochemistry, particularly in terms of REE, of the lunar anorthositic breccias reflects mixing of earlier more fractionated rocks as a result of impact events which produced the ringed lunar basins (Taylor and Jakes, 1974). An equally widely held view is that the same material, though obviously modified by impacts, reflects the primitive composition of an early accreted surface layer on the moon which differed from the deeper source region of mare basalts by its higher Al_2O_3 content, Mg/Fe ratio and content of lithophile trace elements (Hubbard et al., 1974). A similar suggestion has been made for the earth on indirect grounds by Murthy (1976) and on evidence from the volatile content of Archaean high-grade anorthosites by Morgan et al. (1976). Glikson (1976) examined the weight of circumstantial evidence for large-scale impacts contributing to the formation of Archaean supracrustal sequences and suggested that the basic-ultrabasic volcanics of the earliest greenstone belts might represent mantle diapirism triggered by major impacts and their long-term elevation of geoisotherms in the mantle; however, only unequivocal demonstration of structural features or petrographic textures related to impacts could verify this possibility. In this light it seems advisable to conduct a major study on those rocks of anorthositic affinities which hitherto have been merely reported from some of the oldest greenstone terrains (Anheusser et al., 1969; Windley, 1970). They may represent derivatives of pre-impact crust engulfed and transported by basic-ultrabasic magmas.

ACKNOWLEDGEMENTS

The authors wish to thank Dr P. Potts and Mrs Mary Sarre of the Open University for their help with the analyses, and Dr A. Glikson for fruitful discussion. SAD acknowledges a grant from the Open University Overseas Travel Fund which enabled him to visit the Archaean of South India, and SMN and SMH are grateful to the Director, NGRI, who supported their work. Secretarial assistance was provided by Sue Boggie and Joan Waugh at the Open University.

REFERENCES

Anheusser, C.R., Mason, R., Viljoen, M.J. and Viljoen, R.P., 1969. A reappraisal of some aspects of Precambrian shield geology. Bull. geol. Soc. Am., 80: 2175-2200.

Arndt, N.T., 1976. Melting relations of ultramafic lavas (komatiites) at one atmosphere and high pressure. Carnegie Inst. Wash. Yearbk., 75: 555-562.

Arth, J.G., 1976. Behaviour of trace elements during magmatic processes - a summary of theoretical models and their applications. J. Res. U.S. geol. Surv., 4: 41-47.

Condie, K.C., 1976. Trace element geochemistry of Archaean greenstone belts. Earth Sci. Rev., 12: 393-417.

Condie, K.C. and Hunter, D.R., 1976. Trace element geochemistry of Archaean granitic rocks from the Barberton region, South Africa. Earth Planet. Sci. Lett., 29: 389-400.

Glikson, A.Y., 1976. Earliest Precambrian ultramafic-mafic volcanic rocks: ancient oceanic crust or relic terrestrial maria. Geology, 4: 201-205.

Green, D.H., Nichols, I.A., Viljoen, M.J. and Viljoen, R.P., 1975. Experimental demonstration of the existence of peridotitic liquids in earliest Archaean magmatism. Geology, 3: 11-14.

Green, T.H., 1969. High-pressure experimental studies on the origin of anorthosite. Can. J. Earth Sci., 6: 427-440.

Green, T.H., Brunfelt, A.O. and Heier, K.S., 1972. Rare-earth element distribution and K/Rb ratios in granulites, mangerites and anorthosites, Lofoten-Vesteraalen, Norway. Geochim. Cosmoch. Acta, 36: 241-257.

Hawkesworth, C.J. and O'Nions, R.K., 1977. The petrogenesis of some Archaean volcanic rocks from southern Africa. J. Petrol., 18: 487-520.

Henderson, P., Fishlock, S.J., Laul, J.G., Cooper, T.D., Conard, R.L., Boynton, W.V. and Schmitt, R.A., 1976. Rare-earth element abundances in rocks and minerals from the Fiskenaesset complex, West Greenland. Earth Planet. Sci. Lett., 30: 37-49.

Hubbard, N.J., Rhodes, J.M., Wismann, H., Shih, C.-Y. and Bansal, B.M., 1974. The chemical definition and interpretation of rock types returned from the non-mare regions of the moon. Proc. 5th Lunar Sci. Conf., Suppl. Geochim. Cosmoch. Acta, 5(2): 1227-1246.

Morgan, J.W., Ganapathy, R., Higuchi, H. and Krahenbuhls, U., 1976. Volatile and siderphile trace elements in anorthositic rocks from Fiskenaesset, West Greenland: comparison with lunar and meteoritic analogues. Geochim. Cosmoch. Acta, 40: 861-887.

Murthy, V.R., 1976. Composition of the core and the early chemical history of the earth. In: B.F. Windley (Editor), The Early History of the Earth. Wiley, Lond., pp. 21-32.

Nakamura, N., 1974. Determination of REE, Ba, Fe, Mg, Na and K in carbonaceous and ordinary chondrites. Geochim. Cosmoch. Acta, 38: 757-775.

Naqvi, S.M. and Hussein, S.M. (in press). Geochemistry of meta-anorthosite basalts from a greenstone belt in Karnataka, India. Geology.

Nesbitt, R.W. and Sun, S.-S., 1976. Geochemistry of Archaean spinifex-textured peridotites and magnesian and low-magnesian tholeiites. Earth Planet. Sci. Lett., 31: 433-453.

O'Nions, R.K. and Pankhurst, R.J., 1974. Rare-earth element distributions in Archaean gneisses and anorthosites, Godthaab area, West Greenland. Earth Planet. Sci. Lett., 22: 328-338.

Paul, D.K., Potts, P.J., Gibson, I.L. and Harris, P.G., 1955. Rare earth abundances in Indian kimberlite. Earth Planet. Sci. Lett., 25: 151-158.

Phinney, W.C., 1969. Anorthosite occurrences in Keweenawan rocks of north-eastern Minnesota. In: Y.W. Isaachsen (Editor), Origin of Anorthosite and Related Rocks. New York State Mus. Sci. Serv. Mem., 18: 135-143.

Radhakrishna, B.P., 1976. Mineralisation episodes in the Dharwar craton of Peninsular India. J. geol. Soc. India, 17: 79-83.

Ramakrishna, M., Viswanatha, M.N. and Swami Nath, J., 1976. Basement-cover relationships of Peninsular Gneiss with high-grade schists and greenstone belts of southern Karnataka. J. geol. Soc. India, 17: 97-111.

Rama Rao, B., 1940. The Archaean complex of Mysore. Bull. Mysore Geol. Dept., 17: 102 pp.

Taylor, S.R., 1975. Lunar Science: a post-Apollo View. Pergamon, 372 pp.

Taylor, S.R. and Jakes, P., 1974. The geochemical evolution of the moon. Proc. 5th Lunar Sci. Conf., Suppl. Geochim. Cosmoch. Acta, 5(2): 1287-1305.

Viljoen, M.J. and Viljoen, R.P., 1969. Evidence of the existence of a mobile extrusive peridotitic magma from the Komati Formation of the Onverwacht Group. Trans. geol. Soc. S.Afr. Sp. Publ., 2: 87-123.

Watson, J.V., 1969. The Precambrian gneiss complex of Ness, Lewis, in relation to the effects of Laxfordian regeneration. Scott. J. Geol., 5: 269-285.

Windley, B.F., 1970. Anorthosites in the early crust of the earth and moon. Nature, 226: 333-335.

Windley, B.F. and Smith, J.V., 1976. Archaean high-grade complexes and modern continental margins. Nature, 260: 671-675.

Wood, J.A., Dickey Jr., J.S., Ursula, B.M. and Powell, B.N., 1970. Lunar anorthosites and a geophysical model of the moon. Geochim. Cosmoch. Acta, 34: 965-988.

Yoder, H.S., 1968. Experimental studies bearing on the origin of anorthosite. In: Y.W. Isaachsen (Editor), Origin of Anorthosites and Related Rocks. Mem. New York State Mus. Sci. Serv., 18: 13-22.

Yoder, H.S. and Tilley, C.E., 1962. Origin of basalt magmas. An experimental study of natural and synthetic rock systems. J. Petrol., 3: 342-532.

RELICT PLUTONIC TEXTURES IN ARCHAEAN ULTRAMAFIC ROCKS FROM THE FISKENAESSET REGION, SOUTHERN WEST GREENLAND; IMPLICATIONS FOR CRUSTAL THICKNESS

C.R.L. Friend
Department of Geology, Oxford Polytechnic
and
D.J. Hughes
Department of Geology, Portsmouth Polytechnic

ABSTRACT

The field relationships of ultramafic bodies with amphibolitic supracrustal rocks and a later, layered gabbro-anorthosite complex are described. Together these rocks underwent deformation and metamorphism to upper amphibolite facies at c. 2700-2800 Ma. This deformational event was accompanied by the intrusion of large quantities of dioritic to granodioritic material, now converted to gneisses.

The ultramafic bodies contain several boudins in which are preserved various coarse-grained textures as well as mineral and chemical layering. These textures predate the 2700 Ma metamorphism. The textures include symplectic intergrowths of (i) olivine + spinel, and (ii) orthopyroxene + spinel with rare occurrences of (i) overgrown by (ii). The host rock to the boudins is a deformed and recrystallised analogue of the undeformed rock.

Geothermometry carried out on the co-existing olivine-spinel pairs in the coarse-grained relict textures and the deformed host rock suggest differing equilibration temperatures. The deformed rocks equilibrated at c. 700-800°C whilst the undeformed textures equilibrated at c. 1000°C. The former temperature is consistent with the metamorphic environment observed for the surrounding rocks, whilst the latter is far too high.

In the light of experimental work in the literature these textures are interpreted as plutonic igneous textures and the anomalous temperatures of equilibration reflect this event. Preliminary melting experiments performed on these rocks suggest that at 1400°C and 10 kb total water pressure, olivine and spinel are still on the liquidus. This provides minimum conditions which far exceed the observed metamorphic conditions and imply formation under a crust of at least 30 km thick. Since there is an absence of any old sialic crust in the region it is suggested that these rocks reflect conditions in the Archaean upper mantle underlying an oceanic area.

INTRODUCTION

The Fiskenaesset region (Fig. 1) lies some 110 km south of Godthaab in the central Archaean block of West Greenland (Bridgwater et al., 1976). About 75% of the region consists of a series of polyphase, partly syntectonic tonalitic-to-grandioritic gneisses. These contain amphibolites which are predominantly of supracrustal origin (Andersen and Friend, 1973; Andersen, 1974; Friend, 1975), the precursors of which were formed at about 3000 Ma (Burwell et al., in press). The anorthosites of the Fiskenaesset complex were intruded into these amphibolites (Windley et al., 1973; Escher and Myers, 1976). Subsequently the whole suite of rocks was complexly deformed and metamorphosed to upper amphibolite facies conditions and locally to hornblende-granulite facies conditions (Kalsbeek, 1976a,b). The culmination of the granulite facies conditions appears to have occurred at about 2850 Ma (Black et al., 1973; Pidgeon et al., 1976), whilst the amphibolite facies conditions lasted somewhat longer in the southern portion of the region (Pidgeon, 1973; Pidgeon and Hopgood, 1975; Burwell et al., in press), with evidence of open system behaviour continuing in some areas until as recently as 1950 Ma (Pidgeon and Hopgood, 1975).

The amphibolites may be traced throughout the gneisses of the southern part of the Fiskenaesset region (see G.G.U. Rapp. No. 73, Plate 1) and consist of variably preserved pillow lavas, agglomerates, massive lavas or sills and dykes, gabbros and meta-sedimentary rocks. Bodies of ultramafic rocks occur frequently within this sequence. These have been largely converted to metamorphic mineral assemblages but contain occasional relicts of earlier mineralogies. However, bodies with both deformed and relatively anhydrous undeformed assemblages occur at one locality north of Bjørnesund (Fig. 2). Friend and Hughes (1977) have suggested that the textures in the undeformed rocks have a primary igneous origin. It is the purpose of this paper to explain more fully the field relationships of these bodies in relation to the enclosing amphibolites and the relevence of their textures to the metamorphic conditions.

THE AMPHIBOLITES

The general geology of the best preserved unit of amphibolites, the Ravns Storø belt (Fig. 1), was described by Andersen and Friend (1973) and further amplified by consideration of various aspects of their

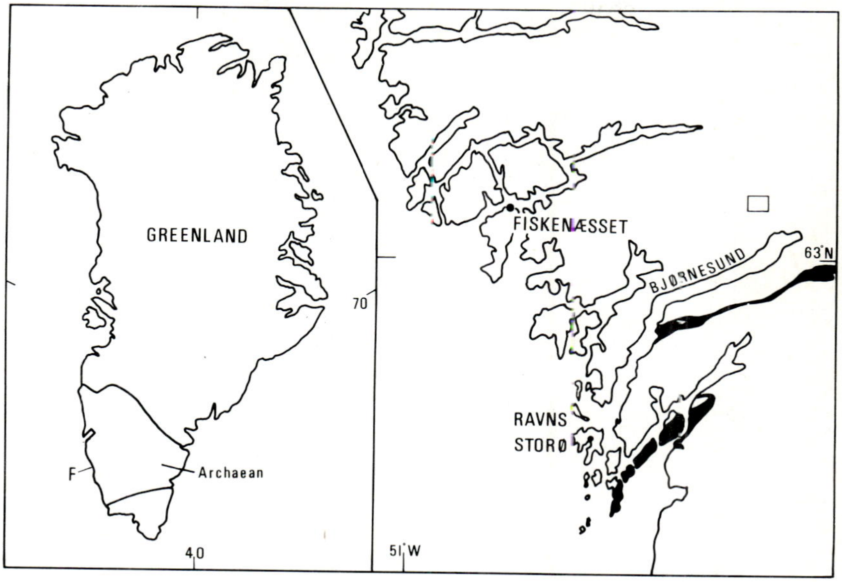

Fig. 1. Location map of the Fiskenaesset region, Ravns Storø and Bjørnesund with respect to the central Archaean block. The box represents the area depicted in Fig. 2. The amphibolites in the vicinity of Bjørnesund are shaded black.

chemistry and structural relations (Andersen, 1974; Friend, 1975). A summary of these results may be found in Bridgwater et al. (1976). Other studies of amphibolites in the Fiskenaesset region include those of Williams (1973) who examined part of the belt occurring along the southern shore of Bjørnesund (Fig. 1) and Rivalenti (1976) who studied amphibolites and ultramafic rocks occurring near Fiskenaesset.

The origin of many of these amphibolites as a series of extrusive, partly sub-aqueous lava flows is now established beyond doubt, since well-preserved pillow structures occur at several localities. Gradations from pillow lavas through pillow breccias to agglomeratic amphibolites may be recognised. Associated with these lithologies are more homogeneous, massive units which are interpreted as penecontemporaneous sills and/or flows. Some of these horizons display compositional layering from ultramafic to a more basaltic composition. Interlayered with the amphibolites are some unusual leucocratic schists which comprise assemblages of: quartz-plagioclase (oligoclase)-orthoamphibole (anthophyllite/gedrite)-cummingtonite$\pm$ biotite$\pm$cordierite$\pm$staurolite$\pm$garnet (Friend, 1976), which are interpreted as weathering products and residues of spilitised basalts and keratophyre associations. Minor meta-sedimentary material is represented as sillimanite$\pm$garnet schists.

Penecontemporaneous, rhythmically layered gabbroic bodies were intruded into this volcano-sedimentary pile (Andersen and Friend, 1973; Andersen, 1974).

Several studies have indicated that mainly isochemical recrystallisation has occurred in the thicker, less deformed units of basic rocks (Windley et al., 1973; Windley and Smith, 1974; Myers, 1975; Rivalenti, 1976). The same situation apparently applies to the Ravns Storø belt (Friend, 1975). Average compositions of groups of selected, homogeneous amphibolites from areas of low total deformation are presented in Table 1 and compared with analyses from Fiskenaesset (Rivalenti, 1976) and other Archaean areas (Viljoen and Viljoen, 1969a,b; Condie, 1976). The conclusion that the basaltic-type precursors of these amphibolites developed by the low pressure fractionation of firstly olivine and later olivine accompanied by clinopyroxene (Friend, 1975) was also reached by Rivalenti (1976) for those amphibolites near Fiskenaesset. However, unlike those amphibolites near Fiskenaesset, where rocks of alkali olivine basaltic affinity are reported (Rivalenti, 1976), no such compositions have been found to occur in the Ravns Storø belt. The close resemblance of many Archaean basic and ultrabasic rocks to modern oceanic basalts and associated rocks has been noted from many areas (Viljoen and Viljoen, 1969c; Condie, 1976, Herrmann et al., 1976; Gunn, 1976). The evidence from the Ravns Storø belt would appear to support these observations (Friend, 1975) and it is concluded that these amphibolites also represent oceanic-type crust.

The amphibolites are the oldest material found in the region (Andersen, 1974; Friend, 1975) and there is no isotopic evidence from the enclosing gneisses which would suggest that there was a pre-existing sialic basement which was later remobilised. In particular the initial ratio $^{87}Sr/^{86}Sr$ value of .702 is far too low to allow this (Pidgeon and Hopgood, 1975; Kalsbeek, 1976c; Moorbath, 1976). Also, it is now known that where older basement has existed, contamination of the lead isotope systems of the younger rocks, both gneisses and anorthosites, has occurred. No such contamination has as yet been discovered in rocks from the Fiskenaesset region

TABLE 1: Comparison of basic rocks from the Ravns Storø area with similar rocks from the Fiskenaesset area, other representative analyses of Archaean komatiites and a modern picrite

	1*	2*	3*	4	5	6	7	8	9
SiO_2	43.95	46.52	48.25	47.81	49.50	51.4	49.7	41.61	43.89
TiO_2	0.40	0.76	1.05	0.92	1.12	1.92	1.0	0.31	1.54
Al_2O_3	8.78	11.04	12.96	7.94	14.43	14.8	14.9	2.70	5.51
Fe_2O_3	13.04	12.96	14.48	2.99	3.24	2.1	2.6	5.63	6.50
FeO	-	-	-	10.11	8.39	8.3	8.8	4.35	7.32
MnO	0.20	0.22	0.24	0.25	0.20	-	-	0.17	0.17
MgO	27.55	15.38	10.72	11.72	6.90	6.7	6.3	30.58	25.53
CaO	5.80	11.60	10.33	12.83	10.55	10.7	9.4	4.29	4.51
Na_2O	0.13	1.34	2.50	2.15	2.78	2.7	2.1	0.15	0.81
K_2O	0.03	0.09	0.19	0.44	0.46	0.18	0.32	0.03	0.59
P_2O_5	0.12	0.09	0.13	0.09	0.14	-	-	0.02	0.28
H_2O+	-	-	-	1.63	1.37	-	-	8.81	2.77
				98.92	99.08	98.80	95.12	98.65	99.42
Cr	-	-	-	1035	366	350	175	-	-
Ni	1091	322	133	341	141	225	100	-	-
Rb	nd	nd	1	16	11	4	10	-	-
Sr	16	79	154	84	121	100	165	-	-
$\frac{CaO}{Al_2O_3}$	0.66	1.05	0.89	1.62	0.73	0.72	0.63	1.59	0.82
$\frac{Na_2O}{K_2O}$	4.33	14.89	19.23	4.89	6.04	15.00	6.56	5.00	1.37
K/Rb	-	-	1577	228	347	350	266	-	-

1,2,3: Average analyses of selected groups of amphibolites from Ravns Storø. 1: 10 ultramafic schists; 2: 22 high-Mg basaltic amphibolites, 3: 11 tholeiitic basaltic amphibolites (Friend, 1975).

4,9: Ultramafic rocks, Fiskenaesset (Rivalenti, 1976).

5,22: Tholeiitic amphibolites, Fiskenaesset (Rivalenti, 1976).

6,7: Depleted and enriched Archaean tholeiites (Condie, 1976).

8: 8 ultramafic komatiites (Viljoen and Viljoen, 1969a).

9: Picrite, Shamandali Hills (Cox et al., 1965).

---

* Fe_2O_3 = total iron, major element analyses summed to 100% on a water-free basis.

(Moorbath, pers. comm). Thus the contention is that the basaltic precursors of the Ravns Storø amphibolites were underlain directly by upper mantle material at the time of their eruption.

THE ULTRAMAFIC ROCKS

Numerous ultramafic bodies, which are often contained within well-defined horizons, are incorporated in the amphibolites. They are typically highly deformed and possess sheared and foliated margins. In many examples they are totally altered to hydrous metamorphic assemblages of serpentine, talc, chlorite, tremolite and anthophyllite. However, some bodies still contain relicts of layering attributable to igneous processes (Friend and Hughes, 1977). Some of these layers are occasionally folded. Rarely, in some of the larger bodies relicts of non-hydrous minerals, such as olivine and green aluminous spinel, are preserved as kernals within hydrous alteration products. Ultramafic bodies are also distributed throughout the trains of amphibolite agmatite and enclaves within the younger gneisses, but in these bodies alteration and reaction are much more intense, usually resulting in an extensive zonation within the body.

Within the amphibolites there is no field evidence to suggest that these bodies originated as the lower, differentiated portions of a larger, layered basic body. They are most frequently restricted to areas of amphibolite which may be interpreted as pillow lavas and associated rocks. Equally it can be demonstrated that the bodies are quite distinct from the Fiskenaesset anorthosite complex since that intrusion often engulfs and cross-cuts a fabric within these bodies. As there is no evidence of any older sialic crust, such as exists in the Godthaabsfjord region (McGregor, 1973), and there are reasonable grounds for considering that the amphibolites in the Ravns Storø-Bjornesund area represent original oceanic-type crust, it is contended that these ultramafic bodies may represent portions of the upper mantle that became incorporated into the amphibolite during deformation.

Bridgwater et al (1974) proposed a model which involves horizontal tectonic movements dominating the early stages of development of the crust of West Greenland, resulting in the intercalation of sheets of older gneisses (the Amitsoq gneisses) and younger amphibolites (the Malene supracrustals). This mechanism of superimposing sheets of material may be used to explain the distribution of ultramafic bodies within the amphibolites. During the early stages of deformation of the oceanic crust a sheet of upper mantle material became incorporated into the amphibolites so formed. On further deformation this sheet formed boudins enclosed by a schistose matrix, but still contained within a fairly well-defined horizon. This restricted disposition of ultramafic bodies is well displayed in the Ravns Storø belt and in other amphibolites in the Fiskenaesset region and further north in the Godthaabsfjord region (Friend and Hall, 1977).

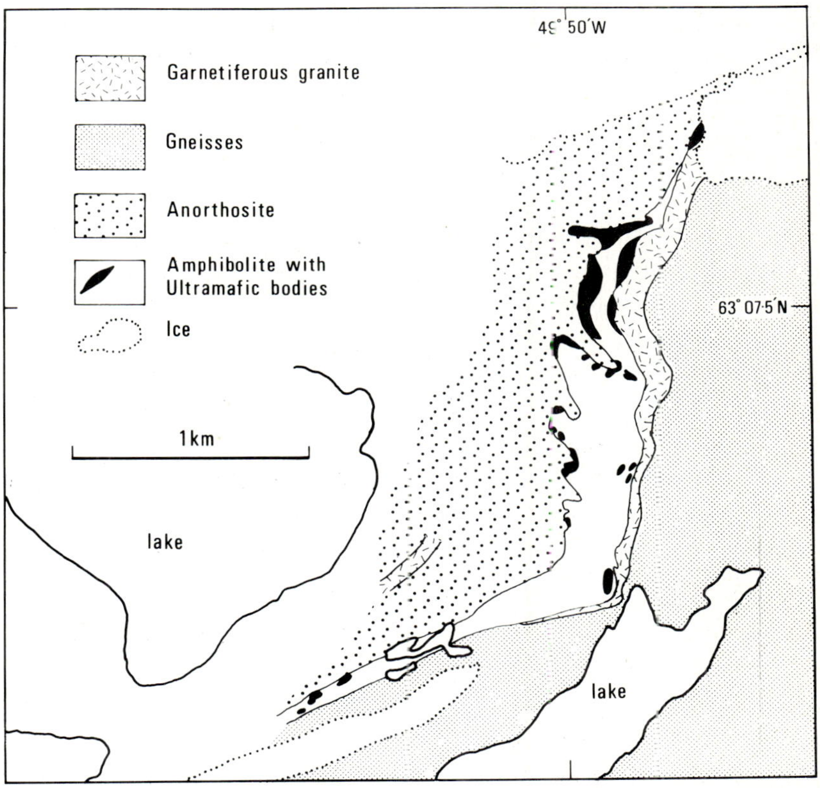

Fig. 2. Sketch map of the geology of the area north of Bjørnesund (see Fig. 1) showing the distribution of ultramafic, amphibolitic and anorthositic rocks in gneisses.

Within the amphibolites north of Bjørnesund (Fig. 2), where a thick, well-preserved unit of the Fiskenaesset anorthosite complex occurs (Walton, 1973), the ultramafic bodies escaped much of the hydration and deformation that occurred during metamorphism (Friend and Hughes, 1977). The bodies comprise deformed spinel-lherzolites, spinel-harzburgites, spinel-dunites and dunites. The tectonic fabric of these bodies contains boudins which preserve undeformed textures (Fig. 3). The relationship of the boudins to the foliated matrix clearly demonstrates the pre-deformational origin of this coarse-grained texture. It is this tectonic fabric which is cut by the anorthosite, which thus separates the two rock units.

An understanding of the origin of these bodies is therefore probably only attainable at this one locality where the pretectonic mineralogy and textures are preserved. The interpretation of the textures hinges upon the reconstruction of the layering which is present. The rocks are layered from magnetite + aluminous green spinel at the base, through assemblages of forsterite + aluminous green spinel, to bronzite + aluminous green and blue spinels at the top. The relationships of the spinel-lherzolites in the vicinity is impossible to ascertain since they are totally recrystallised to foliated, lineated masses in which no coarse-grained textures are preserved.

Within the forsterite and bronzite-bearing portions size-grading of the crystals is evident (Fig. 3). Symplectic intergrowths of forsterite and aluminous green spinel (Fig. 3), which may attain dimensions up to 5 cm in diameter, are contained within portions of the layering that reach pegmatitic proportions. Higher up in the sequence symplectites of bronzite and pale-blue spinel up to 10 cm diameter are found (this occurrence is described in detail in Friend and Hughes, 1977).

DISCUSSION

Symplectites and intergrowths of olivine, pyroxenes and spinels have recently come under scrutiny with the discovery of such structures in ultramafic nodules occurring in kimberlites (Dawson and Reid, 1970; Gurney et al., 1973; Wyatt et al., 1975; Smith et al., 1976). These textures have also been studied experimentally in an attempt to elucidate their origin (Wyatt et al., 1975; Wyatt, 1977). Similar symplectic intergrowths have been discovered in lunar samples (Roedder and Weiblen, 1971, 1972; Gooley et al., 1974; Bell et al., 1975). All of these workers concluded that the symplectites and intergrowths were the result of crystallisation at high pressures, either in the upper mantle in the case of the nodules from kimberlites or in the deep crust in the case of the lunar examples. The conclusions of Wyatt et al. (1975) are extremely relevant in the following discussion in that they suggest that the textures were the

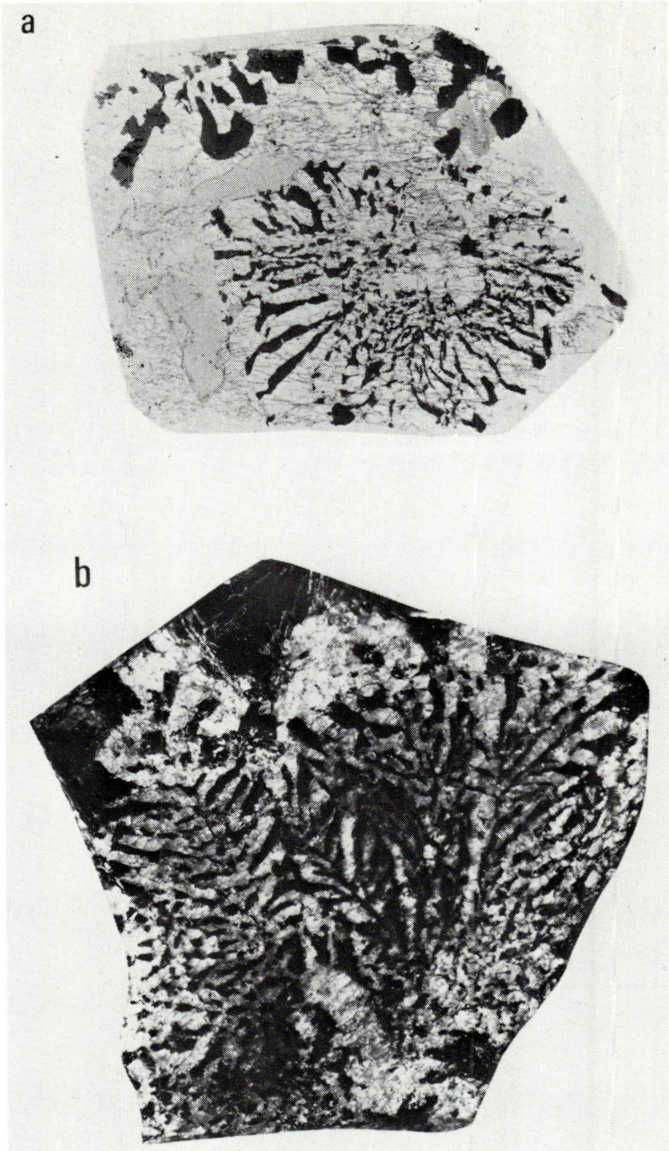

Fig. 3a. 183431b, symplectic intergrowth of green aluminous spinel and forsteritic olivine. Width of photo is about 20 mm.

Fig. 3b. 183431a, symplectic intergrowth of green aluminous spinel and forsteritic olivine. The branching nature of the spinel is clearly displayed. Width of photo is about 20 mm.

result of eutectic crystallisation.

The study of Archaean ultramafic rocks has led to the formulation of several models regarding the evolution of the earth's crust and upper mantle (Viljoen and Viljoen, 1969c; Green, 1974; Green et al., 1974; McIver and Lenthall, 1974; Hart and Brooks, 1977). The present studies were begun because of the unusual nature of the textures preserved, with the thought that they may have some bearing upon the problems of early Archaean magmatic processes.

Average compositions and structural formulae for the spinels and the co-existing olivines are presented in Table 2. The spinel compositions are plotted in Fig. 4 where their highly aluminous nature is apparent. They fall into the aluminous portion of the field of spinels occurring in ultramafic xenoliths from kimberlites and alkali basalts (Basu and McGregor, 1975).

Irvine (1965) produced, and Jackson (1969) refined, a geothermometer which utilises the distribution of Mg and Fe^{2+} between co-existing olivines and spinel pairs. Irvine's (1965) equation:

$$K_D \begin{matrix} \text{Mg-Fe}^{2+} \\ \text{Ol-Sp} \end{matrix} = \frac{X_{Mg}^{Ol} \cdot X_{Fe}^{Sp}}{X_{Fe}^{Ol} \cdot X_{Mg}^{Sp}} \qquad (1)$$

was modified by Jackson (1969) since K_D is not a constant at fixed temperature and pressure because it is also dependent upon amounts of Cr, Al and Fe^{3+} present in the spinels. By substituting free energies into the expression for $K_D \begin{matrix} \text{Mg-Fe}^{2+} \\ \text{Ol-Sp} \end{matrix}$ Jackson (1969) obtained the following equation:

$$T= \frac{(5580Y_{Cr} + 1018Y_{Al} - 1702Y_{Fe^{3+}} + 2400)}{(0.9Y_{Cr} + 2.56Y_{Al} - 3.08Y_{Fe^{3+}} + 1.987\, l_n(K_D)^{Mg-Fe^{2+}}_{Ol-Sp})} \qquad (2)$$

where Y_{Cr}, Y_{Al} and $Y_{Fe^{3+}}$ are the mole fractions of those cations present in the co-existing spinel.

Using this equation for the co-existing olivine-spinel pairs in a recrystallised lherzolite (GGU 120471), values ranging from 713-803°C are obtained for the temperatures of equilibration. Similar temperatures for the upper amphibolite facies conditions have been obtained by a study of corona structures within anorthosite northeast of Fiskenaesset (Myers and Platt, 1977). Applying the same formula to the undeformed textures temperatures in the range 856-907°C are obtained. Thus it is possible that there is a real difference

TABLE 2: Average compositions of spinels and co-existing olivines in the ultramafic rocks

GGU No:	120471 green spinel cores	120471 green spinel margins	183431 green spinel symp	183431 green spinel int	183450 blue spinel		120471 olivine cores	120471 olivine margins	183431 forsterite	183450 forsterite
Al_2O_3	55.6	54.5	67.2	67.5	67.9	SiO_2	38.8	38.6	41.2	41.4
Cr_2O_3	3.43	4.18	0.04	nd	0.09	FeO	23.0	23.3	7.50	3.80
Fe_2O_3*	6.18	6.36	2.90	1.40	0.29	MgO	38.3	37.9	51.1	53.4
FeO	22.6	23.3	6.40	6.60	3.70	CaO	0.02	0.02	na	na
MgO	11.9	11.2	22.9	23.6	25.1	TiO_2	0.02	0.02	na	na
NiO	na	na	nd	nd	0.07	NiO	0.22	0.24	0.05	nd
TiO_2	na	na	nd	nd	0.02					
	99.71	99.44	99.44	99.10	97.17		100.36	100.08	99.85	98.60
			32(O)					4(O)		
Al	14.368	14.256	15.683	15.754	15.901	Si	1.007	1.007	1.000	1.000
Cr	1.072	1.120	0.007	-	0.014	Fe	0.500	0.505	0.152	0.077
Fe^{3+}	0.568	0.680	0.432	0.209	0.043	Mg	1.481	1.474	1.848	1.923
Fe^{2+}	4.136	4.320	1.060	1.093	0.616	Ca	0.001	0.001	-	-
Mg	3.880	3.592	6.757	6.963	7.430	Ti	tr	tr	-	-
Ni	-	-	-	-	0.011	Ni	0.005	0.005	0.001	-
Ti	-	-	-	-	0.004					
R^{3+}	16.008	16.056	16.116	15.963	15.962	Fo	74.8	74.5	92	96
R^{2+}	8.016	7.912	7.817	8.056	8.057	Fa	25.2	25.5	8	4
$\frac{Cr}{Cr+Al+Fe^{3+}}$	0.067	0.070	0.043	-	0.088					
$\frac{Mg}{Mg+Fe^{2+}}$	0.486	0.449	86.44	86.43	92.34					

na = not analysed, nd = not detected, tr = trace, symp = symplectic spinet, int = interstitial spinel

* Iron analysed as FeO, Fe_2O_3 calculated assuming stoichiometry

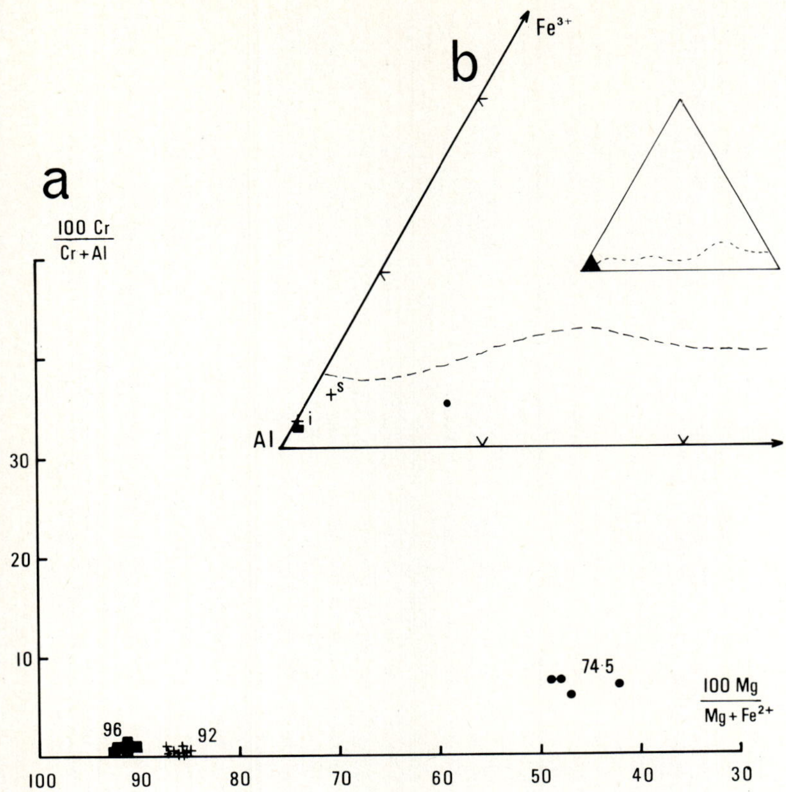

Fig. 4a. Plot of the composition of spinel within Cr/(Cr+Al) and Mg/(Mg+Se^{2+}). Numbers indicate the position (Mol%Fo) of the co-existing olivine.

Fig. 4b. Position of the spinel within Al:Cr:Fe^{3+}. The area below the dashed line is the field of Al-Cr spinel from ultramafic xenoliths (Basu and McGregor, 1975).
Symbols: Dots G.G.U. 180471, crosses 183431, separated in (b) as s, symplectic and i, interstitial types, squares 183450.

between the equilibration temperatures of the recrystallised (re-equilibrated) and undeformed (perhaps non-re-equilibrated) assemblages.

Because of uncertainty in some results obtained from this formula, Evans and Frost (1975), Medaris (1975) and Basu (1977) used an alternative method of plotting the results in a three-dimensional figure with axes which represent $l_n\ (K_D)^{Mg-Fe}_{Ol-Sp}$, Y^{SP}_{Cr} and $Y^{SP}_{Fe^{3+}}$ derived from the formula:

$$l_nK_{DOl-Sp}^{Mg-Fe} = l_nK_b + Y_{Cr}^{Sp}l_nK_e + Y_{Fe^{3+}}^{Sp}l_nK_f, \quad (3)$$

where Y_i^{Sp} is the amount of trivalent cation in the spinel and K_n is the equilibrium constant for the following exchange reactions:

b) $\frac{1}{2}FeSiO_4 + MgAl_2O_4 = \frac{1}{2}MgSiO_4 + FeAl_2O_4$

e) $MgCr_2O_4 + FeAl_2O_4 = FeCr_2O_4 + MgAl_2O_4$

f) $FeAl_2O_4 + MgFe_2O_4 = MgAl_2O_4 + Fe_3O_4$ (Irvine, 1965).

A plot of these calculations applied to the olivine-spinel pairs projected onto the face Y_{Cr}^{Sp} at $Y_{Fe^{3+}}^{Sp} = 0$ and assuming a value of 4.0 for l_nK_f (Irvine, 1965; Evans and Frost, 1975; Basu, 1977) is presented in Fig. 5. It is evident that the values for 120471, which lie close to the 700°C isotherm, correspond fairly well with the values obtained by the first method, and probably confirm re-equilibration of the assemblage under metamorphic conditions. However, the values obtained for the undeformed textured rocks plot towards the 1200°C isotherm and are slightly higher than the range of values obtained by use of the alternative method. This increases the possibility that the assemblage might not have completely re-equilibrated under the prevailing metamorphic conditions. Further, as there is no evidence for an earlier, high temperature granulite facies event it is concluded that the values indicate that the assemblages had a primary, igneous origin.

CONCLUSIONS

The results of the experimental work carried out by Wyatt (1977) in the system ilmenite-clinopyroxene suggest that the minimum conditions for symplectite formation in an anhydrous state were about 1300°C and at 20 kb. The results supported the earlier work (Wyatt et al., 1975) which indicated that the symplectites originated by eutectic crystallisation rather than any other mechanism. Green et al. (1974), using a silicate system of komatiitic composition, produced bladed textures together with elongate cumulate olivines by quenching a melt from 1650°C at 10 kb. These experimentally produced bladed textures bear a strong resemblance to the bladed portions of the textures produced by Wyatt et al. (1975) and Wyatt (1977).

All of these high-pressure experiments produced textures which have striking similarities to the symplectic and bladed mineral associations

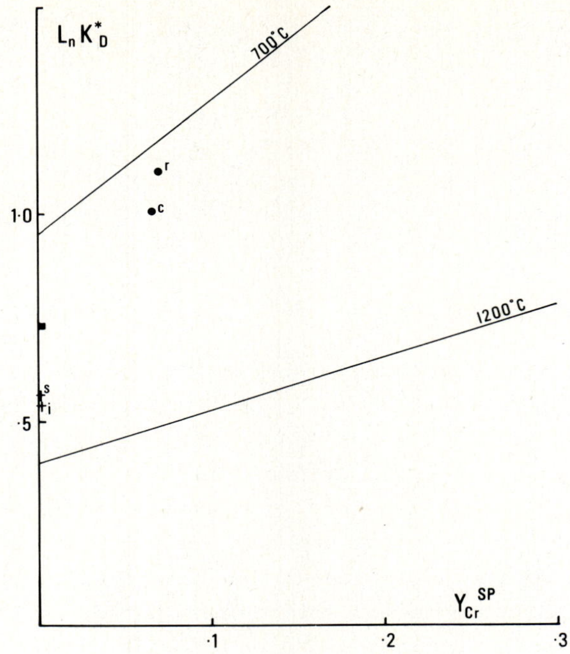

Fig. 5. Plot of the spinels on the face Y_{Cr}^{SP} of the three dimensional figure $Y_{Cr}^{SP}:L_N^{Cr}K_d^*:Y_{Fe3+}^{SP}$ at $Y_{Fe3+}^{SP} = 0$.
Symbols as in Fig. 4, labels rc to 120471 indicate averages for the rims and cores respectively, the 700 and 1200°C isotherms are from Evans and Frost (1975).

found in the naturally occurring ultramafic rocks described here (Fig. 3).

Preliminary melting experiments on a representative of these rocks (149720) suggest that forsterite and aluminous spinel are still on the liquidus at 1400°C and 10 kb total water pressure. This therefore places the melting regime of these spinel-rich ultramafic rocks within the regimes of the experimental systems of Green et al. (1975), Wyatt et al. (1975) and Wyatt (1977). It seems plausible, therefore, to equate the preserved textures with an igneous system with crystallisation temperatures of at least 1400°C at 10 kb. The difference between the calculated equilibration temperatures of the spinel-olivine pairs and the maximum metamorphic temperatures likely in the prevailing metamorphic conditions would appear to support this contention.

To attain pressures of 10 kb implies a crustal thickness approaching 35 km; however there is no evidence from the surrounding gneisses and amphibolites that they have been subjected to such conditions. This would seem to preclude the possibility that the ultramafic rocks have crystallised in their present surroundings. It is therefore suggested that these rocks crystallised at conditions pertaining to the upper mantle and preserve a relict of these conditions in the undeformed boudins which contain the coarse-grained textures. It is hoped that further melting experiments will be carried out on these unusual rocks and which will clarify their origin and conditions of crystallisation.

ACKNOWLEDGEMENTS

This paper is published with the permission of the Director of Grønlands Geologiske Undersøgelse, to whom we extend our thanks for providing the facilities to work in Greenland. Thanks are also due to Dr. J.V.P. Long of the Department of Mineralogy and Petrology, Cambridge University, for use of microprobe facilities. This work was supported by a NERC grant. Dr. C. Ford of Edinburgh University is thanked for the preliminary melting experiments carried out on 149720. Mr. P. Cameron is thanked for producing the photographs.

The award of a Royal Society grant enabling one of us (CRLF) to present this paper is gratefully acknowledged, also Oxford Polytechnic for supplementary assistance.

Our thanks are also due to our various colleagues for reading the manuscript at several stages of preparation.

REFERENCES

Andersen, L.S., 1974. Rapport over geologien mellem Frederikshaabs Isblink og Quvilik fjord - Vest Grønland. Unpub. Thesis, Univ. Copenhagen.

Andersen, L.S. and Friend, C.R.L., 1973. Structure of the Ravns Storø amphibolite belt in the Fiskenaesset region. Rapp. Grønlands geol. Unders., 51: 37-40.

Basu, A.R., 1977. Olivine-spinel equilibria in lherzolite xenoliths from San Quintin, Baja, California. Earth Planet. Sci. Lett., 33: 443-450.

Basu, A.R. and McGregor, I.D., 1975. Chromite spinels from ultramafic xenoliths. Geochim. Cosmochim. Acta, 39: 937-945.

Bell, P.M., Mao, H.K., Roedder, E. and Wieblen, P.W., 1975. The problem of the origin of symplectites in olivine-bearing lunar rocks. Proc. 6th Lunar Sci. Conf., 1: 231-248.

Black, L.P., Moorbath, S., Pankhurst, R.J. and Windley, B.F., 1973. $^{207}Pb/^{206}Pb$ whole-rock age of the granulite facies metamorphic event in West Greenland. Nature Phys. Sci., 244: 50-51.

Bridgwater, D., McGregor, V.R. and Myers, J.S., 1974. A horizontal tectonic regime in the Archaean of West Greenland and its implications for early crustal thickening. Precamb. Res., 1: 179-198.

Bridgwater, D., Keto, L., McGregor, V.R. and Myers, J.S., 1976. Archaean gneiss complex of Greenland. In: A. Escher and W.S. Watt (Editors), Geology of Greenland. Grønlands geol. Unders., Copenhagen, 18-75.

Burwell, A.D.M., Friend, C.R.L. and Hughes, D.J., in press. Geochronology of amphibolites and ultramafic rocks from the Fiskenaesset region, southern West Greenland. Rapp. Grønlands geol. Unders.

Condie, K.C., 1976. Trace-element geochemistry of Archaean greenstone belts. Earth Sci. Rev., 12: 393-417.

Cox, K.C., Johnson, R.L., Monkman, L.J., Stillman, C.J., Vail, J.R. and Wood, D.N., 1965. The geology of the Nuanetsi igneous province. Phil. Trans. R. Soc. Lond., A257: 71-218.

Dawson, J.B. and Ried, A.M., 1970. A pyroxene-ilmentite intergrowth from the Monastery Mine, South Africa. Contrib. Mineral. Petrol., 26: 296-301.

Escher, J.C. and Myers, J.S., 1976. New evidence concerning the original relationships of early Precambrian volcanics and anorthosites in the Fiskenaesset region, southern West Greenland. Rapp. Grønlands geol. Unders., 75: 72-76.

Evans, B.W. and Frost, B.R., 1975. Chrome-spinel in progressive metamorphism - a preliminary analysis. Geochim. Cosmochim. Acta, 39: 959-972.

Friend, C.R.L., 1975. The geology and geochemistry of the Preketilidian basement complex in the Ravns Storø Area, Fiskenaesset Region, southern West Greenland. Unpub. Thesis, Univ. London.

Friend, C.R.L., 1976. Field relationships and petrology of leucocratic schists from the Ravns Storø Group, Fiskenaesset, southern West Greenland. Rapp. Grønlands geol. Unders., 73: 81-85.

Friend, C.R.L. and Hall, R.P., 1977. Field work in the Ivisartoq area inner Godthaabsfjord, southern West Greenland. Rapp. Grønlands geol. Unders., 85: 54-60.

Friend, C.R.L. and Hughes, D.J., 1977. Archaean aluminous ultrabasic rocks with primary igneous textures from the Fiskenaesset region, southern West Greenland. Earth Planet. Sci. Lett., 36: 157-167.

Gooley, R., Brett, R. and Warner, J., 1974. A lunar rock of deep crustal origin: sample 76535. Geochim. Cosmochim. Acta, 38: 1329-1339.

Green, D.H., 1974. Genesis of Archaean peridotitic magmas and constraints on Archaean geothermal gradients and tectonics. Geology, 3: 15-18.

Green, D.H., Nicholls, I.A., Viljoen, M.J. and Viljoen, R.P., 1974. Experimental demonstration of the existence of peridotitic liquids in earliest Archaean magmatism. Geology, 3: 11-14.

Grønlands Geologiske Undersøgelse, 1976. Rapp. Nr. 73.

Gunn, B., 1976. A comparison of modern and Archaean oceanic crust and island arc petrochemistry. In: The Early History of the Earth B.F. Windley (Editor), Wiley, London, 405-417.

Gurney, J.J., Fesq, H.W. and Kable, E.J.D., 1973. Clinopyroxene-ilmenite intergrowths from kimberlites: a re-appraisal. In: Lesotho Kimberlites, P.H. Nixon (Editor), Lesotho National Development Council, 238-253.

Hart, S.R. and Brooks, C., 1977. The geochemistry and evolution of early Precambrian mantle. Contrib. Mineral. Petrol., 61: 109-128.

Herrmann, A.G., Blanchard, D.P., Haskin, L.A., Jacobs, J.W., Knake, D., Koroter, R.L. and Brannon, J.C., 1976. Major, minor and trace element compositions of peridotitic and basaltic komatiites from the Precambrian crust of southern Africa. Contrib. Mineral. Petrol., 59: 1-12.

Irvine, T.N., 1965. Chromian spinel as a petrogenetic indicator. Pt.1. Theory. Can. J. Earth Sci., 2: 648-672.

Jackson, E.D., 1969. Chemical variation in co-existing chromite and olivine in chromite zones of the Stillwater Complex. Econ. Geol., 4: 41-71.

Kalsbeek, F., 1976a. Metamorphism of Archaean rocks of West Greenland In: The Early History of the Earth, B.F. Windley (Editor), Wiley, London, 225-235.

Kalsbeek, F., 1976b. Metamorphism in the Fiskenaesset region. Rapp. Grønlands geol. Unders., 73: 34-41.

Kalsbeek, F., 1976c. Rb-Sr whole-rock age of the Ilivertalik Granite and other rocks from the Fiskenaesset area. Bull. geol. Soc. Denmark, 25: 85-88.

McGregor, V.R., 1973. The early Precambrian gneisses of the Godthaab district West Greenland. Phil. Trans. R. Soc. Lond., A273: 343-358.

McIver, J.R. and Lenthall, D.H., 1974. Mafic and ultramafic extrusives of the Barberton Mountain land in terms of the CMAS system. Precamb. Res., 1: 327-343.

Medaris, L.G., 1975. Co-existing spinel and silicates in alpine peridotites of the granulite facies. Geochim. Cosmochim. Acta, 39: 947-958.

Moorbath, S., 1976. Age and isotope constraints for the evolution of Archaean crust. In: The Early History of the Earth, B.F. Windley (Editor), Wiley, London, 351-360.

Myers, J.S., 1975. Igneous stratigraphy of Archaean anorthosite at Majorqap qava, near Fiskenaesset, South-West Greenland. Rapp. Grønlands geol. Unders., 74: 27 pp.

Myers, J.S. and Platt, R.G., 1977. Mineral chemistry of layered Archaean anorthosite at Majorqap qava, near Fiskenaesset, southwest Greenland. Lithos, 10: 59-72.

Pidgeon, R.T., 1973. Discordant U-Pb isotopic systems in zircons from a granite north-west of Frederikshaabs Isblink, Fiskenaesset region. Rapp. Grønlands geol. Unders., 51: 28-30.

Pidgeon, R.T. and Hopgood, A.M., 1975. Geochronology of Archaean gneisses and tonalites from north of the Frederikshaabs Isblink, S.W. Greenland. Geochim. Cosmochim. Acta, 39: 1333-1346.

Pidgeon, R.T., Aftalion, M. and Kalsbeek, F., 1976. The age of the Ilivertalik granite. Rapp. Grønlands geol. Unders., 73: 31-33.

Rivalenti, G., 1976. Geochemistry of metavolcanic amphibolites from south-west Greenland. In: The Early History of the Earth, B.F. Windley (Editor), Wiley, London, 213-223.

Roedder, E. and Wieblen, P.W., 1971. Petrology of silicate melt inclusions Apollo 11 & 12, and terrestrial equivalents. Proc. 2nd Lunar Sci. Conf., 507-528.

Roedder, E. and Wieblen, P.W., 1972. Petrographic features and petrologic significance of melt inclusions in Apollo 14 & 15 rocks. Proc. 3rd Lunar Sci. Conf., 251-279.

Smith, C.B., McCallum, M.E. and Eggler, D.H., 1976. Clinopyroxene-ilmenite intergrowths from the Iron Mountain kimberlite district, Wyoming. Carnegie Inst. Wash. Yr. Bk., 75: 542-544.

Viljoen, M.J. and Viljoen, R.P., 1969a. The geology and geochemistry of the lower ultramafic unit of the Onverwacht Group and a proposed new class of igneous rock. Geol. Soc. S. Africa Sp. Pub., 2: 55-86.

Viljoen, M.J. and Viljoen, R.P., 1969b. Evidence for the existence of a mobile extrusive peridotitic magma from the Komati Formation of the Onverwacht Group. Geol. Soc. S. Africa. Sp. Pub., 2: 87-112.

Viljoen, M.J. and Viljoen, R.P., 1969c. A reappraisal of the Granite-Greenstone Terrains of shield areas based on the Barberton model. Geol. Soc. S. Africa Sp. Pub., 2: 265-274.

Walton, B.J., 1973. The structure and stratigraphy of the anorthosite complex in the area north of Bjørnesund, near Fiskenaesset. Rapp. Grønlands geol. Unders., 51: 60-64.

Williams, H.R., 1973. The geology of an area to the northeast of Bjørnesund, near Fiskenaesset, West Greenland. Unpub. Thesis, Univ. Exeter.

Windley, B.F., Herd, R.K. and Bowden, A.A., 1973. The Fiskenaesset complex, West Greenland. Pt. 1: A preliminary study of the stratigraphy, petrology and whole-rock chemistry from Qeqertarssuatsiaq. Bull. Grønlands geol. Unders., 106: 80 pp.

Windley, B.F. and Smith, J.V., 1974. The Fiskenaesset complex, West Greenland. Pt. 2: General mineral chemistry from Qeqertarssuatsiaq. Meddr. Grønland, 196, pt. 4: 54 pp.

Wyatt, B.A., 1977. The melting and crystallisation behaviour of a naturally occurring clinopyroxene-ilmenite intergrowth. Contr. Mineral. Petrol., 61: 1-9.

Wyatt, B.A., McCallister, R.H., Boyd, F.R. and Ohashi, Y., 1975. An experimentally produced clinopyroxene-ilmenite intergrowth. Carnegie Inst. Wash. Yr. Bk., 74: 536-538.

MINERALISATION

MINERALOGY AND GEOCHEMISTRY OF A VANADIFEROUS TITANO-MAGNETITE DEPOSITE AND ASSOCIATED COPPER MINERALISATION IN GABBRO-ANORTHOSITES NEAR MASANIKERE, SHIMOGA DISTRICT, KARNATAKA, INDIA

A.S. Ramiengar, N. Chayapathi, K.R. Raghanandan, M.S. Rao
Geological Survey of India

and P. Rama Rao
National Geophysical Research Institute, Hyderabad, India

Abstract

Vanadiferous titano-magnetite deposits, occurring as differentiates of gabbro-anorthosite complexes, are associated with and genetically related to basic magmatic activity in some Archaean greenstone belts. The largest deposit of this kind in Karnataka occurs near Masanikere in the Shimoga district. The Fe-Ti-V-Cu mineralisation is found in a gabbro-anorthosite complex intrusive into the metasediments of the Chitradurga Group of the Dharwar Supergroup. The complex occurs as an intrusion occupying an area of 2.5 km^2. Lithologically it comprises (a) magnetite-gabbro with magnetite bands, (b) gabbro, (c) gabbroic anorthosite, and (d) anorthosite. The Chitradurga Group of metasediments are represented by quartz-chlorite-carbonate schists with interbeds of quartzite. The mineralisation is restricted to the magnetite-gabbro which extends over 1.5 x 0.8 km in the basal part of the complex.

There have been two phases of mineralisation: one oxide and the other sulphide. Ti-magnetite observed in different stages of martitisation is the primary oxide mineral together with exsolved ilmenite and spinel. No discrete vanadium-bearing mineral is observed. Vanadium is present as (a) partial substitution for Fe^{3+} in magnetite, and (b) in the form of VFe_2O_4, similar to $TiFe_2O_4$ (ulvosphinel). The sulphide phase is represented by pyrite and chalcopyrite. Correlation coefficients obtained by statistical analysis indicate a significant positive correlation between Ti and V, and to a lesser extent between Fe^{3+} and V. Minor element data and statistical analysis of element ratios indicate a late-stage crystallisation during magmatic differentiation. The deposit may thus be grouped under the liquid magmatic type of Schneiderhohn with differentiation-crystallisation-segregation. Evidence of minor pneumatolytic activity is also present.

In comformity with the recent trends of correlating influx of some specific metals with different stages of crustal evolution, the present studies show that Fe-Ti-V-Cu mineralisation was characteristic of the early Precambrian in the southern parts of the Indian Shield.

INTRODUCTION

Vanadiferous titano-magnetite deposits are associated with and genetically related to gabbro-anorthosite complexes in some Archaean greenstone belts. Similar deposits occur in Karnataka in a number of places, viz. Nuggehalli Schist Belt in

Hassan district, and Shimoga Schist Belt in Shimoga and North Kanara districts. The largest deposit of this kind in Karnataka occurs near Masanikere in the Shimoga district (Fig. 1).

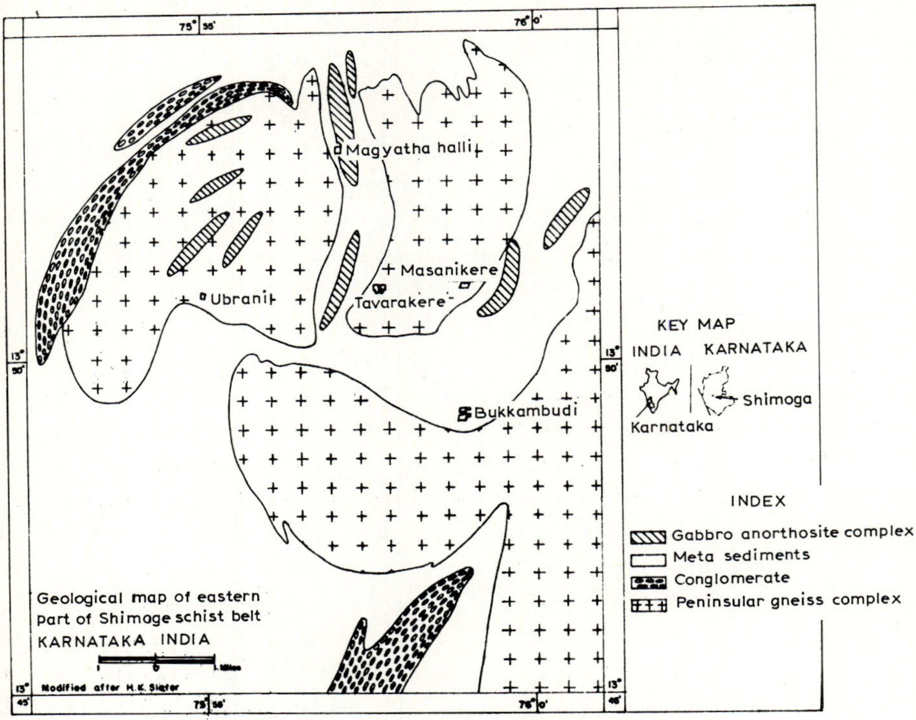

Fig. 1. Geological map of eastern part of Shimoga schist belt.

The mineralogical and geochemical characters of the vanadiferous titano-magnetite and associated copper ore are described for the first time in this paper.

ANALYTICAL METHODS

Major elements were determined by wet chemical methods and flame photometry. Minor elements were analysed on an atomic absorption spectrometer after scanning the samples on a Jarrel-Ash grating spectrograph. Platinum and silver were determined by spark source mass spectrometry in an AEI MS 702 mass spectrometer equipped with an RF spark source. Mass spectra were recorded over a wide exposure range of 0.01-300 nano coloumbs and element concentrations calculated by taking ratios of selected isotopic line intensities, Pt (194, 195, 196) and Ag (107, 109) lines, the ratios of corresponding charge exposures and the preconcentration factor.

GEOLOGICAL SETTING

The eastern part of the Shimoga Schist Belt exposes a number of gabbro-anorthosite bodies of varying dimensions, from less than 1 km^2 to 2.5 km^2. These bodies carry bands and lenses of vanadiferous titano-magnetite. The general stratigraphic succession of this part of the schist belt is given in Table 1.

Table 1. Stratigraphy of the Eastern Part of the Shimoga Schist Belt

	Quartz vein
	Dolerite dyke
INTRUSIVES	Magnetite-gabbro, leuco-gabbro, gabbroic-anorthosite
	Anorthosite
	Ultramafites (serpentinite, talc-chlorite schist)
CHITRADURGA GROUP OF THE DHARWAR SUPERGROUP	Acid Volcanic rocks
	Quartz-chlorite-carbonate schist with quartzite bands
	Polymict conglomerate
	.... UNCONFORMITY ....
	Granitic Gneiss (Peninsular Gneiss Complex)

Granitic gneisses of the Peninsular Gneiss Complex form the basement for the metasediments of the Chitradurga Group. The basal polymict conglomerate consists of pebbles, cobbles and boulders of granitic gneiss, banded gneiss, quartzite, and banded magnetite quartzite, set in a quartz-chlorite matrix. The conglomerate is overlain by a sequence of quartz-chlorite-carbonate schists with interbeds of quartzite and minor flows of acid volcanic rocks. The gabbro-anorthosite complexes were emplaced as narrow elongated bodies (measuring 500 x 100 m to 3 x 0.5 km) within the supracrustal rocks and the basement granitic gneiss.

MASANIKERE DEPOSIT

The basic complex near Masanikere comprises magnetite-gabbro, meta-gabbro, gabbroic anorthosite and anorthosite, showing transitional relationships and auto-hydrative effects. Magnetite bands with minor silicates (chlorite 5 - 10%) grade into magnetite -gabbro (magnetite 25 - 50%) which, on further depletion of magnetite and mafic contents, grades into a normal gabbro and to anorthosite. The metasediments of Masanikere area are folded into a broad synform with a low plunge of 15° - 20° towards NNW. The gabbro-anorthosite complex occupies the core of this synform.

PETROGRAPHY AND MINERALOGY

The magnetite-gabbro is built up of 25 - 50% magnetite in a mesh essentially composed of saussuritised plagioclase (An_{35-38}) epidote, clinozoisite and chlorite. Disseminations of pyrite and chalcopyrite are common. Magnetite-gabbro shows excellent cumulus texture.

The meta-gabbro consists of saussuritised plagioclase and chlorite as common constituents with accessory magnetite and pyrite. The meta-gabbro, with a decrease in the mafic components of mostly chlorite and occasional uralite, grades into gabbroic anorthosite, and finally to anorthosite. Secondary quartz-carbonate veins are common in the complex.

Porphyroblasts of magnetite crystals are developed in quartz-chlorite schist near its contact with the gabbro. These magnetite crystals are due to the recrystallisation of the Fe-oxide in the metasediments. Morphologically and chemically these magnetite crystals are different from those associated with the intrusive complex.

Optical studies indicate two phases of mineralisation, a major oxide and a minor sulphide phase. The ore minerals are magnetite, spinel (ulvospinel?), ilmenite, rutile, hematite, pyrite, chalcopyrite and traces of sphalerite. Geothite and covellite occur as alteration products. The paragenesis of these minerals is shown in Fig. 2.

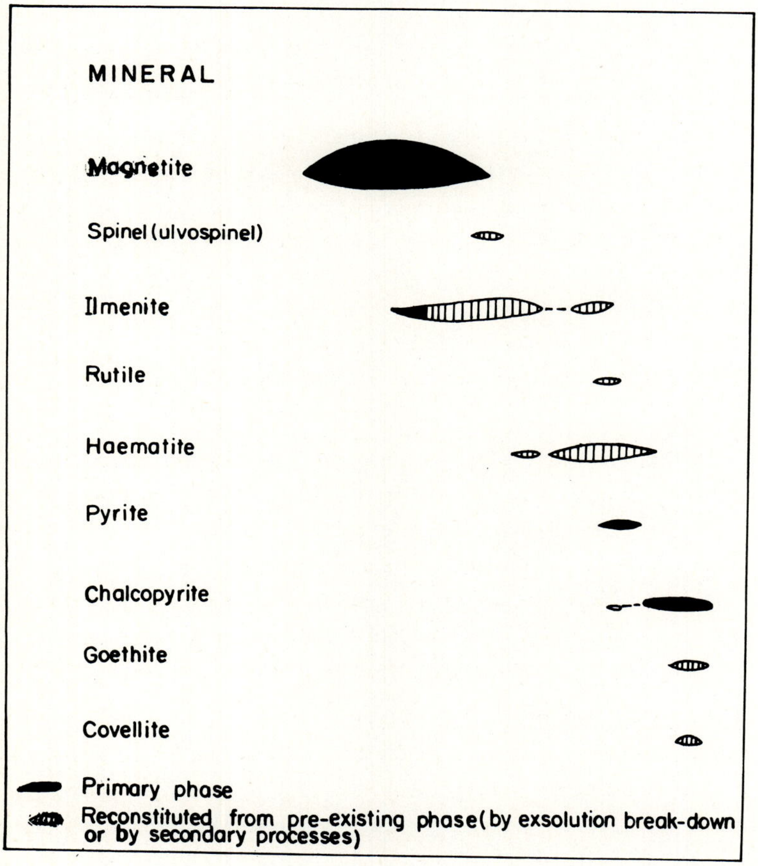

Fig. 2. Paragenesis of ore minerals.

1. The Oxide Phase

Ti magnetite exhibiting different stages of martitisation is the principal primary oxide mineral. It is hypidiomorphic and varies in grain size generally from 200 - 800 μ. The Ti-content of the magnetite is present as exsolved ilmenite of two generations; ilmenite I and II. Ilmenite I occurs as coarse blades or plates particularly along the grain margins of magnetite, while ilmenite II is present as thin exsolution lamellae. In both cases the (0001) lamellae of ilmenite is parallel to the (111) directions of magnetite. Ilmenite I shows polysynthetic twinning in places and has sub-exsolution discs and blebs of hematite parallel to (0001) of ilmenite. Such patches of hematite sub-exsolution are restricted to the central or inner parts of ilmenite.

Martitisation of magnetite starts as thin plates of hematite (hematite II) along the (111) directions of magnetite and invades along grain margins, cracks and twin lamellae. All transitions are observed from a partly to completely martitised magnetite and its original nature is revealed by the relics of ilmenite. Partial breakdown of the ilmenite to rutile is observed in some samples. Exsolution of a spinel (ulvospinel?) parallel to the (100) directions of magnetite has been noticed in some cases.

2. The Sulphide Phase

Pyrite and chalcopyrite are the primary sulphide ore minerals. Pyrite occurs as idiomorphic to hypidiomorphic crystals (40 - 90 μ) showing evidence of cataclasis, the fractures being healed or cemented by chalcopyrite in places. Pyrite usually occurs independently of magnetite. Chalcopyrite occurs in two different forms, perhaps belonging to two generations. Chalcopyrite I occurs as minute droplets and as hypidiomorphic inclusions in the magnetite, this early generation of chalcopyrite seeming to have formed simultaneously with magnetite due to liquid immiscibility. Chalcopyrite II occurs in relatively larger quantities, is xenomorphic and encloses pyrite in places. Fractures in pyrite and to a lesser extent in magnetite are healed by chalcopyrite II. Traces of sphalerite occur invariably as exsolution starlets in chalcopyrite, which indicates a relatively high temperature (Ramdohr, 1969).

3. Secondary Minerals

Supergene replacement of geothite by chalcopyrite as concentric celloform bands is noticed occasionally, as is also secondary covellite replacing chalcopyrite.

GEOCHEMISTRY

1. Major Elements

Samples from different rock units of the gabbro-anorthosite complex were selected from borehole cores and analysed for major elements (Table 2). Analyses of ferrogabbro from Ovre Roddal, Norway, and Bushveld, S. Africa, and gabbro-diorite from

Norlisk II, U.S.S.R. are tabulated for comparison.

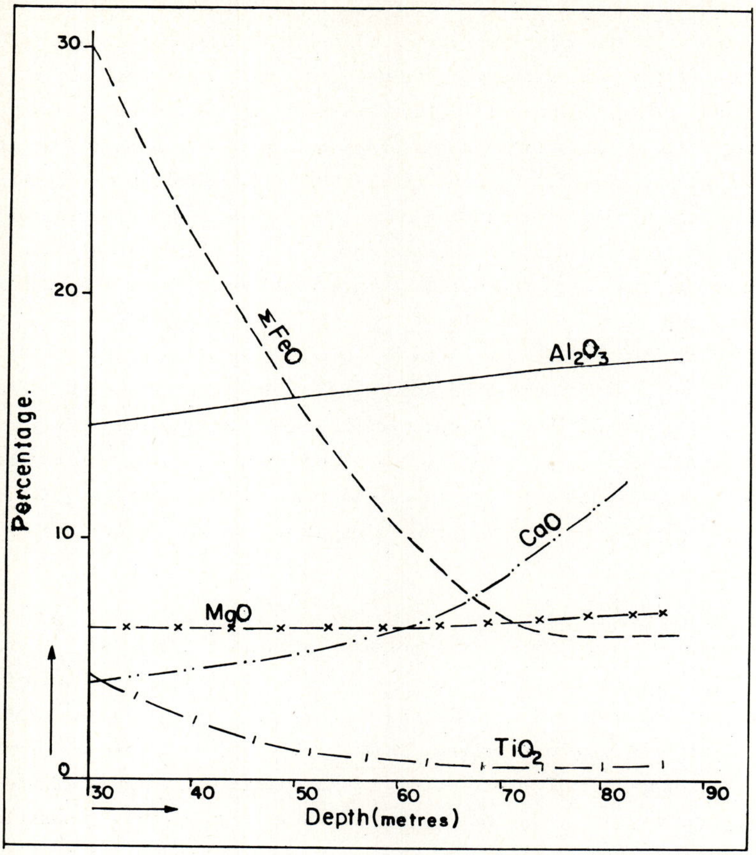

Fig. 3. Variation of some of the major elements in different members of the gabbro-anorthosite complex.

The major element content of the magnetite gabbro from the Masanikere complex compares broadly with the ferro-gabbro from Ovre Roddal, Norway and Bushveld, S. Africa.

The chemical data from the Masanikere complex are interpreted taking into account the auto-hydrative effects the rocks have undergone. The salient deductions are: (a) the FeO : Fe_2O_3 ratio increases with depth reflecting the degree of martitisation, (b) the variation diagram (Fig. 3) for the oxides of elements Fe, Ti, Mg, Al and Ca indicates a steep fall in the total Fe content from magnetite-gabbro to meta-gabbro; (c) the TiO_2 value closely follows the FeO pattern; and (d) CaO and Al_2O_3 show a steady increase towards the anorthosite whereas the MgO content is more or less

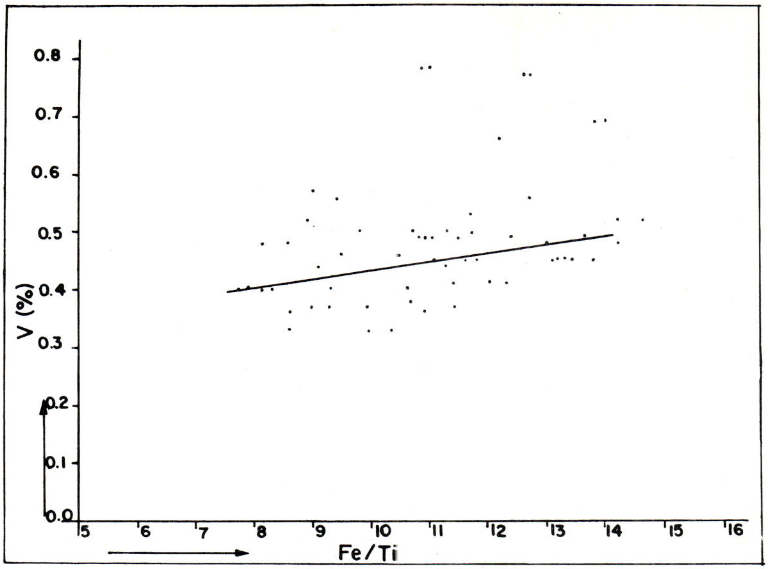

Fig. 4. Variation of V with Fe/Ti.

uniform throughout the complex. A study of the chemical data and elemental variation suggests that the members of the complex are differentiated from an iron-rich basic magma. A high sulphur content and low partial pressure of oxygen might have helped in the formation of iron-rich residual liquids (Turner and Verhoogen, 1961).

The ore band with Ti-magnetite and interstitial silicates (mainly chlorite) shows the following range of composition: Fe:(43.17 - 58.08%), FeO:(1.38 - 10.52%), TiO_2:(5.01 - 12.2%), and V:(0.23 - 0.93%). The Fe : Ti ratio in the ore bands varies from 1.08 - 16.90

Table 3 compares the composition of the oxide minerals of the Masanikere ore with those of other bodies. The high Fe : Ti ratio in the Masanikere deposit is due to a relatively low TiO_2 content, suggesting a late-stage crystallisation, as indicated by Lister (1966) from other areas.

2. Minor Elements

The minor elements determined in the complex are V, Cu, Ni, Co, Cr, Ga and Mo, (Table 4). Vanadium is the dominant minor element in the magnetite-gabbro and magnetite bands; it ranges from 0.23 - 0.93%. Optical studies have not disclosed the presence of any discrete vanadium mineral. Crystal chemistry suggests the possibility of vanadium being present as (a) partial substitution for Fe^{3+} in magnetite, and (b) in the form of VFe_2O_4 similar to $TiFe_2O_4$ (Ulvospinel). The statistically significant positive correlation obtained for the element pairs Ti and V supports the second possibility.

TABLE 2: Major Element Analyses

	1	2	3	4	5	6	7	8	9	10	11	12	13	14
SiO_2	28.98	51.30	52.36	46.25	52.02	34.98	23.83	22.91	25.89	44.05	50.65	56.60	50.70	35.16
TiO_2	4.30	0.70	0.49	0.67	0.72	2.80	4.61	3.75	4.17	1.61	0.21	1.75	1.86	5.13
Al_2O_3	14.54	16.32	17.47	16.85	17.04	15.80	12.08	12.96	12.03	13.34	25.17	13.30	13.80	9.18
Fe_2O_3	18.78	4.05	1.02	0.67	2.02	10.12	17.30	14.80	16.87	4.80	1.63	5.42	5.43	9.00
FeO	13.88	4.88	5.60	6.17	4.40	15.32	22.00	21.44	18.86	11.02	2.94	9.29	8.34	21.53
MgO	6.24	6.39	8.14	5.50	6.81	5.52	6.35	7.33	7.37	6.51	0.35	0.57	5.59	4.24
CaO	3.92	9.42	5.28	14.13	12.05	5.66	3.36	7.63	4.71	10.21	13.30	6.35	10.70	6.97
Na_2O	2.45	3.65	5.80	3.80	3.00	4.20	3.65	1.70	4.35	2.80	4.13	3.51	2.37	2.24
K_2O	0.15	0.05	0.25	0.10	0.15	0.65	0.30	0.15	0.55	0.15	0.05	2.60	0.49	1.67
P_2O_5	0.16	0.10	0.13	0.16	0.11	0.11	0.14	0.12	0.15	0.12	0.13	0.36	0.48	3.24
MnO	0.16	0.41	0.10	Tr	0.30	0.21	0.21	0.23	0.13	0.25	0.13	0.26	0.20	0.73
L.O.I.	4.32	3.63	3.20	3.78	2.02	3.48	3.91	5.47	2.67	5.86	-	-	-	0.57
TOTAL	97.88	100.90	99.84	98.08	100.64	98.85	97.74	98.49	97.75	100.72	98.69	100.01	99.96	100.03

Sample Nos: 1, 6-9, magnetite-gabbro, Masanikere; 2-5, 10, meta-gabbro, Masanikere; 11, meta-anorthosite, Masanikere (Ramakrishnan et al., in press); 12, ferro-gabbro, Bushveld, S.Africa (Hall, 1932); 13, gabbro-diorite, Norlisk II, USSR (Zolutuchin and Vasilev, 1967); 14, ferro-gabbro, Øvre Roddal, Norway (Gjelsvik, 1957).

TABLE 3: Comparison of the composition of oxide minerals of the Masanikere deposit with those of other deposits

Range	Masanikere	Mayurbjanj District, Orissa, India (1)	Main magnetite seam, Bushveld (2)	Duluth gabbro (3)
Fe (%)	43.20 - 58.10	48.00 - 61.00	55.80 - 57.50	48.50 - 55.30
TiO_2 (%)	5.01 - 12.20	11.60 - 18.30	12.20 - 13.90	18.40 - 33.40
Fe:Ti	8.08 - 16.90	5.00 - 7.00	7.00 - 8.00	3.20 - 5.00
V (%)	0.23 - 0.93	0.60 - 0.90	0.84 - 1.00	-

(1) Tugarinov et al. (1974); (2) Willemse (1969); (3) Lister (1966)

Table 4. MINOR ELEMENT ANALYSES

Sample No.	V (%)	Cu (%)	Cr (ppm)	Ga (ppm)	Ni (ppm)	Co (ppm)	Remarks
MSK - 1/10	-	0.18	270	24	485	175	Magnetite-gabbro
36	-	0.15	170	22	420	180	Ditto
75	-	0.10	190	8	195	55	Meta-gabbro free from magnetite
82	-	0.04	140	8	210	70	Ditto
88	-	0.14	165	8	225	75	Ditto
MSK - 6/5	0.51	0.30	230	40	480	190	V-Ti-Magnetite band
12	0.43	0.36	155	38	840	310	Ditto
20	0.62	0.49	85	34	810	300	Ditto
28	0.48	0.15	390	36	390	210	Ditto
31	-	0.12	160	26	450	190	Magnetite-gabbro
40	-	0.42	90	20	330	-	Ditto
72	-	0.15	145	15	-	-	Meta-gabbro free from magnetite
MSK - 7/6	0.56	0.39	160	38	-	-	V-Ti-Magnetite band
10	0.40	0.36	85	36	-	-	Ditto
15	0.39	0.15	120	32	750	285	Ditto
25	0.28	0.15	170	32	1000	225	Ditto
27	-	0.21	185	30	1100	220	Magnetite-gabbro
33	-	0.36	480	26	690	330	Ditto
37	-	0.03	440	10	230	80	Leuco-gabbro
40	-	0.06	240	8	210	95	Ditto

Fig. 4 gives the Fe : Ti ratio and its relation with the vanadium content in the combined oxide minerals. Although the points are scattered on the graph, a general pattern indicating an increase of vanadium with Fe : Ti is noticed. The content of vanadium in the combined oxide minerals from sample to sample within a pluton is highest in those oxide minerals that have the highest Fe : Ti ratio. This relation is an expression of the relative amounts of magnetite and ilmenite and the partition of minor elements between the two minerals (Lister, 1966).

Copper in the complex is concentrated in the magnetite-gabbro and V-Ti-magnetite bands; this suggests a distinct lithologic control for the copper mineralisation. In these rocks it varies from 0.3 - 1.56%, whereas in other rock types it is very low (0.05 - 0.1%). Geochemically, Cu^{2+} is closer in size to Fe^{2+} among the major elements and the Cu : Fe ratios increase with fractionation (Taylor, 1965). In the Masanikere Complex the copper content in the magma must have been progressively concentrated during the fractionation, eventually crystallising in a sulphide phase.

The relatively dominant second generation chalcopyrite is an expression of this late sulphide phase.

Nickel and cobalt are the other two significant minor elements. Nickel values range from 210 - 1100 ppm and cobalt from 55 - 330 ppm. During the magmatic fractionation nickel is depleted at a faster rate, the Ni : Co ratio falls and the Fe : Co ratio tends to rise. The Ni : Co ratio is a very good index of fractionation (Taylor, 1965) and varies from 0.2 - 0.54 in the Masanikere deposit. Chromium content in the various members of the complex varies from 85 - 480 ppm showing no distinct correlation with rock type. Gallium is an important minor element detected in the Masanikere complex. Appreciable concentration of this minor element is noticed in the magnetite-gabbro and magnetite bands in the range of 30 - 40 ppm, and in gabbro and gabbroic anorthosite from 8 to 26 ppm. Ga^{3+} like V^{3+} may substitute for Fe^{3+} due to crystallographic/chemical affinity. It may also, to a lesser extent, take the place of Al^{3+} (Taylor, 1965). The importance and influence of the entry of Ga^{3+} into Al^{3+} and Fe^{3+} is well established in the Skaergaard intrusion, where Ga^{3+} is predominant in magnetite and plagioclase, while it occurs in minor amounts in pyroxene, olivine, ilmenite and apatite. A similar enrichment of gallium in Ti-magnetite has been reported from some Russian deposits (Vlasov, 1968). The Ga content increases from 4 ppm in the early formed olivine to 10 ppm in pyroxenes and 24 ppm in amphiboles, reaching a maximum of 50 ppm in Ti-magnetite. In the Masanikere complex, higher values (30 - 40 ppm) of gallium are associated with magnetite-gabbro rather than with the other rock types and this clearly indicates the preference of Ga^{3+} to Fe^{3+}. Silver and platinum are also noticed as minor elements. Silver is in the range of 6 - 8 ppm and platinum in the range of 1 - 12.5 ppb. Molybdenum values are generally less than 10 ppm.

STATISTICAL CORRELATION OF ELEMENT PAIRS

An attempt was made to find the correlation between element pairs Ti - V, Fe^{3+} - V, total Fe - V, Co - Ni, V - Cr. The correlation coefficients obtained by statistical analysis were subjected to the 't' test of significance at the 95% confidence level. Statistically significant positive correlation has been obtained between Ti and V in three of the six boreholes examined.

CONCLUSIONS

The gabbro-anorthosite complex near Masanikere is a result of magmatic differentiation of an iron-rich basic magma. The mineralogy of the complex indicates the auto-hydrative effects and low-grade metamorphism the rocks have undergone. The high concentration of magnetite in the basal part of the complex suggests gravitational differentiation. The cumulus texture of the Te-Ti oxides indicates the enrichment of Fe and Ti in the residual liquids which might have migrated downwards through the early formed mush of silicates. The major element chemistry corresponds to the different members of the complex; magnetite-gabbro, gabbro, gabbroic-anorthosite

Table 5. Correlation Coefficients for Element Pairs Ti and V and 't' - Test Data

r(Correlation coefficient)	N(No. of points)	't' calculation	't' (from tables at 95% confidence level)	Drill hole number
0.60	23	2.815	2.074	MSK - 1
0.47	27	2.445	2.056	MSK - 6/11
0.67	22	3.149	2.080	MSK - 7

and anorthosite. The vanadium content (0.23 - 0.93%) is thought to be present either as partial substitution for Fe^{3+} in magnetite or as VFe_2O_4 similar to $TiFe_2O_4$ (ulvospinel). Statistically significant positive correlation between Ti and V supports the possibility of vanadium being present as VFe_2O_4. The copper content in the magma which was initially low has progressively built up during fractionation and eventually resulted in a later sulphide phase. This is supported by the fact that Cu-mineralisation is associated with V and Ti magnetite and restricted to a distinct unit of the complex, namely, magnetite-gabbro. Gallium concentration in the higher ranges of 25 - 40 ppm is confined to the Fe-Ti oxide-rich members of the complex and this indicates the preference of Ga^{3+} to Fe^{3+}. The high ratios of Fe : Ti and Cu : Fe, and the appreciable concentration of Ga in the magnetite-gabbro and magnetite bands suggests a late-stage crystallisation.

Similar deposits of titano-magnetite occur in a number of places in the Karnataka craton. They occur in the Nuggihalli and Shimoga Schists Belts and also in the adjoining Peninsular Gneiss. These deposits share many similarities, although they apparently occur in different litho-stratigraphic units. The Ti-magnetite ores are associated with gabbro-anorthosite complexes, have a moderate but consistent vanadium content and extending copper mineralisation is a common association (Radhakrishna et al., 1972). These common features indicate widespread basic magmatic activity in the craton during the early Precambrian, resulting in an Archaean Fe-Ti-V-Cu metallogenic province in the Southern Indian Shield.

ACKNOWLEDGEMENTS

The authors are grateful to the Director General, Geological Survey of India, Calcutta, and the Director, National Geophysical Research Institute, Hyderabad, for permitting the publication of this paper. The authors also thank Professor V.S. Venkatasubramanian, Professor and Head of the Department of Physics, Indian Institute of Science, Bangalore, for the data on platinum and silver.

REFERENCES

Gjelsvik, T., 1957. Geochemical and mineralogical investigation of titaniferous iron ores, west coast of Norway, Econ. Geol., 54: 482-498.

Hall, A.L., 1932. The Bushveld Igneous Complex of the Central Transvaal. Mem. Geol. Surv. S. Africa, 28: 307.

Lister, G.F., 1966. The composition and origin of selected iron-titanium deposits. Econ. Geol., 61: 283, 304, 305.

Radhakrishan, B.P., Pandit, S.A. and Prabhakar, K.T., 1972. Copper mineralisation in the ultrabasic complex of Nuggehalli, Hassan district, Mysore State. Geol. Soc. Ind., 14: 302-312.

Ramakrishnan, M. Viswanatha M.N., Chayapathi, N. and Kutty T.R.N. (in press). Geology and geochemistry of the anorthosites from Karnataka Craton and their tectonic significance.

Taylor, S.R., 1965. Application of trace element data to problems of petrology. In: Physics and Chemistry of the Earth. 6: 171-175.

Tugarinov, A.I., Laktinova, N.V. and Mohanti, R., 1974. Distribution of V, Cr, Ni, Co, and Cu in titaniferous magnetites from some Indian deposits. 1973: 1326 - 1335.

Turner, F.J. and Verhoogen, J., 1961. Igneous and Metamorphic Petrology. McGraw-Hill, New York. p. 307.

Vlasov, K.A., 1968. Geochemistry and mineralogy of rare elements and genetic types of their deposits. Genetic types of rare element deposits. Vol. 3. Israel programme for scientific translations, Jerusalem.

Willemse, J., 1969. The vanadiferous magnetic iron ore of the Bushveld igneous complex. Econ. Geol. Mono., 4: 1 - 22.

Zolutuchin and Vasilev, 1967. Particular characteristics of formation of some trap intrusives of the southeast Siberian platform. Nauka, Moscow. 231pp.